ASC 中国建筑设计作品选 2013–2017

中国建筑学会
《建筑学报》杂志社
编著

CHINESE ARCHITECTURE: A Selection 2013–2017

同济大学出版社
TONGJI UNIVERSITY PRESS

《中国建筑设计作品选 2013—2017》

策　划
修　龙　仲继寿

编辑委员会
王建国　王骏阳　王　辉　王　澍　孙一民
朱小地　刘东卫　刘克成　朱剑飞　庄惟敏
邵韦平　李兴钢　张伶伶　沈　迪　张　颀
李振宇　张　雷　张鹏举　孟建民　周　恺
赵元超　柳亦春　钱　方　常　青　梅洪元
崔　愷　曹嘉明　韩冬青

主　编
崔　愷

副主编
黄居正　李晓鸿

编　辑
范　雪　田　华　孙凌波
陈佳希　孙晓峰　夏　楠

英　文
刘亦师

光 明 城

LUMINOCITY

CHINESE ARCHITECTURE:

A Selection

2013–2017

前　言

　　始于40年前的改革开放，为沉寂有年的中国社会释放出了巨大的活力，由此激发的政治、经济、文化、技术的急剧变化为中国当代建筑的发展提供了前所未有的动力。对于中国的建筑师而言，这是一个成长和磨砺的时代。大量实践的机会蜂拥而至、国际设计思潮不断涌入、市场经济语境下资本利益诉求强势、政绩观引导下的决策随意都驱使着建筑设计呈现一种匆忙应对、盲目追随的状态。加之21世纪初众多国际事务所的进入更使中国建筑师在重大项目创作中被迫处于从属地位，失去了创作话语权。

　　平庸、粗糙的建筑充斥着城乡环境，间或出现的山寨和奇葩的庸俗建筑更成为社会诟病的焦点。但近几年来情况有了变化，一批有着职业理想和人文关怀的建筑师，立足于本土的现实条件，潜心探索着中国建筑文化发展的各种可能性，逐渐摆脱了向西方的蹒跚学步，更深入地思考和研修中国文化的基因谱系，并将之应用到创作实践中去，形成了某种新的实验性和探索性的趋向。

　　该趋向反映在建筑学内部，通过重置结构、空间、建构、材料等本体性议题，逐步形成了属于中国本土建筑学自生性的内在逻辑；而在建筑学外部，国家倡导的文化复兴、城市更新、美丽乡村建设，种种回归常态的决策方向，也急迫地催促着中国建筑师面对当下现实，自觉地建构起有着强烈问题意识的理论框架和实践机制。

　　时代的群体意识和积极追求，催生出了一大批优秀的建筑作品，它们不仅赢得了国内专业媒体和大众媒体的广泛关注，也得到了海外学术研究机构和媒体的青睐，反映出中国建筑创作的新面貌。《建筑学报》作为中国建筑学会以及行业内最重要的核心期刊，也及时报道了这股新的创作风向，受到同行们的关注和认可。

　　但是由于杂志的篇幅所限，以及分期报道的离散性，很难全面完整地集中呈现中国建筑创作的总体风貌，加之学会在与国际相关学术机构和组织的频繁交往中，也一直缺少一本可以反映我国建筑实践发展状况、包含中英文和图像资料的著作可供交流。因此，2017年初在故宫召开的第七届编委会上，编委们提议由《建筑学报》组成编辑组，承担《中国建筑设计作品选》系列图书的编撰出版任务。

　　2017年6月，编辑部开始启动这一项目，经过认真准备，初步确定了作品遴选的方法、范围以及基本原则：① 虽然作品选具有年鉴性质，但鉴于是系列图书的第一辑，所选作品限定于2013–2017年间竣工完成（极少量重要作品可放宽至2013年）；② 作品落成地点不限于中国内地（大陆），适当兼及港澳台地区，中国建筑师在国外完成的作品，也予纳入；③ 作品不再进行大规模全国范围的海选，而从《建筑学报》登载过的数百个作品，以及中国建筑学会建筑创作奖、WA中国建筑奖获奖作品（排除重复的作品）中初选出首批项目，交由《建筑学报》编委会审定。

编辑部根据编委们审定的结果和反馈意见，进一步优化遴选方法和标准，比如要求对一名建筑师入选作品的数量限定不应超过5个，而对部分地区大型设计机构的优秀设计作品进行了补充提名，同时，编辑部把遴选范围拓展至2017年度全国优秀工程勘察设计行业奖的作品、中国威海国际建筑设计大奖赛获奖作品，从而基本达到了编委们提出的体现代表性、典范性和全局性的编辑出版目标。

在最终所收录的204个设计作品中，国有大型设计集团公司与民营设计机构各占一半。其中个人或合伙人制小型设计事务所贡献的作品数量甚为可观，他们在开放的设计市场中，各因才性，各擅胜场，立一端之说，骋一偏之长，在建构中国当代建筑文化的过程中，逐步拥有了不可小觑的话语权。

然而，综观本书仍不难发现，市场经济体制改革的不断深入和注册建筑师制度的长久实施，使得优秀建筑师无论身处于何种体制中，大多都不再满足于单纯提供技术服务的社会角色，而谋求向具有更高层次的文化创造性活动的身份转化，以确立其文化的主体意识。因此，在设计过程中他们更强调充分、自由地表达个人的价值观，释放出更强烈的创作欲望，从而成就了本书所呈现出的中国当代建筑文化的丰富多彩和多元格局。

就作品分布的地域而言，华东地区无可争议地拔得头筹，占50%；华北次之，占18.7%；华中占8%；西北和西南各占6.5%；东北占4.2%；华南地区意外地仅占3.3%，似乎与其蓬勃的经济极不相埒；港澳台地区和国外则各占1.4%。

全书收录的204个作品按建筑类型划分成五个大类，包括公共建筑、建筑遗产修复及改造、居住建筑、生产设施、城市设计及其他。其中公共建筑类较为复杂，数量达131项之多，因此不得不又将之细分为五个小类：文化/体育建筑、观演/博览建筑、教育/科研建筑、办公/商业建筑和医疗/交通建筑；修复及改造则有38项；城市设计及其他，共计16项；生产设施较少，有6项；令人疑惑的是，居住建筑仅占13项，与持续火爆的中国房地产市场反差极大。

在13个居住建筑项目中，有胡同民居的改造，有为灾民援建的震后居所，也有乡村建设大潮中涌现的民宿，当然，还有城市集合住宅。那些规模较小的民宅村居项目,无论是材料的选择、空间的营造,还是对居住者行为的引导和组织,大都体现出了建筑师的独具匠心;而在量大面广的集合住宅这一居住类型中,虽然有少数项目给人耳目一新之感,体现出建筑师寻求打破惯常平庸的平面布局,赋予居住生活空间以活力的努力,但或许受房地产市场发展阶段的局限,具有前瞻性和探索性的设计作品还是太少,与其建设总量不成比例。

在本书的编辑体例上,每个作品均占2页,按类型分类,按竣工时间排序,并在书的最后为每个作品附录了"索引",提供了更加详细的工程信息及作品所在位置的地理信息,倘若读者想深入了解某作品,可据此按图索骥。

本书的编撰是一次初步的尝试,挂一漏万在所难免,期望读者朋友不吝指正,以便在以后的编撰工作中补遗填漏,使这一系列图书成为一部记录中国当代建筑完整且持续更新的档案。

文/崔愷 黄居正
2018年2月8日

前言　Preface

Preface

The reform and opening-up policy adopted some 40 years ago has freed gigantic energy of Chinese society repressed for decades, and the sea change in politics, economics, culture and technology has become an unprecedented driving force for the development of contemporary Chinese architecture. For Chinese architects, this is the age of growth and empowerment. In a time when ample opportunities of practice emerge along with constant introduction of foreign design ideas, Chinese architectural design has fallen victim hastily and blindly to the interest of capital in the context of market economics and decision-making process predominated by political performance. To make it worse, the fact that a large number of foreign design firms flooded into China around the turn of 21st century has subordinated Chinese architects to a lesser place in crucial projects, losing the power of design discourse.

Mediocre and coarse buildings are crammed in the built environment, with strangely vulgar and replicating ones as the focus of criticism. However, in recent years the situation has been changing: a band of architects with professional and humane concerns have sedated themselves to explore various possibilities of the development of Chinese architectural culture, getting away from the pattern learned from the West but rooting more in local and realistic conditions to consider and study the genetic pedigree of Chinese culture. They have tried to apply what they learned to practical design, and the prospective trend of experimentation and exploration has been more visible.

Inside the field of architecture, ontological issues such as structure, space, construction and materials are rearranged to produce the autogenetic logic undergirding Chinese architecture in the local context. External circumstances such as cultural revival as advocated by the state, urban renewal, and construction of beautiful countryside have amended the direction of decision, urging Chinese architects to confront the realities with strong consciousness of improving theoretical framework and practicing mechanism.

Group consciousness and positive pursuit of the times have become the driving force to the emergence of a large amount of architectural masterpieces. They have not only attracted attention of domestic professional and mass media, but also been recognized by research institutions and media abroad, reflecting the new face of Chinese architectural practice. As the official magazine of Architectural Society of China and the most authoritative Chinese journal of this field, Architectural Journal has timely reported this new trend with extensive impact and recognition in China.

However, due to restrictions of length and discreteness of each issue, it is not easy to offer a full picture of the overall situation of Chinese architectural design. It is also expected that a book in both English and Chinese that can represent the development of Chinese architecture be compiled to facilitate frequent exchange between the Architectural Society of China and international academic institutions and organizations. As a result, in the Seventh Editing Board Meeting of Architectural Journal convened in the Forbidden City in early 2017, it is proposed that the Journal would organize a panel to compile and publish the series of "Selected Projects of the Architectural Society of China".

The editing board initiated this project in June 2017, and made preliminary decision on procedures, scope and basic principles for selection of projects: ① despite a yearbook by nature, it is the first of its kind that includes projects completed between 2013 and 2017, and a very few were built in 2013; ② the geographic sites are not restrained within the mainland but extends to Hong Kong, Macau, Taiwan and foreign countries where the projects produced by Chinese architects were built; ③ the first band of 164 projects was not selected on a national scale, but came either from hundreds of excellent projects published on previous issues of Architectural Journal, or from the winning projects of Architectural Society of China Design Awards and award-winning works of WA China Building Awards, which have undergone review by the editing board of Architectural Journal.

Based on the outcome of review and consequent feedback of the members of the board, the methods and standards of selection were optimized. For example, a maximum of five projects was set for the same architect(s), while supplementary nomination of excellent projects of large design institutions in certain regions was made. In the meantime, the editing board extended the scope of selection to the 2017 China Survey and Design Industry Awards and International Architectural Design Awards of Weihai, China. As a result, the objective of publication of representativeness, universality and overall importance as decided by the members of the board has generally been reached.

All the 204 projects finally selected in this book are equally split into the state-owned institutes and private firms. Particularly, small individuals or partnership design firms contribute to a remarkable portion in all projects, due to distinctive individuality and uniqueness in an open market. They have gained power of architectural discourse that cannot be ignored in the process of contemporary Chinese architectural cultural construction.

It is apparent that with the continuous deepening of market economic reform and implementation of the mechanism of registered architects, preeminent architects, regardless of institutions they are affiliated to, are no longer satisfied with the social role of offering technical service but seeking for a transformation of status of cultural creation on a higher level to construct their cultural identity. As such, they have emphasized more on a sufficient and free expression of individual cultural value, releasing stronger desire of creation, hence the richness and diversity of contemporary Chinese architectural culture as manifest in this book.

In regard to geographical distribution of the buildings selected, the majority of 50% are built in East China, second to which are those of 18.7% in North China. Those built in the Northwest and Southwest are 6.5% respectively, the Northeast is 4.2%. It is unexpected the percentage of South China is as low as 3.3%, strangely disproportionate to the prosperity of economic development of this region. Those in Hong Kong, Macau and Taiwan are 1.4%, the same as overseas.

All 204 projects selected in this book are grouped into five architectural categories: Public Architecture, Renovation/Heritage Preservation, Housing, Production Facilities, and Urban Design/Others. Public Architecture are more complicated than others, so the total amount of 131 projects in this category is sub-divided into five genres: Cultural Architecture/Sports Architecture, Theaters/Museums, Educational Architecture/Scientific Research Architecture, Office/Commercial Architecture, and Medical Architecture/Transportation Architecture. There are 38 projects for the category of preservation and renovation, 16 for urban design, urban landscape and others. Only 6 projects appear in industrial facilities. It is bizarre that the category of residential buildings only includes 13 projects, a sheer contrast to the boom of real estate in contemporary China.

The 13 projects consist of regeneration of hutong alleyway houses, shelters built in the aftermath of earthquakes, homestay hostels as part of rural construction, and certainly urban collective apartments. For the small-scaled projects in countryside, the way architects select materials, manage spatial organization, and encourage activities in space is unique. In contrast, in collective apartment buildings as a most widely seen architectural type with the largest quantity in the city, very few are refreshing that attempt to break through mediocre layout and invigorate living space, while most are constrained by the current development of real estate market with less vision and experimentation, hence the disproportion is produced.

In terms of format, all projects are categorized based on different architectural types and sorted chronically according to completion dates, with two pages for each. At the end of the book an index with more information such as engineering and geographical details for each project is provided, which readers can use as a guide to explore more if interested.

This book pioneers an attempt that entails inevitable mistakes and expects amends from readers, in a way that we as editors can improve the future work of the kind as a constantly upgrading archival series of complete records on contemporary Chinese architecture.

CUI Kai, HUANG Juzheng
2018/02/08

目 录

Contents

公共建筑
Public Architecture

001

263

改造及修复
Renovation/Heritage Preservation

264

341

居住建筑
Housing

342

369

生产设施
Production Facilities

370

383

城市设计及其他
Urban Design/Others

384

417

项目索引
Index

418

440

Public Architecture 公共建筑

001

文化 / 体育
Cultural Architecture
Sports Architecture
002 — 055

- 002 马岔村村民活动中心
Macha Village Center
- 004 张家界国家森林公园游客中心
Zhangjiajie National Forest Park Visitor Center
- 006 天津泰达居民之家
Residents' Homes of TEDA, Tianjin
- 008 华夏河图国际艺术家村
River Origins International Artists' Village
- 010 广州市从化区图书馆
Conghua District Library of Guangzhou
- 012 上海棋院
Shanghai Chess Academy
- 014 海边图书馆
Seashore Library
- 016 东湖国家自主创新示范区公共服务中心
Donghu National Innovation District Civic Center
- 018 重庆桃源居社区中心
Chongqing Taoyuanju Community Center
- 020 罕山生态馆和游客中心
Hanshan Ecological Hall and Tourist Center
- 022 大厂民族宫
Da Chang Cultural Center
- 024 街子古镇梅驿广场
Jiezi Ancient Town Meiyi Square
- 026 天津大学新校区综合体育馆
Gymnasium of New Campus of Tianjin University
- 028 盛乐遗址公园游客中心
Tourist Center of Shengle Heritage Park
- 030 南矶湿地访客中心
Nanji Wetland Reserve Visitor Center
- 032 南京牛首山景区游客中心
Tourist Center in Niushou Scenic, Nanjing
- 034 武汉光谷国际网球中心网球馆
Wuhan Optics Valley International Tennis Center, The Tennis Court
- 036 码头书屋
Library on the Quay
- 038 2014 青岛世界园艺博览会天水、地池综合服务中心
Tianshui/Dichi Service Center of International Horticultural Exposition 2014
- 040 南京万景园小教堂
Nanjing Wanjing Garden Chapel
- 042 无锡阳山田园综合体 I 期 田园生活馆
The 1st Phase of Yangshan Rural Life Complex in Wuxi, Rural Life Pavilion
- 044 康巴艺术中心
Kangba Art Center
- 046 盛泽文化中心
Shengze Cultural Center
- 048 第十三届全国冬季运动会冰上运动中心
Ice Sports Center of the 13th National Winter Games
- 050 黄河口生态旅游区游客服务中心
Visitor Center of Ecological Tourism Area at Delta of Yellow River

目录 / Contents

052	淮安市体育中心 Huai'an City Sports Center		106	北京菜市口输变电站综合体（电力科技馆） Beijing Caishikou Power Transformer Substation Complex (Electric Power Science and Technology Museum)
054	天津武清文化中心 Wuqing Cultural Center, Tianjin		108	中国版画艺术博物馆 China Scratchboard Art Museum
056	遵义市娄山关红军战斗遗址陈列馆 The Site Museum of Loushanguan Battle, Zunyi	观演 / 博览 Theaters/Museums 056 — 131	110	玉树州博物馆 Yushu Museum
058	上海世博会博物馆 Shanghai Expo Museum		112	国家会展中心（上海）Shanghai National Exhibition and Convention Center
060	苏州非物质文化遗产博物馆 Suzhou Intangible Cultural Heritage Museum		114	又见五台山剧场 Encore Wutai Mountain Theater
062	延安大剧院 Yan'an Grand Theatre		116	盘锦城市文化展示馆 Panjin City Cultural Exhibition Center
064	宜昌规划展览馆 Yichang Planning Exhibition Hall		118	上海嘉定保利大剧院 Jiading New Town Poly Grand Theater
066	2015 米兰世博会中国馆 China Pavilion for Expo Milano, 2015		120	玉树文成公主纪念馆 Memorial of Princess Wencheng, Yushu
068	银川当代美术馆 Museum of Contemporary Art Yinchuan		122	绩溪博物馆 Jixi Museum
070	张家界博物馆 Zhangjiajie Museum		124	贾平凹文化艺术馆 Jia Pingwa Culture & Art Gallery
072	商丘博物馆 Shangqiu Museum		126	南京博物院改扩建工程 Nanjing Museum
074	侵华日军南京大屠杀遇难同胞纪念馆三期扩容工程 The Memorial Hall of the Victims in Nanjing Massacre by Japanese Invaders, Phase III		128	乌镇剧院 Wuzhen Theater
076	哈尔滨大剧院 Harbin Opera House		130	金陵美术馆 Jinling Art Museum
078	云南省博物馆新馆 New Yunnan Province Museum	教育 / 科研 Educational Architecture Scientific Research Architecture 132 — 209	132	南开大学新校区核心教学区 Core Teaching Quarter of the New Campus of Nankai University
080	木心美术馆 Muxin Art Museum		134	上海德富路初中 De Fu Junior High School
082	金陵大报恩寺遗址博物馆 Site Museum of Jinling Grand Bao'en Temple		136	苏州实验中学原址重建项目 Reconstruction Project of Suzhou Experimental Middle School
084	临安体育文化会展中心 Lin'an Sports and Culture Center		138	山西兴县 120 师学校 Instruction Building of the 120th Division School in Xing County, Shanxi Province
086	毓绣美术馆 Yu - Hsiu Museum of Art		140	清华大学海洋中心 Tsinghua Ocean Center
088	银川韩美林艺术馆 Yinchuan Han Meilin Art Museum		142	天颐湖儿童体验馆 Tianyi Lake Children's Edutainment Mall
090	蚌埠博物馆及规划档案馆 Bengbu Museum, Urban Planning Exhibition and Archive Centre		144	苏州湾实验小学 Suzhou Bay Experimental Primary School
092	刘海粟美术馆 Liu Haisu Art Museum		146	北京华为环保园 J 地块数据通信研发中心 Beijing Huawei R&D Center
094	龙美术馆（西岸馆）Long Museum (West Bund)		148	FAST 工程观测基地综合楼 Comprehensive Building of FAST Observation
096	无锡阖闾城遗址博物馆 Helv City Historic Site Museum		150	岱山小学 岱山幼儿园 Daishan Primary School and Kindergarten
098	柯力博物馆 Keli Museum		152	中衡设计集团研发中心 The Design and Research Building of ARTS Group Co., Ltd.
100	范曾艺术馆 Fan Zeng Art Gallery		154	华东师范大学附属双语幼儿园 East China Normal University Affiliated Bilingual Kindergarten
102	乌海市黄河渔类增殖站及展示中心 Wuhai Yellow River Fishing Station and Exhibition Center			
104	浙江湖州梁希纪念馆 Huzhou Liangxi Memorial Hall			

156	中福会浦江幼儿园 Pujiang China Welfare Institute Kindergarten	
158	上海国际汽车城研发港 D 地块 Plot D, The R&D and Innovative Port of Anting International Automobile City	
160	芭莎·阳光童趣园 BAZAAR · Sunshine Playhouse	
162	天津大学新校区图书馆 Library on the New Campus of Tianjin University	
164	浙江音乐学院 Zhejiang Conservatory of Music	
166	清华大学南区学生食堂 Central Canteen of Tsinghua University	
168	杭州师范大学仓前校区 Hangzhou Normal University Cangqian Campus	
170	苏州科技城实验小学 Experimental Primary School of Suzhou Science and Technology Town	
172	清控人居科技示范楼 THE-Studio	
174	武汉理工大学南湖校区图书馆 Wuhan University of Technology Nanhu Campus Library	
176	东北大学浑南新校园风雨操场 Hunnan Campus of Northeastern University Gymnasium	
178	北京四中房山校区 Beijing No.4 High School Fangshan Campus	
180	松江名企艺术产业园区 Songjiang Art Campus	
182	中新生态城滨海小外中学部 Binhai Xiaowai High School, Sino-Singapore Tianjin Eco-City	
184	寒地建筑研究中心 Cold Region Architecture Research Center	
186	北京工业大学第四教学楼组团 A Complex of the Teaching Facilities at Beijing University of Technology	
188	同济大学浙江学院图书馆 Zhejiang Campus Library, Tongji University	
190	浙江科技学院安吉新校区 New Campus of Zhejiang University of Science and Technology in Anji City	
192	北京育翔小学回龙观学校 Beijing Yuxiang Primary School Huilongguan School	
194	上海嘉定桃李园实验学校 Shanghai Jiading Tao Li Yuan Experimental School	
196	北京大学光华管理学院西安分院 Peking University Xi'an Branch of Guanghua School of Management	
198	大连华信（国际）软件园 Dalian Hi-Think (International) Software Park	
200	同济大学嘉定校区留学生宿舍及专家公寓 The Foreign Students' Dormitory and Experts' Apartment in Jiading Campus of Tongji University	
202	瓦山——中国美术学院象山校区专家接待中心 Tiles Hill: New Reception Center for the Xiangshan Campus, China Academy of Art	
204	南京三宝科技集团物联网工程中心 Networking Engineering Center, Nanjing Sample Sci-Tech Park	
206	大连理工大学辽东湾校区 The Liaodong Bay Campus of Dalian University of Technology	
208	国电新能源技术研究院 Guodian New Energy Technology Research Institute	

目录 / Contents

办公 / 商业
Offices
Commercial Architecture
210 — 253

210	大乐之野庾村民宿	Lostvilla Boutique Hotel in Yucun
212	湖南城陵矶综合保税区通关服务中心	Hunan Chenglingji Free Trade Zone Customs Clearance Service Center
214	湖上村舍	The House by the Lake
216	石塘互联网会议中心	Shitang Village Internet Conference Center
218	上海中心大厦	Shanghai Tower
220	天赐新能源企业总部	Headquarters of Zhejiang TCI Ecology & New Energy Technology
222	南昌绿地紫峰大厦	Jiangxi Nanchang Greenland Zifeng Tower
224	大舍西岸工作室	Atelier Deshaus Westbund
226	滨江休闲广场商业用房	Binjiang Leisure Plaza and Commercial Housing
228	外滩 SOHO	Bund SOHO
230	水西工作室	Shui Xi Studio
232	北京绿地中心	Beijing Greenland Center
234	宝龙城市广场集装箱售楼处	Baolong Qingpu Plaza Container Sales Office
236	中国商务部驻印尼商务馆舍	Office Building of Chinese Ministry of Commerce in Indonesia
238	苏州科技城国家知识产权局苏州中心办公楼	Patent Examination Cooperation Jiangsu Center of the Patent Office, SIPO
240	北方长城宾馆三号楼	NO.3 Building of Northen Great Wall Hotel
242	玉树藏族自治州行政中心	Yushu Tibetan Autonomous Prefecture Administrative Center
244	西村大院	West Village
246	成都远洋太古里	Sino-Ocean Taikoo Li, Chengdu
248	凌空 SOHO	Sky SOHO
250	尤努斯中国中心陆口格莱珉乡村银行	Yunus China Center Lukou Grameen Village Bank
252	凤凰中心	Phoenix Center

医疗 / 交通
Medical Architecture
Transportation Architecture
254 — 263

254	上海虹桥国际机场 T1 航站楼改造及交通中心工程	Shanghai Hongqiao International Airport T1 Renovation & Traffic Center
256	上海东方肝胆医院	Shanghai Oriental Hepatic Hospital
258	昂洞卫生院	Angdong Hospital
260	太原南站	Taiyuan South Railway Station
262	都江堰大熊猫救护与疾病防控中心	Dujiangyan Giant Panda Conservation and Disease Control Centrer

Renovation Heritage Preservation

改造及修复

264

266	七园居 The Hotel of Septuor	310	徐家汇观象台修缮工程 Refurbishment of L'Observatoire de ZI-KAWEI, Xuhui, Shanghai
268	五龙庙环境整治工程 The Environmental Upgrade of the Five Dragons Temple	312	上海油雕院美术馆及咖啡厅 SPSI Art Museum & Chimney Cafe
270	四叶草之家 Clover House	314	桐庐莪山畲族乡先锋云夕图书馆 Tonglu Librairie Avant-Garde, Ruralisation Library
272	乌镇北栅丝厂改造 Renovation of Beizha Silk Factory in Wuzhen	316	广元千佛崖摩崖造像保护建筑试验段工程 Experimental Protective Structure for Thousand Buddha Cliff
274	五原路工作室 Wuyuan Rd. Studio	318	摩梭家园 Homeland of Mosuo
276	西浜村昆曲学社 The Society of Kun Opera at Xibang Village	320	上海电子工业学校六号楼 / 学生浴室 Block 6 of Shanghai Electronic Industry School/Student Shower Block
278	乡宿上泗安 Shangsi'an Cottage	322	牛背山志愿者之家 Cattle Back Mountain Volunteer House
280	池社 Chi She	324	西河粮油博物馆及村民活动中心 Xihe Cereals and Oils Museum and Villagers' Activity Center
282	四行仓库修缮工程 Protection and Restoration of the Joint Trust Warehouse	326	箭厂胡同文创空间 Arrow Factory Media & Culture Creative Space
284	谢店村传统村落保护与再生规划设计 Planning of Xiedian Traditional Village Protection and Regeneration	328	上海国际时尚中心 Shanghai International Fashion Center
286	天目湖微酒店 Tianmu Lake VIP Club	330	南京下关电厂码头遗址公园 Relics Park for the Coal Dock of Xiaguan Power Plant
288	隐居江南精品酒店 Seclusive Jiangnan Boutique Hotel	332	天津拖拉机厂融创中心 Sunac Center of Tianjin Tractor Factory
290	北京民生现代美术馆 Minsheng Museum of Modern Art	334	吉兆营清真寺翻建工程 Jizhaoying Mosque Renovation
292	首都电影院装修改造 Preservation and Reconstruction of the Capital Cinema	336	天仁合艺美术馆 T_Museum
294	居住集合体 L Housing L	338	回酒店 Hui Hotel
296	微园 Wei Yuan Garden	340	南京愚园修缮与重建 Restoration and Reconstruction of Nanjing Yu Garden
298	陈化成纪念馆移建改造 Removal Renovation of Chen Huacheng Memorial		
300	竞园 22 号楼改造 Jingyuan No.22 Transformation		
302	富丽服装厂改造 Renovation of Fuli Clothing Factory		
304	上海联创国际设计谷 Shanghai UDG International Design Valley		
306	上海延安中路 816 号修缮项目——解放日报社 Renovation Project of 816# Middle Yan'an Road: the Jiefang Daily Office		
308	大理慢屋·揽清度假酒店 Dali Munwood Lakeside Resort Hotel		

Housing 居住建筑

342

344	乌镇·雅园	Wuzhen Graceland
346	泰康之家·燕园	Taikang Yanyuan Community
348	船长之家改造	Renovation of the Captain's House
350	上海龙南佳苑	Shanghai Longnan Garden Social Housing Estate
352	齐云山树屋	The Qiyun Mountain Tree House
354	随园嘉树养生中心	The Health & Longevity Center of the Suiyuan Jia Shu Project
356	深深·深宅	Deep³ Courtyard
358	新青年公社	New Youth Commune
360	浙江山地老宅	The Old Curtilage in the Mountains of Zhejiang
362	退台方院	Stepped Courtyards
364	苏州阳山敬老院	Suzhou Yangshan Nursing Home
366	广州南湖山庄 C、D 区	Villa South Lake in Guangzhou-District C and D Residence
368	拙政别墅	The Humble Administrator's Villa

Production Facilities 生产设施

370

372	唐山乡村有机农场	Tangshan Rural Organic Farm
374	松阳樟溪红糖工坊	Songyang Zhangxi Brown Sugar Workshop
376	雅昌艺术中心	Artron Art Center
378	松风翠山茶油厂	Song Feng Cui Camellia Oil Plant
380	爱慕时尚工厂	Aimer Lingerie Factory
382	武夷山竹筏育制场	Wuyishan Bamboo Raft Factory

Urban Design Others

城市设计及其他

INDEX

项目索引

_384_

_418_

_386_	常德老西门综合片区城市更新 Urban Renewal of Old West City Gate Region	
_388_	驿道廊桥改造 Lounge Bridge Renovation	
_390_	天府新区公安消防队站 Fire Station of Tianfu New District	
_392_	风雨桥 Wind and Rain Bridge	
_394_	隐庐莲舍 Lotus Tea House	
_396_	六边体系 HEX-SYS	
_398_	衢州鹿鸣公园 Quzhou Luming Park	
_400_	长沙滨江文化园 Riverside Cultural Park in Changsha	
_402_	天空之桥 Sky Bridge	
_404_	上海廊下郊野公园核心区景观 Veranda in Shanghai	
_406_	华山绿工场 Green Factory	
_408_	松鹤墓园接待中心 Reception Center for Songhe Cemetery	
_410_	格萨尔广场 Gesar Square	
_412_	林建筑 Forest Building	
_414_	太阳公社竹构系列 Bamboo Design in Taiyang Farming Commune	
_416_	苏仙岭景观瞭望台 Observation Deck in Suxianling	

_420_ 项目索引
Index

Cultural Architecture /Sports Architecture

Theaters/Museums

Educational Architecture /Scientific Research Architecture

Offices/Commercial Architecture

Medical Architecture/Transportation Architecture

Public
Architecture

文化／体育
观演／博览
教育／科研
办公／商业
医疗／交通

公共
建筑

P001

P263

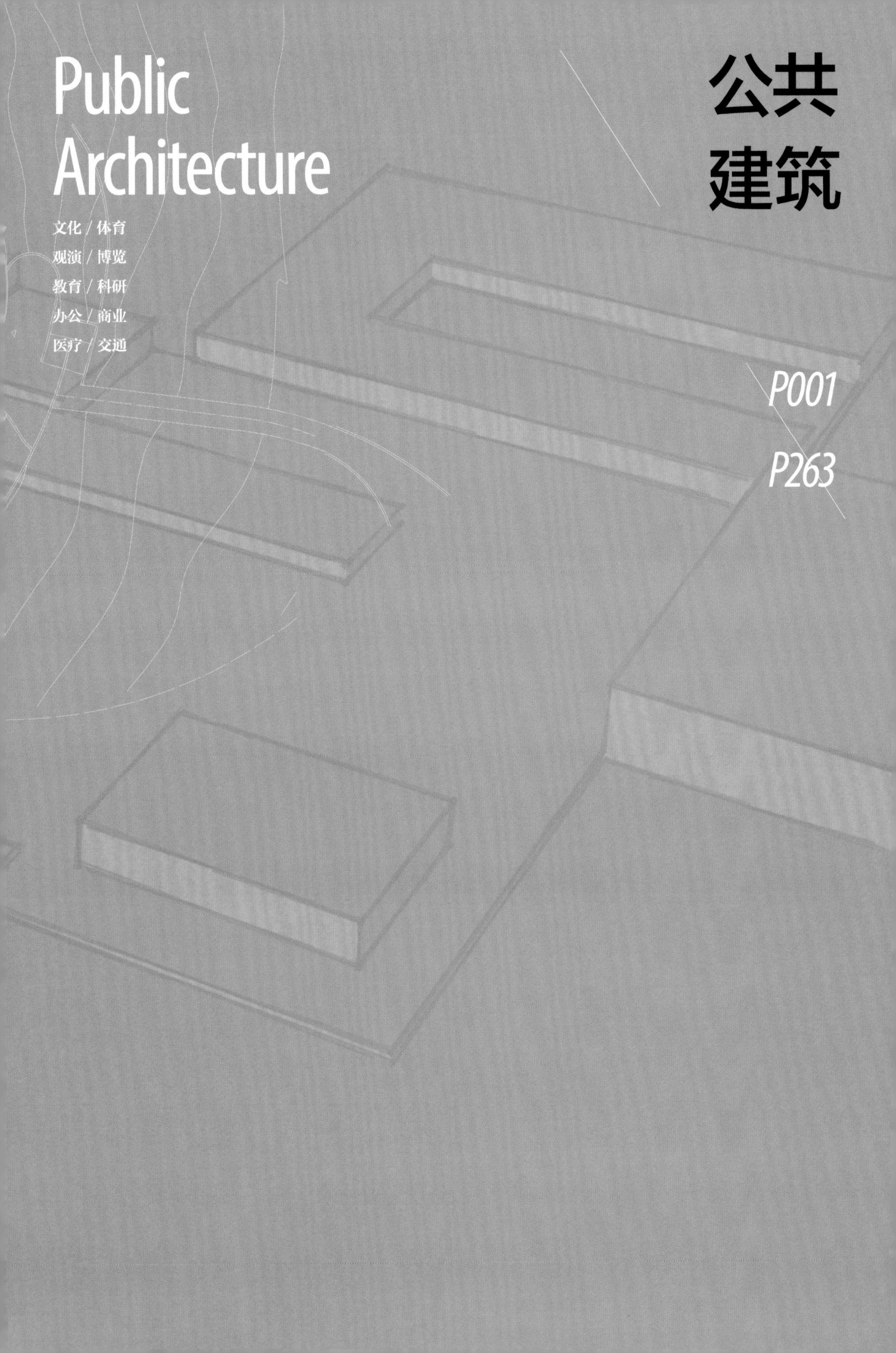

Public Architecture | Cultural Architecture / Sports Architecture

全景
场院

托儿所外墙

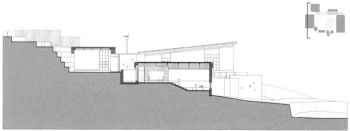

剖面 0 2 5 10m

马岔村村民活动中心
Macha Village Center

建筑设计
土上建筑工作室
蒋蔚 穆钧 周铁钢

甘肃 会宁 | 竣工时间 2016 年 | 摄 影 蒋蔚 李强强 王正阳

Architects
Onearth Architecture
JIANG Wei, MU Jun, ZHOU Tiegang
Location Huining, Gansu
Completion 2016
Photo JIANG Wei, LI Qiangqiang, WANG Zhengyang

托儿所室内

东侧外景

　　该中心是住建部现代夯土民居研究与示范项目中的一项重点内容，由无止桥慈善基金出资并组织当地村民与志愿者共同建造完成。功能包括：多功能厅、商店、医务室和托儿所。除满足马岔村民日常公共生活服务需求外，也是该地区推广现代生土建造技术的培训基地。中心的建设是按当地传统的施工组织模式进行的一次现代夯土建造实践。施工过程本身也是对当地工匠和村民的技术示范与培训。中心所处的甘肃省会宁县马岔村为干旱的黄土高原沟壑区，土资源极其丰富。建筑在空间组合方式上借鉴了当地民居传统的合院形式，并结合山地现状，将若干土房子设置于山坡上不同的标高，围合出一个三合院，开口面向山谷，建筑师希望这几个土房子就像在地里生出的土块，自然地融入到当地的空间景观之中。

场院一角

The center plays a key role in the research and exhibiting demonstration projects of the modern rammed earth residential architecture led by the Ministry of Housing and Urban-Rural Development (MOHURD). It is funded by Wu Zhi Qiao (Bridge to China) Charitable Foundation that organizes the local villagers and volunteers for construction. The center consists of a multi-purpose hall, shops, an infirmary and a nursery. It not only meets the needs of daily public service of the villagers at Macha, but also provides a training base for the promotion of modern earth construction technology. The construction of the center is a new practice in modern rammed earth architecture based on the local traditional construction and organization methods. In addition, it is also a technical demonstration of training for the local craftsmen and villagers through the construction process itself. Ma Cha Village center locates in Huining County, Gansu Province. Because of the arid loess plateau gully area, the resources of earth are extremely rich. The house draws on space form of the traditional courtyard of the local vernacular architecture. Combined with hilly topography, a series of houses at different heights to enclose a courtyard are arranged, open to the valley. The architects hope these earth houses are like the clods grown from the ground that can be naturally integrated into the local landscape.

1 多功能室
2 阅览室
3 场院
4 托儿所
5 厨房
6 沙坑
7 医务室
8 儿童滑梯
9 戏台
10 商店

上层平面　0 1 2　5m

东北侧外景
主入口夜景

张家界游客服务中心位于世界文化与自然双重遗产景区张家界国家森林公园内,一般游客要乘坐山下的"百龙天梯",垂直上升近400米的高度才能到达此地。因为选址的这一特殊性,所以场地周围几乎皆是风景。游客中心的功能以餐饮为主,建筑师试图将多个档次的餐厅和场地的高差结合起来进行设计,同时其体量是逐层递减的关系,这样就会在每一层形成较大的露台,各层露台通过室外的大台阶彼此相连,拾级而上如同山的意象,建筑由此自然地融合在四周的风景当中。游客中心在造型设计中吸收了湘西民居吊脚楼的独特之处,由石材砌筑的基座部分、中间木结构主体部分以及青瓦覆盖的坡屋顶部分三部分构成,在风景中进行"湘土"的重构。

Zhangjiajie National Forest Park Visitor Center is located in Zhangjiajie National Forest Park, which is recognized as both world natural heritage and world cultural heritage scenic site. The visitor center sits on the 400m high "base" with fantastic views, so most visitors have to take the "One-Hundred-Dragon Steps" to approach to it. As catering is the main function of the center, the architects combined the dining halls on various levels of height, and reduced the mass of buildings as they ascend, forming terraces on different stories that connects to each other by the outdoor steps, merging into the scenery just like the mountains. Local stilted buildings provide a unique reference to the creation of the building form. The three parts of the building—the stone base, the main part with wood and the pitched roof with grey tiles, represent the "local Western Hunan architecture" in the scenery.

张家界国家森林公园游客中心
Zhangjiajie National Forest Park Visitor Center

建筑设计
东南大学建筑设计研究院有限公司艺筑工作室
杨志疆 杨程 周妍琳

Architects
Atelier Art & Architect, Architects & Engineers Co., Ltd. of Southeast University
YANG Zhijiang, YANG Cheng, ZHOU Yanlin

湖南 张家界 | 竣工时间 2016年 | 摄 影 夏强

Location Zhangjiajie, Hunan
Completion 2016
Photo XIA Qiang

西北侧夜景
室外露台

三楼餐厅内景

1 专卖店
2 VIP 休息区
3 等候区
4 旅游纪念品商店
5 门厅

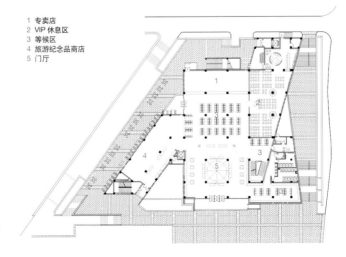

一层平面　0 2 5 10m

入口水院　　　　　　　　　　　从阅览室看庭院
　　　　　　　　　　　　　　　阅览区室内

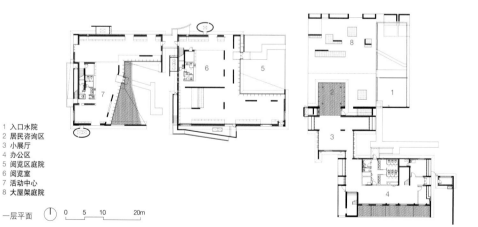

1 入口水院
2 居民咨询区
3 小展厅
4 办公区
5 阅览区庭院
6 阅览室
7 活动中心
8 大屋架庭院

一层平面　　0　5　10　　20m

天津泰达居民之家
Residents' Homes of TEDA, Tianjin

建筑设计
天津华汇工程建筑设计有限公司
周恺

Architects
Huahui Architectural Design & Engineering Co., Ltd.
ZHOU Kai
Location Tianjin
Completion 2016
photo ZHOU Kai

天津　　竣工时间 2016 年　　摄　影　周恺

项目位于居住小区的东侧，既是居民的公共活动中心，也是小区的步行主入口。作为居民之家，建筑涵盖了阅读、展示等文化活动区，咖啡、茶馆等休闲等候区，以及洽谈、管理等办公服务区。在设计上，将建筑化整为零，根据不同的功能及其相互关系，用"加法"的方式，将不同的小尺度区块组合起来。区块之间，着意布置了绿化水景，营造出与自然亲密接触的空间体验。建筑与环境之间不刻意分隔，而是用建筑的方式，或延伸墙体，或挖空屋顶，用一块块充满自然元素的灰空间模糊建筑与环境的界限。在建造方面，采用了形式纯净的清水混凝土作为主要的建筑语汇，并辅以木作装饰。两种材质一硬一软，一刚一柔，协调地组织在一起，为居民创造出清雅、纯粹、宜人的社区交流氛围。

A residents' home, the building consists of areas of cultural activities such as reading and exhibiting space, and leisure and waiting areas such as spaces for coffee and tea along with office service area for negotiation and management. In design, the building is divided according to different functions and inter-relationships. The different small-scale spaces were put together according to the rule of addition. The architects deliberately arranged the green water features between the regions, creating the spatial experience of proximity to nature. Buildings and the environment is not separated by intention, but through construction such as extending the wall or carving out the roof. The boundaries of the building and its surroundings are blurred by a piece of gray space filled with natural elements. In construction, bare concrete was used as the main building material, supplemented by wooden decorations. Hard and soft qualities of the two materials are coordinated together to create an elegant, pure, pleasant atmosphere for social communication in the community.

入口水院
大屋架庭院
模型

夜景

华夏河图国际艺术家村位于银川市郊区，西北面近邻鸣翠湖，远眺贺兰山脉。这是一片沉稳、内敛、质朴并充满力量感的"原生"建筑群落，设计语言来自黄土高原的黄、穆斯林的白、西海固的拱，辅以锈蚀钢板深沉的红、金属构件温和的灰、以及高大通透的玻璃。设计坚持可持续发展的原则，外墙由160万块回收的旧黏土砖组成，保证视觉上不仅如黄土般淳朴自然，在光影下更加呈现出强烈的雕塑感。最终呈现的建筑群整体壮阔而亲切，体形方正中寓变化，平静中包含自由，为艺术的创造留下更多的想象空间。

Facing the serene Lake Mingcui on the outskirts of Yinchuan, the River Origins International Artists' Village is located on the border between lush wetlands and arid desert divided by the Yellow River, overlooking the Helan Mountains from afar. This building group of calmness, pristine and intrinsic sense of power is regionally oriented, using vernacular colors of yellow from the Loess plateau, white from Muslim, the arch form from Xihaigu region and the modern language such as the rusted steel with red stains, soft grey metals and transparent glass. Hinging on the design principle of sustainable development, the Artists' Village has recycled 1.6 million pieces of used clay bricks on facades, in which a rustic, vigorous sense and the sculptural volume with shadows under the light composed impressive forms. The final presentation of the whole complex imprints a kind of geniality with magnificence, peace with freedom, leaving more imagination space for artwork.

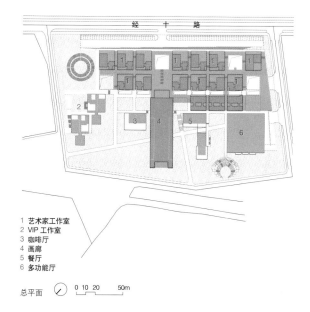

1 艺术家工作室
2 VIP工作室
3 咖啡厅
4 画廊
5 餐厅
6 多功能厅

总平面 0 10 20 50m

华夏河图国际艺术家村
River Origins International Artists' Village

宁夏 银川 | 竣工时间 2016年

建筑设计
北京市建筑设计研究院有限公司王戈工作室
北京科可兰建筑设计咨询有限公司
北京中天建中工程设计有限责任公司

王戈 李阳 于宏涛

摄影 杨超英 李洁苒

Architects BOA STUDIO, Bejing Institute of Architectural Design; KKL Partnership Architects ; Beijing Zhongtian Jianzhong Engineering Design Co., Ltd.
WANG Ge, LI Yang, YU Hongtao
Location Yinchuan, Ningxia
Completion 2016
photo YANG Chaoying, LI Jieran

雪景
咖啡厅东望

画廊室内
餐厅走廊

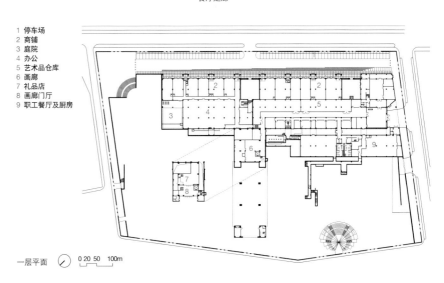

1 停车场
2 商铺
3 庭院
4 办公
5 艺术品仓库
6 画廊
7 礼品店
8 画廊门厅
9 职工餐厅及厨房

一层平面　0 20 50 100m

全景

主楼

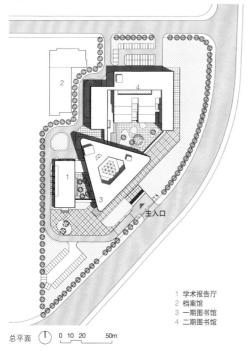

1 学术报告厅
2 档案馆
3 一期图书馆
4 二期图书馆

总平面 0 10 20 50m

项目紧邻穿越城市的流溪河绿道，自然环境优越。设计分为一二两期，一期以旋转的等边三角形平面呼应河流带来的城市网格变化，主体阅览区的三角形体量轻轻浮起两层，越过高起的河堤，最大化室内景观视野。二期与一期形成半围合关系，东南角的开口以及层层退让的露台，使外部河流景观自然过渡到内部庭院并向上延展，形成建筑与自然环境的对话关系。设计秉持低成本被动式节能设计策略，将岭南传统建筑中骑楼、冷巷、庭院、水面等特色气候空间加以发展应用，并结合具有时代特点的退台、立体绿化、立面遮阳等建筑手法，形成一整套综合的绿色建筑语汇，从规划布局、空间灵活性及气候空间设计等方面，对地域性建筑的可持续性问题作出一定的思考和尝试。

Next to the greenbelt of the river winding through the city, Liuxihe River, the project takes the advantage of the natural environment and is developed for two phases. The shape of the building of Phase 1 is laid out as an equilateral triangle rotated to adjust to the city grid cut by the River. The main part of the building: the reading area is raised two stories above the riverbank, to maximize the view of landscape from the interior. The buildings of Phase 2, along with that of Phase 1 form a semi-enclosing layout. The opening at the southeastern corner and the terrace descending by stories allow external river landscape naturally transitioning to internal courtyard and extending upwards, which forms the conversation between the building and the nature. The design hinges on the principle of low-cost passive energy efficiency, which develops from the distinguishing climate spaces of traditional Lingnan architecture, e.g. arcaded space, cold lane, courtyard, water, etc. and also combines contemporary architectural forms including descending terraces, vertical gardens, facade shading, etc., contributing to a whole set of comprehensive green architecture concepts. The design demonstrates some contemplations and attempts on regional architecture, including layout planning, flexible space arrangements, climate space design, etc.

广州市从化区图书馆
Conghua District Library of Guangzhou

广东 广州   竣工时间 2016 年

建筑设计
东意建筑工作室
华南理工大学建筑设计研究院
肖毅强 刘穗杰 齐百慧

摄 影 陈中 王艮 吴嗣铭 林力勤

Architects Atelier Y; Architectural Design and Research Institute of South China University of Technology
XIAO Yiqiang, LIU Suijie, QI Baihui
Location Guangzhou, Guangdong
Completion 2016
Photo CHEN Zhong, WANG Gen, WU Siming, LIN Liqin

庭院及退台
主楼与报告厅之间的连桥

中庭 | 二期连桥

沿街外景

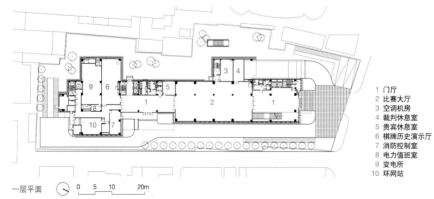

一层平面

1 门厅
2 比赛大厅
3 空调机房
4 裁判休息室
5 贵宾休息室
6 棋牌历史演示厅
7 消防控制室
8 电力值班室
9 变电所
10 环网站

上海棋院
Shanghai Chess Academy

建筑设计
同济大学建筑设计研究院（集团）有限公司
曾群　吴敏　汪颖

Architects
Tongji Architectural Design (Group) Co., Ltd.
ZENG Qun, WU Min, WANG Ying
Location Shanghai
Completion 2016
Photo ZHANG Yong

上海　　竣工时间 2016 年　　摄　影　章勇

办公入口

项目地处上海市繁华商业区南京西路,基地为南北向狭长地块,南北长约 140 米,东西最窄处约 40 米。将室内和室外的虚实空间交错布局,以墙围院,以院破墙,从而在狭小的用地内争取外部空间。院与墙的结合融合了中国传统建筑的精髓,以现代的手法体现传统空间。建筑整体形态完整统一,以安静祥和的姿态出现在充满商业意味的南京西路,与周边建筑形成强烈的对比和反差,从而突出建筑的文化形象。

Situated at the bustling West Nanjing Road in Shanghai, Shanghai Chess Academy sits on a long and narrow site: 140m long from north to south and 40m wide from east to west. The void and real spaces are staggered together from the inside to the outside. Gardens are separated by walls, while breaching the walls at the same time, striving for the exterior space within the limited site. The essence of Chinese traditional architecture is marked by fusing gardens and walls and the architecture recreates traditional space with modern approach. The existence of gardens fills the whole building with Chinese implication while creating an integral form. Presenting itself quietly and peacefully at the busy West Nanjing Road with an obviously commercial flavor, it segregates itself from the surrounding buildings and highlight its own cultural image.

主入口
东南透视

剖面 0 5 10 20m

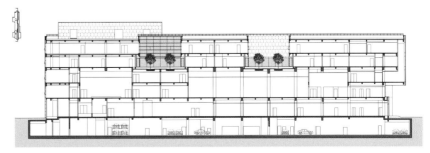

Public Architecture　　　Cultural Architecture / Sports Architecture

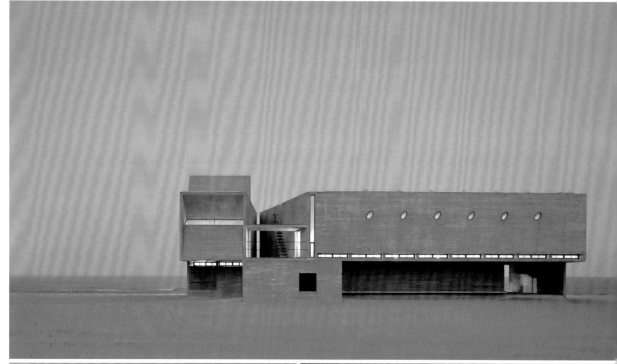

西立面
西南视角

阅读空间

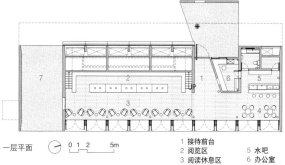

一层平面　　0 1 2　5m

1 接待前台
2 阅览区
3 阅读休息区
4 休息区
5 水吧
6 办公室
7 室外空间

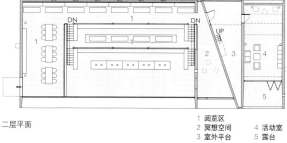

二层平面

1 阅览区
2 冥想空间
3 室外平台
4 活动室
5 露台

海边图书馆
Seashore Library

建筑设计
直向建筑
董功

河北 秦皇岛　　竣工时间 2015 年　　摄　影　苏圣亮 夏至 何斌

Architects
Vector Architects
DONG Gong
Location Qinhuangdao, Hebei
Completion 2015
Photo SU Shengliang, XIA Zhi, HE Bin

阅读空间

该设计的主要理念在于探索空间的界限、身体的活动、光氛围的变化、空气的流通以及海洋的景致之间的共存关系。图书馆东侧面朝大海，在春、夏、秋三季服务于西侧居住区的社区居民，同时免费向社会开放。设计是从剖面开始的，图书馆由一个主要的阅读空间、一个冥想空间、一个活动室和一个小的水吧休息空间构成。建筑师依据每个空间功能需求的不同来设定空间和海的具体关系，来定义光和风进入空间的方式。海，气象万千，随着季节的交替和时光的流动不断演变，像是一出以自然为主题的戏剧。于是建筑师把最重要的阅读空间理解为一个"看台"，逐渐升起的阶梯平台会让空间中不同位置的人更不受阻拦地看到海的景象。弧线的屋顶朝海的方向张开，暗示着空间的主题。

The key points of the design focused on exploring the co-existing relationship of space boundary, the movement of human body, the shifting light ambience, the air ventilation through, and the ocean view. The library faces the ocean to its east. During seasons of spring, summer and fall, it not only serves the community residents at west, but opens to the public as well. The design began with section. The library houses a reading area, a meditation space, activity room, a drinking bar and a resting area. The distinctive relationship between space and the ocean, and how light and wind enter into each room are defined according to the functional need of each space. The Ocean, an ever-changing character continues to alter from season to season, and from morning to night. It is like a drama play of nature. As if giving the stage to this character of ocean, the architects pile up seating platforms raised toward the back, so that everyone has an unblocking view to the stage. The half-arched roof opens toward the sea and implies the main subject of the space.

从冥想空间洞口看向阅读空间
冥想空间

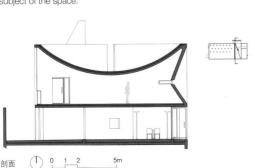
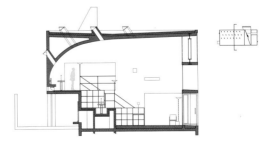

剖面

围合院落夜景

融入环境的建筑群

整个建筑群落围绕1个主广场和3个院落展开。主广场是建筑群落的中心，它依次和西院、北院和东院相通。建筑形成连续的半围合面，向广场和院落展开。院落的半围合状态和高宽比，保证了在此休息和交流的人的舒适度。建设地块是一处背靠山峦的景观用地，内有水塘，植被丰富。为最大限度地保护山峦天际线和基地自然肌理，将服务中心设计成一个由独立的低层单体组成的建筑聚落。各单体通过联接和围合，形成系列的院落空间。而在垂直维度，建筑的架空让院落景观和基地的山水在水平向得以交融。

The entire group of buildings surrounds a main square and three courtyards. The main square is the center of the group, interlinked with the rest of sections on west, north and east. The building forms a continuous semi-enclosed surface, extending towards the square and courtyard. The semi-enclosing condition and ratio of height to width of the courtyard ensure the comfort of people resting and communicating here. The site is a landscape land backing to the mountains, dotted with ponds and vegetation. In order to best preserve the skyline and the natural texture of the site, the service center is designed as a cluster of individual buildings of low profile. Each unit is connected to form a series of enclosed courtyard spaces. The overhanging part of the building allows the courtyard landscape and the site landscape to be integrated in the horizontal dimension.

东湖国家自主创新示范区公共服务中心
Donghu National Innovation District Civic Center

湖北 武汉	竣工时间 2015 年	摄 影 姚力 章勇

建筑设计
中信建筑设计研究总院有限公司
这方建筑师事务所

汤群 杨勇凯 赵仲贵

Architects
CITIC General Institute of Architectural Design and Research Co.,Ltd.;
Architects Zephyr(US) Architects P.C.
TANG Qun, YANG Yongkai, ZHAO Zhonggui
Location Wuhan, Hubei
Completion 2015
Photo YAO Li, ZHANG Yong

东侧立面和景观
围合院落

外墙细部

二层平面　0 5 10 20m

1 服务大厅
2 办公
3 门厅
4 档案馆
5 展厅
6 会议厅
7 食堂
8 镜面水池
9 下沉庭院
10 风雨连廊
11 健身中心

绿化广场鸟瞰
覆土屋顶夜景

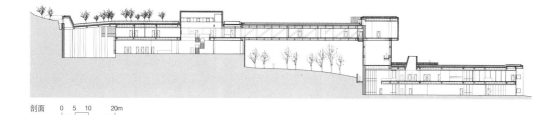

剖面　0　5　10　20m

重庆桃源居社区中心
Chongqing Taoyuanju Community Center

建筑设计
直向建筑
董功

Architects
Vector Architects
DONG Gong
Location Chongqing
Completion 2015
Photo SU Shengliang

重庆　　　竣工时间 2015 年　　　摄　影　苏圣亮

从绿化广场看向体育中心
从景观连廊看向绿化广场

混凝土雨篷之下的坡地花园

社区中心位于重庆市桃园公园半山腰上的一块洼地，四周被起伏的山形围合。设计所面对的首要问题是新的空间建造以什么样的态度介入到原始地貌之中。非同于"在山上盖一幢房子"的通常做法，建筑师希望塑造一个建筑整体趋势和山体相融合的景观意象。新的空间形体依顺着山势而规划，它们的轮廓线和原来的山形融汇相映。整体建筑采用绿色植被屋顶和局部的垂直绿化墙体，进一步强化了建筑与自然山体共存的想法。社区中心包括文化中心、体育中心、社康中心三个基本功能。建筑整体呈环状布局，两个庭院空间被围合于其中，一个是坡地花园，另一个是可以容纳社区生活与集会活动的绿化广场。两个庭院与建筑四周有多条线路相连接，并通过大尺度的洞口与架空，从视线和流线两个层面，将内外空间紧密联系在一起。

The community center is located in the mountains of Taoyuan Park in Chongqing. The starting point is attempting to merge new building outline with the existing wavy topography. Instead of building an "object" in the field, the architects hope to create an imagery of fusing architectural form and hilly landscape together. Green roof and green walls assist to blend the volume into its natural environment, and enhance the thermal co-efficiency of building envelope. The relationship of the "in and out" of architecture spaces is an important aspect in our design as well. Cultural center, athletic center and public health center are the three major programs. A continuous roof connects the three independent buildings into one unified volume. There are multiple paths connecting two courtyards and perimeter of the building. They relate the inside and outside of architecture closely in both visual association and physical connection by large openings and spans.

1 室外平台
2 入口大厅
3 大堂
4 休息区
5 灌溉区
6 实验室
7 教室
8 舞蹈室
9 音乐室
10 体育馆
11 多功能室
12 更衣室
13 羽毛球场
14 餐厅

二层平面　0　10　20　50m

从场地入口看建筑全景
从游客中心看生态馆

罕山生态馆和游客中心
Hanshan Ecological Hall and Tourist Center

内蒙古 通辽 竣工时间 2015 年

建筑设计 内蒙古工大建筑设计有限责任公司 张鹏举

摄影 张广源 张鹏举

Architects Inner Mongolian Grand Architecture Design Co., Ltd.
ZHANG Pengju
Location Tongliao, Inner Mongolia
Completion 2015
photo ZHANG Guangyuan, ZHANG Pengju

项目位于内蒙古通辽市北部罕山林场的入口处，建筑按功能分为博物馆和游客中心两个体量，它们前后错位，分置于两个小山坡上。该地寒冷，阳光和风成为设计的核心要素，同时保护自然环境是设计的切入点。综合应对这些要素和目标，自然平实地生成了建筑形态的基本策略：体量背坡面阳，后部埋入坡内，形体沿等高线顺山体层层而退，表皮材料则是挖方后析出的碎石。由此，建筑融入了场地，保护了自然，节约了造价。

Located at the entrance of Hanshan Forestry Range in the north of Tongliao, Inner Mongolia, the project is divided into two parts: a museum and a tourist center. The two parts are separated on two hill slopes side by side. Due to the cold climate in this region, sunlight and wind prevention become two priorities in design, and environmental preservation is another factor considered. Taking all these factors into consideration, a unique design plan is made. While allowing the sunshine to go through the front of the building and burying the back under the slope, the design enables the building shape to suit the contours of the hill. The materials on surface are made from the crushed stones collected from the foundation work. In this way, the building becomes an organic part of nature which is protected at low cost.

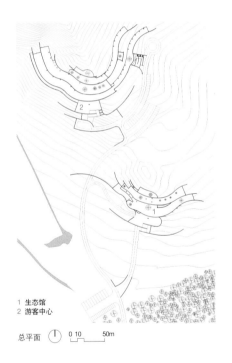

1 生态馆
2 游客中心

总平面 0 10 50m

生态馆局部
游客中心夜景
生态馆展厅

游客中心全景

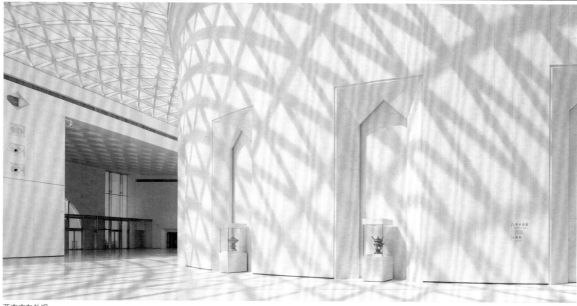

西南方向外观
阳光中庭

大厂县是邻近北京的一个回族聚居地。大厂民族宫是一个文化综合体，包括剧场、展览、会议、社区中心等功能。建筑以传统的清真寺为原型，通过新的材料和技术，以微妙的方式来演绎清真寺的空间结构：四周环绕的拱券从下到上逐渐收分形成优雅的弧线。当整栋建筑倒映水中时，弧形的花瓣形拱券更显清晰灵动，散发出优雅气质。建筑的穹顶不是简单复制传统的符号，而是一种抽象和转译。以一系列花瓣状的壳体构成穹顶，创造性地将内部空间变为半室外的屋顶花园，融合了阳光、空气和绿色。

Dachang County is the settlement for ethnic Hui people near Beijing. Da Chang Muslim Cultural Center was designed as a cultural complex putting functions of theater, exhibition, conference and community center together. Prototyped on a traditional mosque, the building subtly interprets the spatial structure of the mosque with new materials and technologies. The surrounding arches tapered along elegant curves from the base, while the arc petal-shaped arches with reflection in water look vivid and graceful. The dome is used as a translation and abstraction of cultural symbols rather than simple mimicking. The architects constitute the dome with petaloid shells and creatively transform the interior space into a semi-exterior roof garden with sufficient sunshine, fresh air and vegetation.

大厂民族宫 / Da Chang Cultural Center

建筑设计 华南理工大学建筑设计研究院
何镜堂 郭卫宏 盘育丹

河北 廊坊 竣工时间 2015 年 摄影 姚力

Architects Architectural Design & Research Institute of South China University of Technology
HE Jingtang, GUO Weihong, PAN Yudan
Location Langfang, Hebei
Completion 2015
Photo YAO Li

公共建筑 文化/体育

南立面

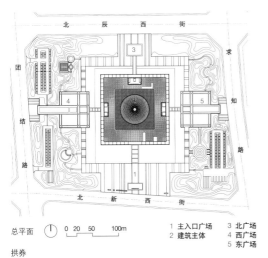

总平面　0 20 50 100m

1 主入口广场
2 建筑主体
3 北广场
4 西广场
5 东广场

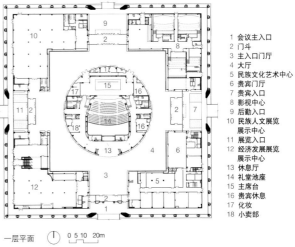

一层平面　0 5 10 20m

1 会议主入口
2 门斗
3 主入口门厅
4 大厅
5 民族文化艺术中心
6 贵宾门厅
7 贵宾入口
8 影视中心
9 后勤入口
10 民族人文展览展示中心
11 展览入口
12 经济发展展览展示中心
13 休息厅
14 礼堂池座
15 主席台
16 贵宾休息
17 化妆
18 小卖部

拱券

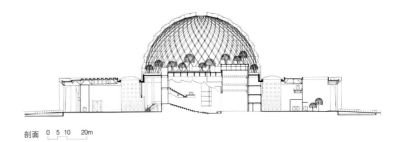

剖面　0 5 10 20m

梅驿广场侧影

咨询服务厅室内

一层平面 0 10 20 50m

1 停车场
2 元丘
3 环道
4 景观水面
5 竹岛
6 办公
7 商店
8 西广场入口
9 快餐店
10 咖啡厅
11 亲水木平台
12 断桥
13 码头
14 瑞龙桥

街子古镇梅驿广场
Jiezi Ancient Town Meiyi Square

四川 崇州 | 竣工时间 2015 年

建筑设计
北京华清安地建筑设计有限公司成都分公司
蒲建聿 刘伯英 蒲兵

摄 影 存在建筑

Architects
Beijing Huaqing An-design Architects Co., Ltd., Chengdu Branch
PU Jianyu, LIU Boying, PU Bing
Location Chongzhou, Sichuan
Completion 2015
Photo Arch-Exist Photography

断桥入口

　　街子古镇地处成都市西57公里的群山之麓，自五代建置至今一千多年历史。梅驿广场，是古镇的中心。梅驿广场由游客中心及配套功能建筑围合形成。总平面根据环境、地形和建设体量，顺应场地条件采取围合、台级式布局。建筑设计取古镇传统民居形式为意向，以山、水、古镇的相邻关系和新城区的城市设计为尺度和背景。建筑以山水为宗，以内观的空间自制力和形式语言的简净，寻求谦和虚澹，与环境气象相和。竹，作为本地区最日常的景观和生产、生活要素，以质朴、平实的材料气质，体现在建筑肌理的主题中。景观设计以水面、元丘和梅林，体现场所人文情怀的清静内敛、灿然成象。

Located at the foot of mountains some 57km west of Chengdu, Jiezi ancient town has a history of more than one thousand years since first establishment. Meiyi Square is the center of the historic town.
Meiyi Square is surrounded by the tourist center and supporting functional buildings. According to the environment, terrain and construction volume, the plan is arranged in line with site conditions and an enclosed and terraced layout is adopted. The traditional residential houses in the historic town are taken as the design concept, and the adjacent relationships between mountains, waters and the town and design of new urban area are studied for scale and background. The theme of the buildings hinges on mountains and waters, seeking humility by internal self-controlled space and simplicity of architectural languages to be harmonious with the environment. Bamboo, as the region's most common landscape and a production and livelihood element, is embodied in the theme of architectural texture with plain and simple temperament. Water, hills and plum forests are featured in landscape design, reflecting the introverted humanities and brilliance.

游客中心通廊
西入口柱廊

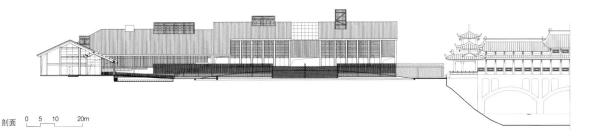

剖面　0　5　10　20m

游泳馆室内

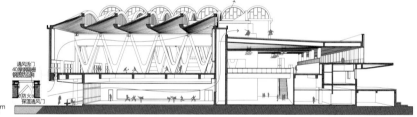

东西剖透视　0　2　5　10m

天津大学新校区综合体育馆，位于校前区北侧，包含体育馆和游泳馆两大部分，以一条跨街的大型缓拱形廊桥将两者的公共空间串通。各类室内运动场地依其对平面尺寸、净空高度及使用方式的不同要求紧凑排列，并以线性公共空间迭加、串联为一个整体。一系列直纹曲面、筒拱及锥形曲面的钢筋混凝土结构，带来大跨度空间和高侧窗采光，在内明露木模混凝土筑造肌理，在外形成沉静而多变的建筑轮廓。极限运动区通过不规则铺展的室外台阶看台，可以一直延伸到公共大厅波浪形渐变的直纹曲面形屋面。东侧长达140米的室内跑道，为大厅带来凸显屋面形状的自然光线和向远处延伸的外部景观。如此成为一个室内与室外、地面与屋面联为一体的"全运动综合体"。

The Gymnasium on the New Campus of Tianjin University is located on the north side of the front part of the campus. The public space consists of a gymnasium and a natatorium, connected by a huge slight-arched corridor. The various kinds of indoor sports venues are compactly arranged according to different sizes, clearance height and usage, and strung up and overlapped by linear public spaces. A series of rulesurfed, barrel-vaulted and conformed reinforced concrete structure result in the large-span space and clerestory daylighting, exposing the texture of concrete made of timber-formwork in the interior and presenting the silent and dynamic silhouette on the outside. The extreme sports area spreads from the irregular exterior steps and terraces to the gradually changed rulesurf of the corrugated roof of public hall. The 140m indoor runway brings in the daylight portraying the roof curve, and also shows the outdoor scenery that extends to the afar. This is how the indoor and outdoor, the ground and roof join in a "Comprehensive Sports Complex".

天津大学新校区综合体育馆
Gymnasium of New Campus of Tianjin University

建筑设计 中国建筑设计院有限公司
李兴钢 张音玄 闫昱

Architects China Architecture Design Group
LI Xinggang, ZHANG Yinxuan, YAN Yu
Location Tianjin
Completion 2015
Photo SUN Haiting, ZHANG Qianxi, ZHANG Guangyuan

天津　　竣工时间 2015年　　摄影　孙海霆 张虔希 张广源

西南侧鸟瞰

篮球场直纹曲面屋顶
游泳馆入口中庭
公共大厅波浪形屋面

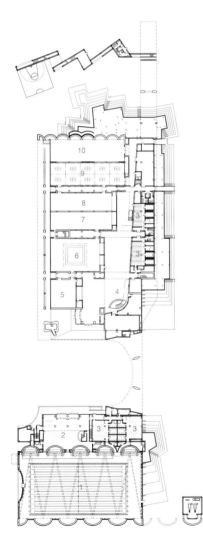

1 游泳馆　　6 跆拳道馆
2 游泳馆门厅　7 体操馆
3 更衣室　　8 健身馆
4 门厅　　　9 乒乓球馆
5 多功能厅　10 武道馆

一层平面　　0 5 10 20m

公共建筑　文化／体育

建筑东侧入口形态

项目位于内蒙古盛乐古城博物馆已经建成的场地中。游客中心位于场地东北角，与入口一侧的附属用房以一种抱合的方式组织布局，设计顺势打散体量，取得与既有建筑尺度上的统一，进而利用场地的高差将建筑嵌埋其中。新建筑与既有建筑拉开间距，在解决采光通风问题的前提下，形成线状分布且有高差的内院。最终，建筑外形方正简明，与原博物馆取得性格上的一致，建筑内院则丰富生动，充满市井生活的气息，延伸了人的游园体验，照应了游客中心这一主体功能。

This building is located on the northeast corner of the site of Inner Mongolia Shengle Museum, embraced by auxiliary buildings at the entrance. The design is made to reduce the mass by making the best use of the existing conditions. In this way, the dimensions of different buildings can complement one another and conceal the visitor center among other tall buildings. The distance between this new building and the existing buildings is enlarged to facilitate the effect of ventilation and illumination, forming a courtyard with a liner plan taking full advantage of height differences. With rich clues of daily life, the new building obtains the same character of the old museum and extends the experience of a museum to the visitor center, with attention paid to its main function.

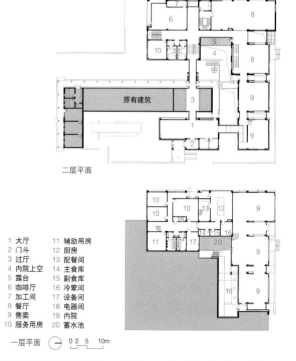

二层平面

1 大厅　　　11 辅助用房
2 门斗　　　12 厨房
3 过厅　　　13 配餐间
4 内院上空　14 主食库
5 露台　　　15 副食库
6 咖啡厅　　16 冷荤间
7 加工间　　17 设备间
8 餐厅　　　18 电器间
9 售卖　　　19 内院
10 服务用房　20 蓄水池

一层平面　　0 2 5　　10m

盛乐遗址公园游客中心
Tourist Center of Shengle Heritage Park

	建筑设计 内蒙古工大建筑设计有限责任公司 张鹏举	**Architects** Inner Mongolian Grand Architecture Design Co., Ltd. ZHANG Pengju **Location** Hohhot, Inner Mongolia **Completion** 2015 **Photo** ZHANG Guangyuan
内蒙古 呼和浩特 ｜ 竣工时间 2015 年	摄　影 张广源	

庭院由西向东局部

外部院落围合关系

局部透视
北侧入口室内

外景

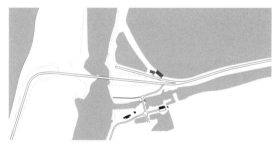

位置示意

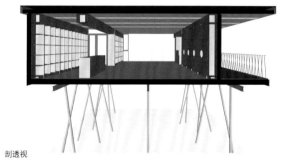

剖透视

南矶湿地访客中心位于鄱阳湖主湖区南部，处在赣江北支、中支和南支三大支流汇入鄱阳湖开放水域冲积形成的三角洲前缘。受鄱阳湖夏洪冬枯的水文节律影响，访客中心建设地点平均每两年有一年被水淹没，少则0.5米，多则2～4米不等，淹没时间一般在数周到两个月不等。针对这一独特的自然地理条件，访客中心建筑采用了钢管架空式基础，确保建筑主体不受洪水侵害之外，也在枯水期提供一个公众停留休憩的独特场所。建筑主体采用大型木制楼板及屋面板吊装，以轻巧结构建造出150平方米的开敞展厅，所有屋面板由特制的家具柜单元组合支撑，实现建筑整体坚固稳定的同时，也提供了各式实用的收纳展示空间。屋顶板组合成的屋面平台整体高出湖床约10米，为访客提供了一个登高望远的理想场所。

The Nanji Wetland Visitor Center sits in a nature reserve on the front edge of where Poyang Lake and three tributaries of Gan River meet. Affected by Poyang Lake, the construction site is inundated by tidal water regularly from half a meter to 2-4m every other year. Each flooding period lasts from several weeks to 2 months. The visitor center is lifted up from the ground by steel—pipe stilts that not only protect the building from flooding, but also create a public space for rest. The main building is composed of large wooden insulation sandwich panels, providing an exhibition space of 150m². The large roof panels are supported by specifically designed furniture modules, which integrate structural purpose with storage function together. The roof is designed to provide a flat platform. The platform rises up from the lake at 10m, providing an ideal place of fantastic views.

南矶湿地访客中心
Nanji Wetland Reserve Visitor Center

建筑设计
深圳元远建筑科技发展有限公司
香港中文大学建筑学院
朱竞翔 吴程辉 韩国日

Architects
Unitinno Architectural Technology Development Co., Ltd.;
School of Architecture, The Chinese University of Hong Kong
ZHU Jingxiang, WU Chenghui, HAN Guori
Location Nanchang, Jiangxi
Completion 2015
Photo WU Chenghui

江西 南昌 | 竣工时间 2015年 | 摄 影 吴程辉

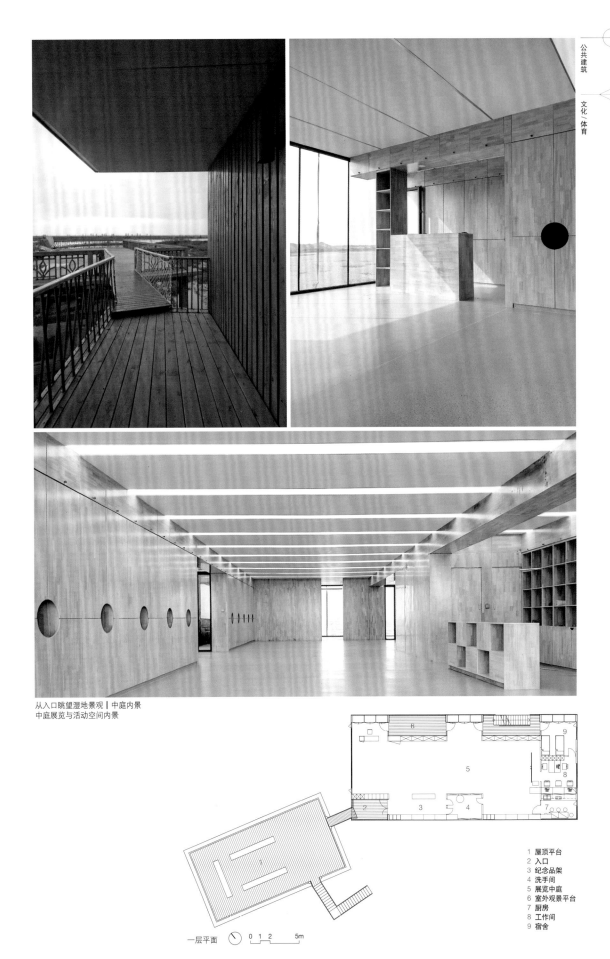

从入口眺望湿地景观 | 中庭内景
中庭展览与活动空间内景

一层平面

1 屋顶平台
2 入口
3 纪念品架
4 洗手间
5 展览中庭
6 室外观景平台
7 厨房
8 工作间
9 宿舍

抬升的建筑和场地
局部鸟瞰

南京牛首山景区游客中心
Tourist Center in Niushou Scenic, Nanjing

建筑设计
东南大学建筑设计研究院有限公司
王建国 朱渊 吴云鹏

江苏 南京　　竣工时间 2015年　　摄　影 许昊皓

Architects
Architects & Engineers Co., Ltd. of Southeast University
WANG Jianguo, ZHU Yuan, WU Yunpeng
Location Nanjing, Jiangsu
Completion 2015
Photo XU Haohao

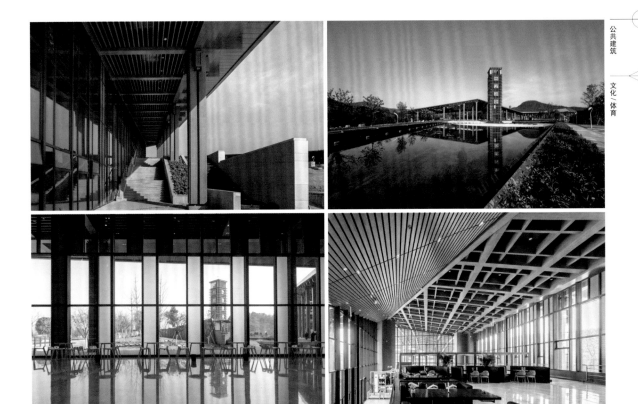

檐下空间
室内局部

水池与建筑
茶室内景

牛首山景区是南京市"十二五"期间的重大文化项目，以长期安奉释迦牟尼佛顶骨舍利闻名。本项目是牛首山东麓入口处的标志性建筑，既是景区接待量最大的游客中心，也作为公共广场为城市服务。

设计根据场地地形标高变化，采用两组在平面上和体型上连续折叠的建筑体量布局，高低错落、虚实相间。起伏的屋面和深灰色钛锌板的使用，是对山形的呼应和江南建筑气质的演绎。设计在审美意象上考虑了佛祖舍利和牛首山佛教发展的年代属性，总体抽象撷取简约唐风，并在游客的路线设计上融入禅宗文化要素，回应了公众心目中所预期的集体记忆。建筑功能包括售票、电瓶车换乘、展览、小型放映、售卖、办公及停车库等。两组建筑围合出的公共空间从城市道路延伸至景区内部，不同层次的场所设计兼顾了参禅人流的礼仪性空间和市民休闲的亲和性空间。

The Niushou Scenic Area is one of the major cultural projects of Twelfth Five-Year Plan in Nanjing, renowned for long for the placement of Buddhist relics. This building is a landmark at the east entrance of the Niushou Scenic, both as a tourist center for most visitors and as a public plaza serving the city.

According to the variance of height on site, two groups of builds continuously folded both in layout and form are arranged. The corrugated roof and the use of dark grey titanium-zinc panels echo to the profile of mountains and the spirit of exquisite Jiangnan architectures. Considering the age of the Buddhist relics and Buddhist history in this region, the design adopts a simplified Tang dynasty style, and combines cultural elements of Zen in designing tourist routes, a respond to the collective memory of the public. The functions of the building include ticket selling, electric car transfering, exhibitions, video halls, souvenir shops, offices and parking. The public space guarded by the two groups of building extend from the city road to the inner space of the scenic. Site planning on different height levels takes both religious worshiping and normal public leisure into account in the same time.

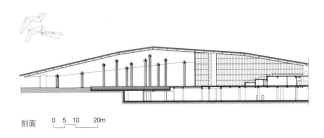

剖面   0 5 10   20m

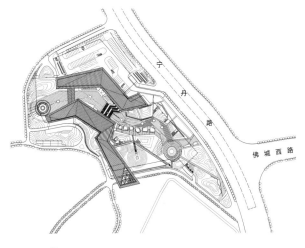

总平面   0 20 50   100m

鸟瞰

武汉光谷国际网球中心 15 000 座网球馆作为武汉国际网球公开赛"WTA 超五巡回赛"中心场馆，是华中地区首个跨度最大、可开合屋盖网球馆。其总体规划利用现有地形高差将建筑融入自然环境中。建筑外表皮由 64 根空间弯扭构件自下向上扭曲倾斜，似飞速旋转的网球，并使"气旋"钢结构、玻璃幕墙、遮阳与照明系统结合，室内空间纯粹通透，体现"旋风球场"的魅力。最先进的可开启屋面和充裕的内场尺寸使之成为全天候多功能场馆。通过采用先进的节能和雨水回用系统、大范围的预制构件使建筑对环境的影响降至最低，同时与全三维参数化设计和工业化预制紧密结合，实现较高的建筑完成度。

As the main venue of Wuhan international tennis tournament for "WTA Super Five Tour", the Wuhan Optics Valley International Tennis Center of 15 000 seats is the first tennis stadium with retractable roof of large span in the Central China. The overall planning uses the existing terrain characteristics to blend the building with the natural environment. The outer surface of the building consists of 64 spatial curved shafts twisted clockwise like a tennis ball rotating at full speed, making a perfect combination of the air-twisted steel structure, glass curtain wall, sun shading system and lighting system. As a result, the pure space reflects the charm of the "whirlwind" of stadium. The most advanced system of the retractable roof and the ample indoor space make the stadium an all-weather multi-purpose stadium. Prefabrication for most parts minimizes negative impacts on the environment, and a combination of the advanced full three-dimensional parametric design and industrial prefabrication guarantees the completion of the construction on an extreme high level.

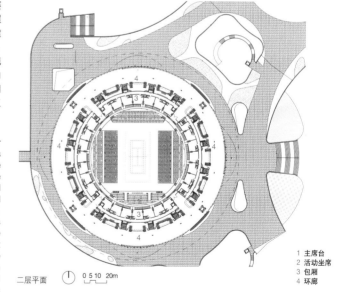

二层平面

1 主席台
2 活动坐席
3 包厢
4 环廊

武汉光谷国际网球中心 网球馆
Wuhan Optics Valley International Tennis Center, The Tennis Court

湖北 武汉 | 竣工时间 2015 年

建筑设计
中信建筑设计研究总院有限公司
陆晓明 叶炜 姜瀚

摄 影 李扬 章勇

Architects CITIC General Institute of Architectural Design and Research Co., Ltd.
LU Xiaoming, YE Wei, JIANG Han
Location Wuhan, Hubei
Completion 2015
Photo LI Yang, ZHANG Yong

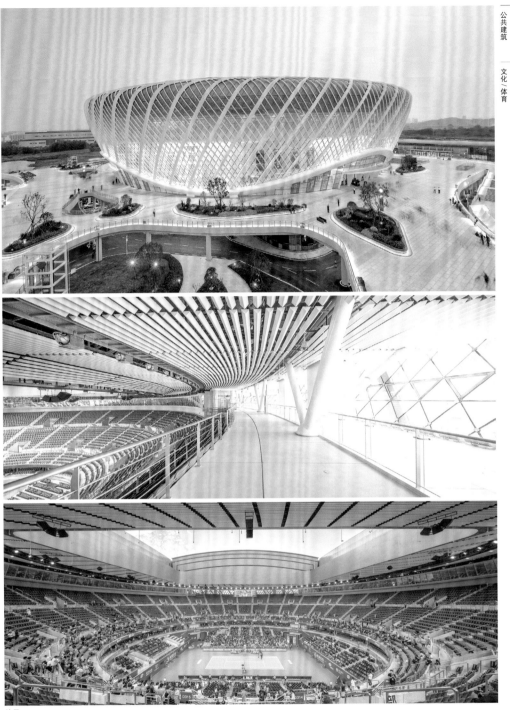

外景
环廊
比赛场地与看台

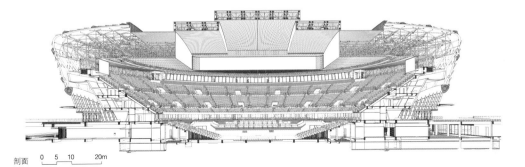

剖面　0　5　10　20m

码头之上的书屋

本项目是利用江边的一座废弃码头建造一个书屋。书屋是工业码头改建而成的滨江公园的重要组成部分，其目的不仅是为市民提供阅读书籍和观景的场所，同时也保留人们对场所的历史记忆。码头采用毛石和混凝土砌筑，长宽约40m×14m，设计中32m×14m的书屋主体钢结构框架被6根钢筋混凝土柱子支撑在码头上，大尺度的结构出挑既形成了强烈的空间张力，也切实降低了对码头的破坏程度，同时避免建筑受江水袭扰。架空建筑形体的底层为原有码头空间，中层是书屋的阅览空间，顶层是观江平台。粗糙的毛石、硬朗的钢、温暖的竹三者之间的平衡关系表达出当下中国人的自然观。

This project is a library built on an abandoned quay by the riverside. The library is an important part of the riverside park which was converted from an industrial wharf. The purpose of the library is not only to provide a place to read books and viewing for people, but also to preserve the historical memory of the place. The quay was constructed of rubble and concrete. It is 40m long and 14m wide. The architects designed a 32m×14m meters steel structure which is supported by 6 reinforced concrete columns on the quay. At the bottom of the building, there is the quay space. In the middle of the building, there is the reading room of the library. On the top of the building, there is the platform that could see the whole riverside. The rough rubble, hard steel and soft bamboo which reflect the Chinese's natural view at this moment.

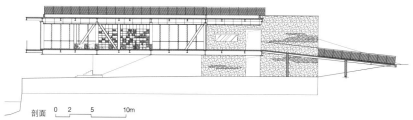

剖面 0 2 5 10m

码头书屋
Library on the Quay

安徽 铜陵 | 竣工时间 2015年

建筑设计
东南大学建筑设计研究院有限公司建筑技术与艺术（ATA）工作室
李竹 王嘉峻

摄 影 钟宁

Architects
Architecture Technology and Art (ATA) Studio, Architects & Engineers Co., Ltd of Southeast University
LI Zhu, **WANG** Jiajun
Location Tongling, Anhui
Completion 2015
Photo ZHONG Ning

外立面材料组合
栈道入口

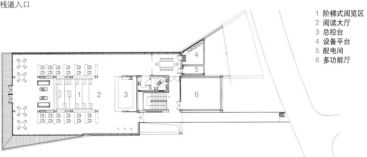

1 阶梯式阅览区
2 阅读大厅
3 总控台
4 设备平台
5 配电间
6 多功能厅

一层平面　0　2　5　10m

台阶阅览区

天水屋顶平台
二层西侧外观

地池二层商业局部

"天水"和"地池"是百果山上的两个原有湖泊，两个服务中心也因为分别坐落在两个湖边而得名。作为园博会园区内的主要建筑，承担着人流集散、活动集聚、餐饮、休闲景观、文化传播、展示等多项功能。由于建筑性质及地理位置等的特殊性，在设计中需要处理好建筑、环境、人的关系。通过合理地利用地形高差，将建筑与环境作为一个整体设计，功能按照不同标高分区设置，尽量减小建筑体量的同时获得最佳的景观朝向。最大限度地保留原有地形地貌和原有植被。

为了使游客在方便到达的同时拥有丰富的观览体验，设计中应用了多路径游览系统。

"Tianshui" and "Dichi" are the two existing lakes on Baiguo Mountain, and the two service centers are named because of their lakeside locations. As the primary buildings of the Expo area, they contain multiple functions including hub of circulations, activity center, restaurant, recreational landscape, cultural communication, exhibition, etc. Considering its distinctiveness by nature and its geographical location, the relations of architecture, the environment and human should be well dealt with. Taking advantage of the height differences, the architect combine buildings and the environment in design as a whole, setting several functional areas according to various heights to obtain best views with least architectural volume. Original terrain and vegetation have been preserved as much as possible. A system of multiple routes is adopted to enrich tourist experience and enhance easy accessibilities.

2014青岛世界园艺博览会天水、地池综合服务中心
Tianshui/Dichi Service Center of International Horticultural Exposition 2014

山东青岛　　竣工时间 2014 年

建筑设计
华汇设计（北京）
青岛北洋建筑设计有限公司
王振飞　王鹿鸣　李宏宇

摄　影　王振飞　济南多彩摄影

Architects HHD_FUN; Beiyang Design Group
WANG Zhenfei, WANG Luming, LI Hongyu
Location Qingdao, Shandong
Completion 2014
Photo WANG Zhenfei, Duocai Photo

地池西北角下沉广场
地池鸟瞰（北侧）

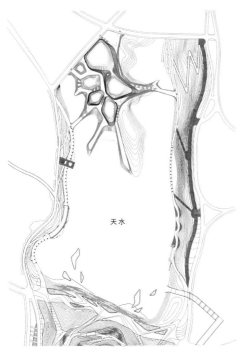

"天水"总平面　　0 10　50m

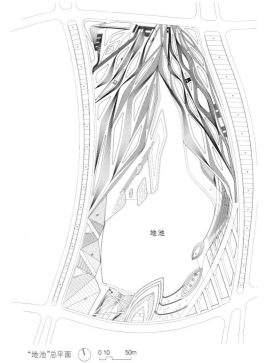

"地池"总平面　　0 10　50m

入口夜景

项目位于南京滨江风光带万景园段内,是一个面积仅200平方米的小教堂,由南京金陵协和神学院的牧师主持,满足信众的聚会、婚礼等功能。这个钢木结构的小教堂具有平和的外形与充满神秘宗教力量的内部空间、质朴的材料和精致的构造逻辑,设计周期仅一个月,并在45天内完成建造。

The project—a 200m² small chapel, is located in Wanjing Garden along Nanjing's riverfront. Hosted by the priests from Nanjing Union Theological Seminary, it houses religious activities such as worshiping and wedding services. Made of simple materials and ingenious construction logics, the chapel in wood and steel structure has an ordinary appearance and mysterious interior space of religious power. It took only one month for design, and it has been built in 45 days.

大厅室内

南京万景园小教堂
Nanjing Wanjing Garden Chapel

江苏 南京 | 竣工时间 2014 年

建筑设计
张雷联合建筑事务所
南京大学建筑规划设计研究院有限公司
张雷

摄影 姚力

Architects
AZL Architects; Institute of Architecture Design and Planning Co., Ltd, Nanjing University
ZHANG Lei
Location Nanjing, Jiangsu
Completion 2014
Photo YAO Li

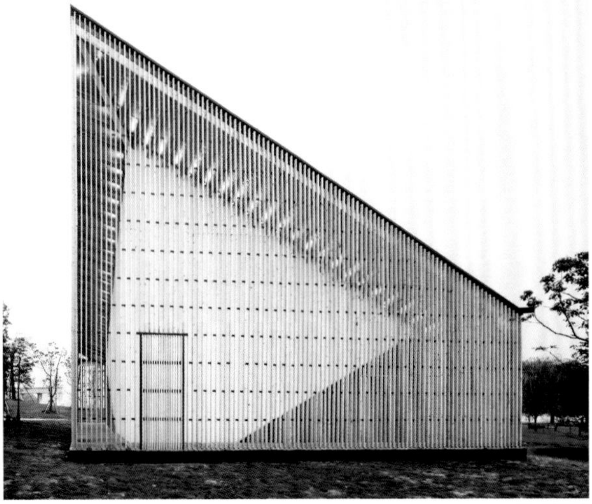

东北立面｜回廊
沿湖外景｜大厅室内

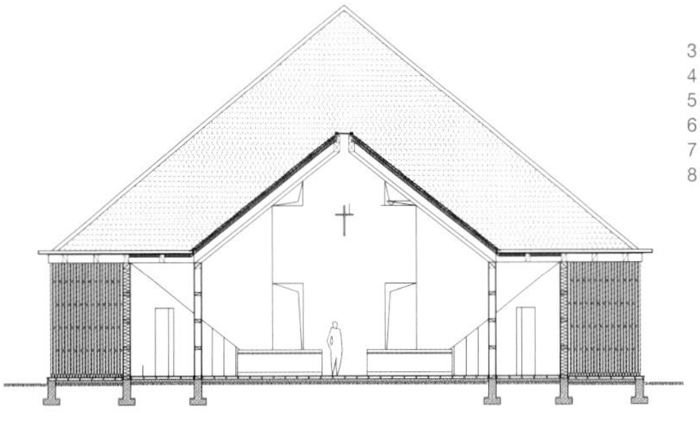

1 38/89mmSPF 木格栅
2 外墙构造：
　15mm 石膏板
　墙体龙骨
　15mm 石膏板
3 38/89mmSPF 木十字架
4 隐框玻璃
5 白色教堂椅
6 白色木地板
7 原色木地板
8 φ70/230mm 黑色壁灯

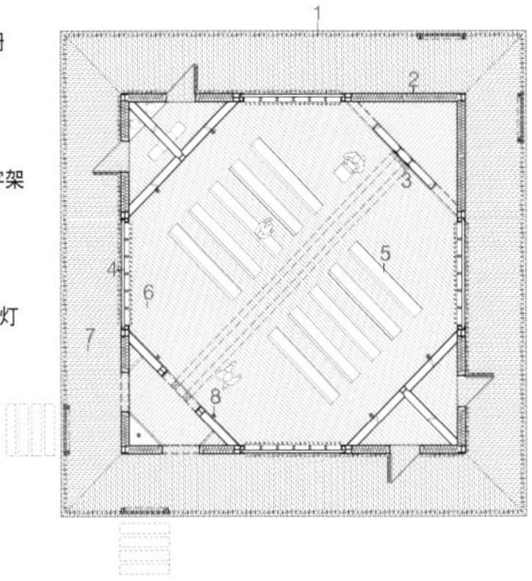

剖面　0 1 2　5m

平面　0 1 2　5m

西侧外观
东南侧外观

项目位于田园综合体一期开发的示范区。田园生活馆的设计目标，是根植于阳山丰富的自然及人文资源，寻求新建筑与原有环境的协调共生，为"当代田园"这一新型生活方式提供与之相适应的空间形态。田园生活馆的空间构成源于传统建筑的空间原型——庭园空间，而围合庭园的建筑南北通透、东西实墙。建筑的东西侧外墙设计提取了地域的象征元素——桃花，烘托"桃文化"主题，带来浓郁的地域文化气息。桃花图案的镂空铝板的运用，使建筑表情丰富，白天与夜晚呈现不同效果。

The pavilion is located in the area of the first phase of Yangshan Rural Life Complex development demonstration area. Seeking coordination and symbiosis between new building and the original environment by rooting in the rich natural and cultural resources of Yangshan, it aims at creating space to adapt to the new life style of contemporary garden. The spatial structure of the Rural Life Pavilion originates from the prototype of space in traditional architecture—the garden enclosed by permeable north-south walls and solid east-west walls. The design of the eastern and western walls is inspired by a symbolic local element—peach blossom, which highlights the theme of "peach culture" with rich cultural implications. Through the application of pierced aluminum panels with peach patterns, the building has various expressions in day and night.

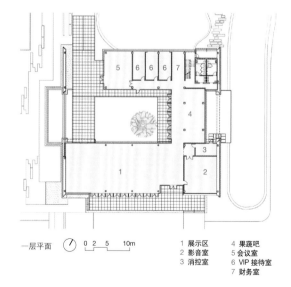

一层平面 0 2 5 10m

1 展示区　4 果蔬吧
2 影音室　5 会议室
3 消控室　6 VIP接待室
　　　　　7 财务室

无锡阳山田园综合体I期 田园生活馆
The 1st Phase of Yangshan Rural Life Complex in Wuxi, Rural Life Pavilion

江苏 无锡　　竣工时间 2014年

建筑设计
联创国际设计集团
东南大学
钱强

摄影 姚力

Architects
United Design Group Co., Ltd.; Southeast University
QIAN Qiang
Location Wuxi, Jiangsu
Completion 2014
Photo YAO Li

西南侧外观
庭园东望
西山墙

文化馆及图书馆侧内街入口
从内街看大剧院

康巴艺术中心
Kangba Art Center

建筑设计
中国建筑设计院有限公司
崔愷 关飞 曾瑞

青海 玉树藏族自治州   竣工时间 2014年   摄影 张广源 关飞

Architects
China Architecture Design Group
CUI Kai, GUAN Fei, ZENG Rui
Location YuShu, Tibetan Autonomous Prefecture, QingHai
Completion 2014
Photo ZHANG Guangyuan, GUAN Fei

大剧院室外院落环廊　　　　　　　　　　　　从康巴艺术中心内街看格萨尔王雕像

文化馆演艺厅

总平面　0 20 50 100m

1 剧场
2 图书馆
3 文化馆
4 剧团
5 电影院
6 剧场前广场
7 停车场
8 唐蕃古道商业街
9 格萨尔广场
10 小学

　　该项目是玉树灾后重建项目，援建资金来自社会援助，功能由玉树州剧场、剧团、文化馆、图书馆组成。设计主要特点：1) 尊重城市文脉：布局自由松散但错落有致，强调与结古寺、唐蕃古道商业街、格萨尔广场等周边要素的呼应，密度上与传统城市肌理吻合。2) 功能复合使用：尝试将州剧场与县剧场合并，将州剧团的辅助功能区与大剧院的后台合并，将排练厅与室外演艺区合并，注重各功能区的通用性。3) 院落和台地：通过院落空间的组合体现藏式建筑的空间精神，并尝试在建筑形态上体现台地特征。4) 材料构造：外墙装饰材料通过混凝土空心砌块自由叠砌，构造上与传统石墙相契合，再现藏式建筑粗犷的表面肌理的同时，减轻了构造难度，并降低了造价。

The Kangba Art Center is a reconstruction project in the aftermath of Yushu earthquake, funded by social assistances. It's a comprehensive building including a theater, a theatrical troupe office, a cultural building and a library. The main characteristics of the design include: 1) Respect of the cultural context of Yushu County: the architectural density of Kangba art center mutually observes the norm of the traditional city, and the layout is free in plan but connected closely to the surrounding landmarks, such as Jiegu temple, Tangbo Gudao commercial street and Sa'er Square, etc.; 2) Composite functions: the prefecture theater is combined with the county theater, merging the service area of theatrical troupe office with the that of performance with an emphasis on common use of each function; 3) Courtyard and Plateau: the arrangement of courtyards bespeaks the essence of Tibetan architecture and embodies the characteristics of plateau buildings through architectural form; 4) Material and Construction: Concrete hollow blocks are used as the main material on the facade, piled up freely by workers, which successfully represents the roughness of traditional Tibetan architecture. In this way, it reduces both the degree of difficulty of construction and construction cost.

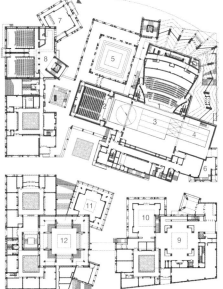

一层平面　0 5 10 20m

1 大剧院
2 多功能剧场
3 主舞台
4 侧台
5 半室外演艺
6 票务厅
7 电影院门厅
8 电影院大厅
9 期刊阅览
10 儿童阅览
11 展厅
12 共享大厅

主入口夜景

盛泽文化中心
Shengze Cultural Center

建筑设计
同济大学建筑与城市规划学院
苏州九城都市建筑设计有限公司
李立

江苏 吴江　　竣工时间 2014 年　　摄　影 姚力

Architects
College of Architecture and Urban Planning, Tongji University;
Suzhou 9 Town Urban Architecture Design Co., Ltd.
LI Li
Location Wujiang, Jiangsu
Completion 2014
Photo YAO Li

西南侧外景
南向外景

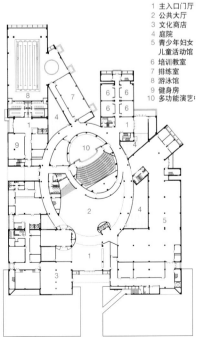

1 主入口门厅
2 公共大厅
3 文化商店
4 庭院
5 青少年妇女儿童活动馆
6 培训教室
7 排练室
8 游泳馆
9 健身房
10 多功能演艺厅

一层平面

0 5 10 20m

中央大厅

盛泽文化中心是包含各类文化设施的文化综合体，涵盖了博物馆、图书馆、科技馆、青少年活动中心、广电中心、体育健身等多种功能，建筑设计的难点在于如何处理复杂功能的不同空间需求并合理构建具有地方特色的建筑形象。建筑总体围绕内部的椭圆形小剧场呈U形环绕布局。建筑主体北侧呈发散状的多个延伸空间，在结构和功能上与建筑主体功能融为一体。建筑外立面的设计灵感来自江南传统砖雕艺术，错落的民居山墙以及经典的小桥流水布置，是江南传统元素在现代建筑形式下的表达，并且各元素都承担着建筑内部的特定功能，是内部空间外在的延伸。

Shengze Cultural Center is a cultural complex containing various cultural facilities, including museum, library, science and technology museum, youth activity center, radio and television center, sports fitness and many other functions. The difficulty of design is how to deal with various spatial requirements of different functions with local characteristics. The architecture surrounds a U-shaped small elliptical theater in the interior. A series of radiated spaces is settled in the north part of the main architecture, integrated with the building's main structure and functions. The design of facades is inspired by the traditional brick art in the lower Yangtze River region in addition to elements in traditional settlements such as scattered gable walls and bridges. The expression of these traditional elements is under the guise of modern architecture, representing specific functions within the building as an extension of the external space.

西北侧鸟瞰

本工程是目前我国最大规模的冰上运动建筑综合体，由速滑馆、冰球馆、冰壶馆及媒体中心、运动员公寓等功能组成。设计从新疆自然环境与历史文脉入手，以天山雪莲与丝绸之路为创作主题，将体育建筑与自然景观有机融合，塑造了鲜明的当代体育建筑地域特色，营造了丝路花谷的建筑群体意象。设计中采用围合式的建筑布局抵御冷风侵袭，利用高效导风的流线型屋盖形态减少屋面雪荷，优化建筑形体实现节能降耗。功能布局与交通流线充分考虑建筑之间的便利联系以及赛事的合理组织，全面提供灵活的赛时及赛后空间环境，创造多样化的活动空间。在建筑、结构、设备、材料等方面充分应用新技术，致力于实现本项目的绿色建筑目标。

The project is currently China's largest ice sports complex building, including speed skating hall, ice hockey hall, curling hall and media center, athletes apartment and other functions. Designing perposals start from the Xinjiang natural environment to historical context. With Tianshan snow lotus and Silk Road the theme of creation, the sports architecture and the natural landscape organicly integrate with each other, shaping the distinctive characteristics of contemporary sports architecture, and creating a Silk Road Valley building groups. The design uses enclosed building layout to resist the cold wind invasion, tand streamline roof to reduce the snow load as well as efficiently guiding the wind the building shape to achieve energy saving. Convenient links between buildings and rational organization of the event are taken into account when it turns to functional layout and traffic flow line design spatial flexibility during and after each match, creating a variety of activities space. In the construction, structure, equipment, materials and other aspects of design, new technologies are fully applied, aiming at achieving the project's green building goals.

1 大道速滑馆
2 冰球馆
3 冰壶馆
4 餐厅宿舍
5 媒体中心及组委会办公
6 足球训练场
7 停车场
8 篮球场
9 广场中心景观

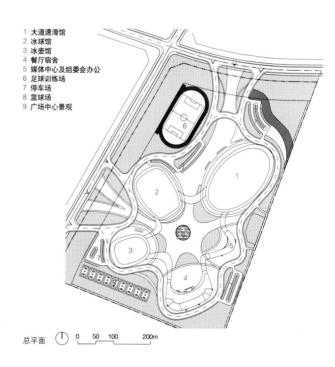

总平面

第十三届全国冬季运动会冰上运动中心
Ice Sports Center of the 13th National Winter Games

新疆 乌鲁木齐 | 竣工时间 2014 年

建筑设计 哈尔滨工业大学建筑设计研究院 梅洪元

摄影 韦树祥 存在建筑

Architects Architecture Design and Reasearch Institute of Harbin Institute of Technology MEI Hongyuan
Location Urumqi, Xinjiang
Completion 2014
Photo WEI Shuxiang, Arch-Exist Studio

整体外观

冰球馆
大道速滑馆入口
大道速滑馆室内

从南向园林看建筑
东北向鸟瞰

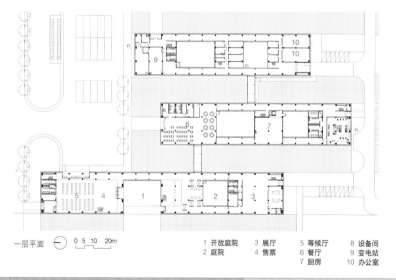

南侧湿地内院

一层平面 0 5 10 20m

1 开放庭院　3 展厅　　5 等候厅　8 设备间
2 庭院　　　4 售票　　6 餐厅　　9 变电站
　　　　　　　　　　　7 厨房　　10 办公室

黄河口生态旅游区游客服务中心
Visitor Center of Ecological Tourism Area at Delta of Yellow River

山东 东营 ｜ 竣工时间 2014 年

建筑设计：同济大学建筑设计研究院（集团）有限公司麟和建筑工作室

李麟学 刘旸 周凯锋

摄影 苏圣亮

Architects Atelier L+, Tongji Architectural Design (Group) Co., Ltd.
LI Linxue, LIU Yang, ZHOU Kaifeng
Location Dongying, Shandong
Completion 2014
Photo SU Shengliang

入口庭院
主入口门厅

从湿地看建筑

　　黄河口生态旅游区游客服务中心力图通过地方气候响应进行建筑自然系统的建构，对于环境要素与能量流动的关注是促成形式、格局、空间、与材料建构的基础。多层夯土材料的建造实验，不仅是环境技术的创新，更是当代建筑自然诗意的呈现。黄河口生态旅游区位于黄河入海口处的黄河三角洲国家自然保护区内，游客服务中心就位于该保护区的入口区域。建筑在体量、景观、公共空间与流线多方面与周边地貌和景观和谐共存。服务中心是集接待、展览、候车、餐饮、办公、会议于一体的综合性服务设施，设计以3个体型平展的建筑错落布置，融入广袤的黄河湿地，6个院落嵌入其中，并通过巨大的开口与地景尺度相呼应。回应地域气候，最大程度减少能耗，这也是该项目的设计目标。

Visitor Center of Ecological Tourism Area at Delta of Yellow River tries to construct a nature-based architectural system through responding to the local climate. The consideration for enviromental elements and energy flow is the motivation of the form, layout, space and materials. The built experiment of multi-stories' rammed earth is not only the enviromental technique creation but also the presentation of natural poetry of contemporary architecture. Yellow River Ecological Tourism Zone is located in the Yellow River Delta State Reserve, at whose entrance sits the Visitor Center. As a complex building consisting of reception, exhibition, bus waiting, restaurants, offices and conference rooms, the center is integrated with the natural surroundings in terms of volume, landscape, public space and circulations. Three horizontally spread buildings blend into the wild wetland of the Yellow River, connected by six courtyards, which corresponds to the scale of landscape of the site by huge openings. Another primary objective in design studies how to respond to the climate and minimize energy consumption.

Public Architecture   Cultural Architecture / Sports Architecture

东南向鸟瞰

体育场内景

1 主体育场　　7 排球场
2 体育馆　　　8 足球训练场
3 游泳馆　　　9 棒球场
4 综合训练馆　10 篮球场
5 热身训练场　11 体育场入口广场
6 网球场　　　12 市民健身广场
　　　　　　　13 停车场

总平面　　0　5　10　20m

　　淮安体育中心位于淮安城市规划中的新城西区，引入体育中心与新城肌理相融入、与未来社区生活相衔接的城市设计理念，建筑师将其定位为多功能混合的运动型城市活力中心。

　　设计避免采用将体育场馆作为"纪念物"的单一做法，而是让体育场与体育馆和游泳馆肩负不同的城市责任。体育场位于场地中心，作为周边街道的视觉焦点，满足市民与决策者对新城区标志性形象的需求。同时，体育馆与游泳馆则作为城市肌理的延伸部分，贴近城市街道布局，以创造面向社区的多元生活界面，力图融入未来的城市街区，更好地满足社区化的服务需求。体育馆与游泳馆整合成综合体，并通过共用功能用房减少了总投资，同时将用于体育运动的设施面积最大化。建筑形象构思取"水波"之意，并与结构形式相契合，结合屋面层叠造型设置天窗，为室内提供自然采光通风，节约建筑运营能耗。

Huai'an City Sports Center (HCSC) is located in the west part of the planned Huai'an New Town. Based on the concept of the sports center blending in the context of new town land linking up with future community life, the project aims as a urban lively center with mutiple functions integrated.

A didactic approach to monumentalizing sports facilities is avoided. Instead, the main stadium, the gymnasium and the aquatic center are designed to play different urban roles respectively. The stadium sits at the center of the site, becoming a visual focus and landmark of the new town. The key of the design is that the gymnasium and aquatic center serve as the extension of urban fabric, adjacent to streets gently and creating dynamic interfaces with communities. As such, the design integrates with future urban communities and better matches requirements of local communities. The gymnasium and aquatic center are designed as a whole complex, sharing auxiliary rooms to reduce construction cost and maximize the areas for sports facilities. The design is inspired by the image of "wave" in accordance with structural performance. Louvers are placed in combination with construction layers on the roof, providing natural lighting and ventilation to reduce energy consumption.

淮安市体育中心
Huai'an City Sports Center

建筑设计
华南理工大学建筑设计研究院
孙一民 陶亮 叶伟康

江苏 淮安　　竣工时间 2014 年　　摄影 谢光源 王祥东

Architects
Architectural Design and Research Institute of South China University of Technology
SUN Yimin, TAO Liang, YE Weikang
Location Huai'an, Jiangsu
Completion 2014
Photo XIE Guangyuan, WANG Xiangdong

体育中心外景
体育场内景

体育场内景

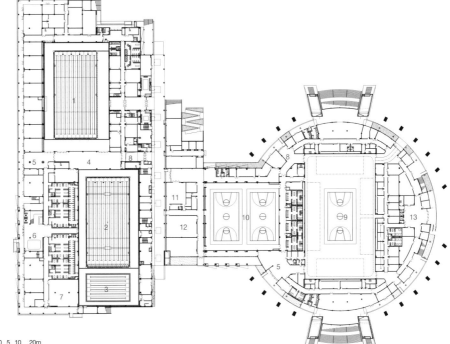

1 比赛池
2 训练池
3 儿童池
4 赛时检录区
5 运动员及随队官员门厅
6 运动员及日常运营门厅
7 健身房
8 新闻媒体门厅
9 比赛场地
10 热身训练馆
11 新闻发布厅
12 健身房
13 贵宾门厅

体育馆游泳馆首层平面　　0 5 10 20m

在本案中，建筑师借助简洁的立方体盒子进行堆叠，希望能够尽量淡化建筑形象，突出市民广场。从使用方面来看，建筑可以从任何方向进入。建筑底层在沿广场一侧布置商业店铺和服务设施等，方便市民使用。此外大部分底层空间是架空的，将场地还给市民。人们既可以利用架空空间遮阳避雨、组织活动，也可以从广场散步到建筑的庭院中来。设计选择金属铝板和玻璃作为不同盒子的材质，建筑主体的二层和三层由穿孔铝板整体围合，柔和地强化了整体感和体量感，加上首层的架空设计，建筑的轻盈感自然而生。铝板的穿孔形式是通过计算机辅助生成流水的肌理，最终效果若隐若现，恰到好处。

In this case, the architect uses the simple cube to pile up boxes, hoping to downplay the image of the building and highlight the public square. Visitors can enter the building from any direction. At the bottom of the building, commercial shops and service rooms are arranged on the ground floor for easy access, and most of the ground floor is open to the public with stilted structure. People can use this space for protection from rain, for other activities, or as a path of stroll to the courtyard. Aluminum panel and glass are selected as enclosing materials for boxes. The second and third floors of the main building are surrounded by perforated aluminum plates, strengthening the overall sense of entity and volume in a gentle way. Combined with the overhead space on the ground, the lightness of the building is obvious. The form of perforation of the aluminum plate is generated by the computer-assisted technology imitating the texture of flowing water, which gives the architecture looming temperament perfectly.

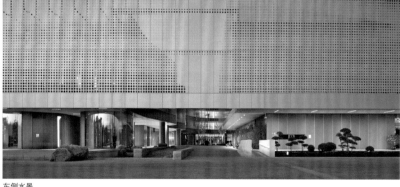

东侧水景
建筑南向入口

天津武清文化中心
Wuqing Cultural Center, Tianjin

建筑设计
天津华汇工程建筑设计有限公司
周恺

天津 | 竣工时间 2014 年 | 摄 影 周恺 魏刚

Architects
Huahui Architectural Design & Engineering Co., Ltd.
ZHOU Kai
Location Tianjin
Completion 2014
Photo ZHOU Kai, WEI Gang

公共建筑 文化／体育

南侧外景
内庭水景

图书馆门厅

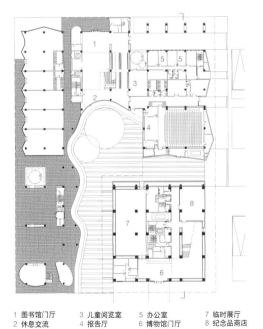

1 图书馆门厅　3 儿童阅览室　5 办公室　7 临时展厅
2 休息交流　　4 报告厅　　　6 博物馆门厅　8 纪念品商店

一层平面　　0 2 5　10m

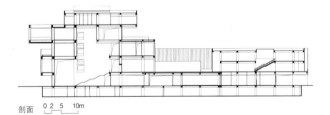

剖面　0 2 5　10m

内院

鸟瞰

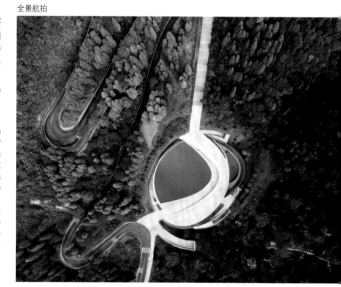

全景航拍

位于娄山关景区入口处的陈列馆既是一个独立的展馆，又是游客步行进入娄山关景区的必经之路。建筑的大部分埋入大地，仅以一道简洁有力的弧墙置于群山之中，限定出建筑的场所。延续自场地的迂回转折的缓坡将场地入口至上山步道的整个路径串联起来。应对山区潮湿多雨的气候，在场地的西南侧设置浅池收集雨水，周边的苍山倒映在水中，如镜的水面恬静温柔，与周围若斧似戟的大山形成鲜明对比，塑造出安宁而肃穆的纪念场所气氛。

Sited at the entrance of Loushanguan Scenic Area, the exhibition hall is not only an independent building but also an inevitable path for visitors to enter the whole area. Most of the building is buried beneath the ground, only a concise and powerful arc wall is placed amongst the mountains, indicating the perimeters of the site. The gentle slopes winding through the site connects the entire path form the site entrance to the hiking trail. In response to the local humid and rainy weather, a shallow pool on the southwest side of the site is placed to collect rainwater and reflect the surrounding mountains. The surface of the water is as smooth as a mirror, quiet and mild, in sharp contrast to the mountains as if carved by axes, creating a peaceful and solemn memorial atmosphere.

遵义市娄山关红军战斗遗址陈列馆
The Site Museum of Loushanguan Battle, Zunyi

贵州 遵义　　竣工时间 2017 年

建筑设计　同济大学建筑设计研究院（集团）有限公司　任力之

摄影　章勇

Architects Tong Ji Architectural Design (Group) Co., Ltd. REN Lizhi
Location Zunyi, Guizhou
Completion 2017
Photo ZHANG Yong

公共建筑 观演／博览

主入口坡道水景
入口空间外景

1 入口门厅
2 贵宾室
3 室外平台
4 室外楼梯
5 下沉庭院
6 售票
7 主入口室外广场
8 景观水池
9 室外广场
10 柴发机房

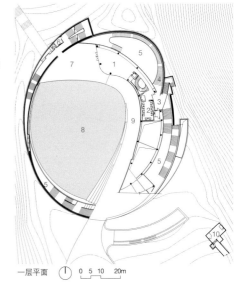

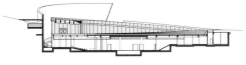

剖面 0 5 10 20m

一层平面 0 5 10 20m

下沉庭院步道　　　　　　　　　　　　　　　下沉庭院

东向鸟瞰

世博会博物馆由上海市政府和国际展览局合作共建，是世博专题博物馆。世博会博物馆将全面综合地陈列展示世博会160多年历史、2010年世博会盛况，以及2010年以后各届世博会的情况。

方案设计以"承载欢乐记忆的容器"为主题，将历史的河谷与欢庆之云两个不同色彩、材质及功能的形体进行组合，形成丰富的半室外空间，拓展博物馆的外部展场区域，形成与世博会主题相符的特质空间。"欢庆之云"造型取云的形态所代表的未来、开放、瞬间的寓意，整体形态简洁流畅，在建筑底部以三条云柱扭转收分，形成逐级上升的动势，具有强烈的未来感。在建筑中部，云柱扭转展开，并相互联系，逐渐于建筑顶部形成一个完整的曲线流畅的"欢庆之云"轮廓。"历史河谷"造型取河谷的形态所代表的历史、冥想、永恒的寓意，整体形态纯粹有力。主体建筑外观为矩形，通过切削，形成不确定的建筑形式，中部的峡谷将建筑内部形式多样化，打破固定的空间模式，具有强烈的雕塑感。

Established by joint efforts of Shanghai municipal government and the Bureau of International Expositions, World Expo Museum is a unique world expo thematic museum that comprehensively displays the history of world expositions in the past 160 years, spectacular events of the World Expo 2010 and the future expositions after 2010.

This project is designed with the theme of "a container carrying all happy memories of mankind", combining two parts of "Historical Valley" and "Jubilant Cloud" of different colors, materials and functions to form an abundant semi-outdoor space and extend outdoor exhibition areas echoing to the main theme of the Expo. "Jubilant Cloud" imitates the shape of clouds, which represents the meaning of future, openness and moment. The overall form is simple and smooth, and the bottom of the building is twisted by three cloud-like columns to form the momentum of rising, which has a strong sense of future. In the middle of the building, the cloud columns turn around and connect with each other, gradually form a complete curved line of 'Jubilant Cloud'. 'Historical Valley' imitates the shape of valley, symbolizing the meaning of history, meditation and eternality. The whole form is pure and powerful, and the profile of the main building is rectangular with an uncertain building form through cutting the mass. Central canyons diversify the internal area of the building, which breaks the fixed spatial pattern with a strong sense of sculpture.

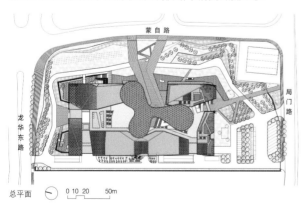

总平面 0 10 20 50m

上海世博会博物馆
Shanghai Expo Museum

上海 竣工时间 2017年

建筑设计
华建集团华东建筑设计研究总院
汪孝安 杨明 刘海洋

摄影 邵峰

Architects
East China Architectural Design & Research Inistitute
WANG Xiao'an, YANG Ming, LIU Haiyang
Location Shanghai
Completion 2017
Photo SHAO Feng

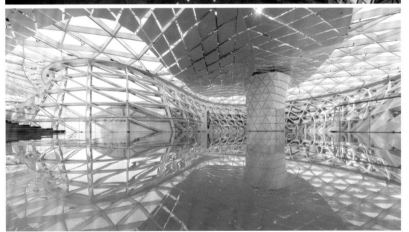

南向入口
公共区室外庭院
云厅室内

剖面 0 2 5 10m

从亲水平台看向连廊和庭院

苏州非物质文化遗产博物馆坐落于苏州吴中区园艺博览会会址东侧。这个地点周围相对开阔的自然环境和苏州本地的文化及空间意象给设计带来了最初的启发。设计策略是按照功能的区别把此项目的综合体量拆分，并置入不同的院落，然后利用风雨连廊让他们彼此联接。在江南多烟雨的气候条件下，人们可以通过连廊在不同的院落和体量之间随意移动而不必担心多变天气的影响。此外，每个院落各自的主体性空间都拥有不同的功能和特征。建筑师利用覆土遮掩住了博物馆的建筑体量，进一步强化了建筑与自然的融合。苏州非物质文化遗产博物馆在给游客展示大量文化知识的同时，也给予他们丰富的互动体验。它附带的三个小庭院除了能引入自然采光和通风，也会指引游客登上屋顶的展览平台。人们在享受户外美景的同时，也可以在此举办各种活动。

Suzhou Intangible Cultural Heritage Museum is located at east part of the Suzhou Horticultural Exposition Park. The site is surrounded by river on its three sides, and the design is inspired by its natural environment and the culture of the city—Suzhou. The strategy is connecting different scattered functions by courtyards and outdoor corridors in order to generate the spatial experience of local building types. In consideration of local climates, it allows visitors to walk through those outdoor corridors between different courtyards when raining. The main spaces located in the courtyards are designed with identities. Meanwhile, the majority of the volume is covered by earth, facilitating the integration with nature. To meet the program requirements and reduce the negative impact on natural environment, the majority of the volume is covered by green roof. It amplifies the theme of blending the boundary between architecture and nature. The Suzhou Intangible cultural heritage museum offers people large amount of information and interactive experiences during their visit. Three atriums create natural ventilation and lighting, and direct visitors to the roof exhibition platform. It forms a public park with various types of vegetation where outdoor performances, dining events as well as educational and interactive experience can take place.

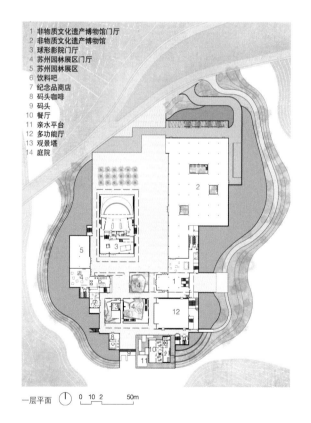

1 非物质文化遗产博物馆门厅
2 非物质文化遗产博物馆
3 球形影院门厅
4 苏州园林展区门厅
5 苏州园林展区
6 饮料吧
7 纪念品商店
8 码头咖啡
9 码头
10 餐厅
11 亲水平台
12 多功能厅
13 观景塔
14 庭院

一层平面

苏州非物质文化遗产博物馆
Suzhou Intangible Cultural Heritage Museum

江苏 苏州　　竣工时间 2016 年　　建筑设计 直向建筑　　董功　　摄影 陈颢 Eiichi Kano

Architects Vector Architects
DONG Gong
Location Suzhou, Jiangsu
Completion 2016
Photo CHEN Hao, Eiichi Kano

公共建筑 观演／博览

覆土屋顶公园

观景塔南立面

餐厅楼梯

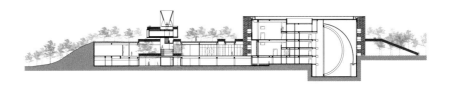
剖面　0　5　10　20m

剧院鸟瞰

延安大剧院建于延安北部新城。把延安传统建筑元素窑洞作为大剧院的主要设计元素，赋予其地域特色；建筑以简洁的方形体量相互穿插生成：两侧各11道拱门如厚重基石承托起屋顶，它们通过轻盈的玻璃体量衔接，形体间虚实结合。21米的大悬挑空间自然形成主入口。大剧院的核心公共空间是7道大拱组成的公共大厅，在功能上承接剧场三大厅，在空间上作为主入口的大拱门的延续，无需太多人工装饰，便产生令人震撼的感觉。窑洞符号合理的再生赋予了建筑更高层次的价值。建筑外立面主要应用艺术混凝土、玻璃、金属3种材质：艺术混凝土有着黄土大地的味道，肌理贴近自然，如同沟壑，使建筑如从大地中生长出来；屋顶的造型轻盈，凭添现代开放气息。

The project is located in the Northern New Area of Yan'an. The traditional building element of this region—cave dwelling is used as the key element for the design. The architectural form is generated by spatial organization: 11 arches on both sides support the roof as a heavy base, with an effect of a combination of the void and the solid by light glass connecting the two parts. The 21m main entrance comes into existence naturally due to huge cantilever space. The central space of the grand theater is a public hall consisting of 7 arches as a continuation of the three theaters functionally, and an extension of the main entrance spatially. A striking impression is achieved with no excessive artificial ornaments. The reasonable regeneration of the symbol of cave dwelling endows value of a higher level to the building. Artistic concrete, glass and metal are used on the facades. Artistic concrete has the flavor of the earth with natural texture as ravines, which makes the building look as if arising from the earth. The shape of the light and graceful roof also adds a modern and open flavor to the building.

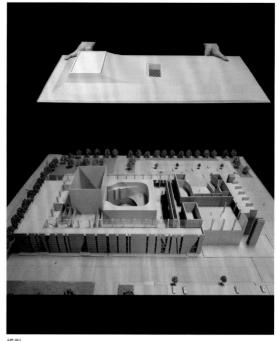

模型

延安大剧院
Yan'an Grand Theatre

陕西 延安　竣工时间 2016年

建筑设计
中国建筑西北设计研究院有限公司
赵元超 李强 李彬

摄　影 陈溯

Architects
China Northwest Architecture Design and Research Institute Co., Ltd.
ZHAO Yuanchao, LI Qiang, LI Bin
Location Yan'an, Shaanxi
Completion 2016
Photo CHEN Su

公共建筑 观演／博览

剧院主入口
拱形公共大厅

音乐厅

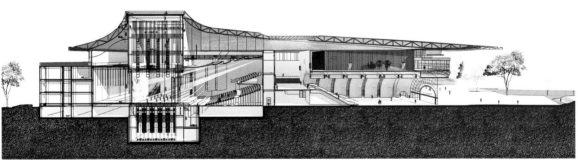

剖面

1 拱形大厅
2 门厅
3 戏剧厅
4 临时展厅
5 大剧场
6 候场
7 贵宾接待

一层平面　0 5 10 20m

主入口夜景

宜昌规划展览馆结合周边山体环境，充分利用场地长度和宽度，将形体展开呈起伏之势，在呼应延绵山体的同时，更增加了空间的丰富性和流线的变化。随着标高的不同，分别可以达到室内庭院、大展厅、二层平台和屋顶花园。

建筑形态与周边地貌相呼应，仿佛层峦叠嶂的山体。景观设计延续山体形象这一母题，用大面积斜面草坡与建筑呼应。游人行走于建筑内外，犹如爬山观景，别有一番趣味，由此形成人与建筑、建筑与环境的对话。建筑主体呈现出银白色光泽，俯瞰时宛若墨绿色山地之中烘托的玉石，从而体现出"行走宜昌，夷陵拾玉"的设计主题。

The design of Yichang Planning Exhibition Hall takes the surrounding mountain environment into account and takes advantage of the length and width of the site to stretch out its form in response to continous mountains, enriching spatial diversities and circulation changes. On different height levels, visitors can approach to the interior courtyard, the large exhibition hall, the platform on the second floor and a roof garden.

The architectural form echoes to the surrounding landscape as if overlapped peaks of the mountain. Landscape design continues with the motif of an image of the mountain, using a large area of grass and building ramps to speak to the building. Visitors walk in and outside the building are like climbing and sightseeing, hence dialogues between human and architecture, and architecture with the environment are formed. The main building gives a silver-white luster, like a jade within the dark green mountains, indicating the theme of "walking in Yichang, collecting jaded in Yiling".

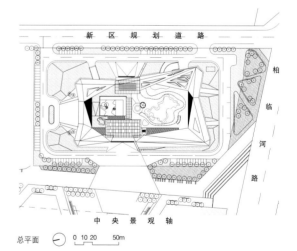

总平面

宜昌规划展览馆
Yichang Planning Exhibition Hall

建筑设计
华东建筑设计研究院有限公司
华东都市建筑设计研究总院
孙晓恒 丁蓉 张冉

湖北 宜昌　竣工时间 2016 年　摄影 胡义杰

Architects
East China Architectural Design & Research Institute Co., Ltd.
East China Urban Architectural Design & Research Institute
SUN Xiaoheng, DING Rong ZHANG Ran
Location Yichang, Hubei
Completion 2016
Photo HU Yijie

东立面远景
内庭院

百叶细部

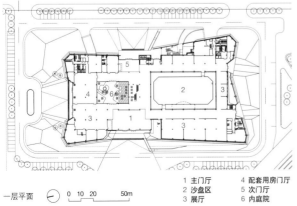

一层平面　0 10 20 50m

1 主门厅　　4 配套用房门厅
2 沙盘区　　5 次门厅
3 展厅　　　6 内庭院

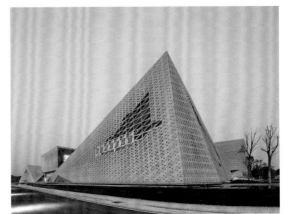

西北角外景

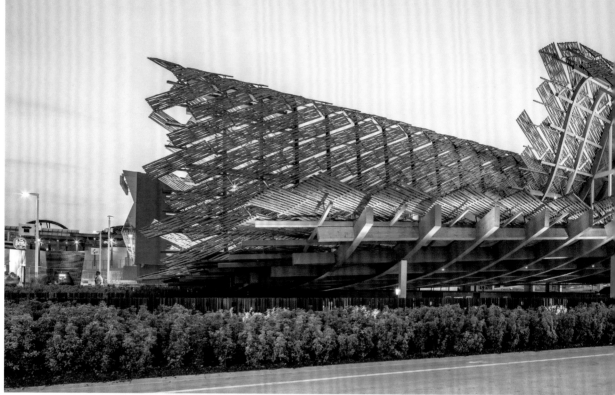

东南侧视角

2015米兰世博会中国馆是中国首次以独立自建馆的形式赴海外参展的世博会场馆，其设计理念来源于对本次世博会主题"滋养地球、为生命加油"及中国馆主题"希望的田野，生命的源泉"的理解和思考。建筑师在面对场地南侧主入口和北侧景观河的两个主立面分别拓扑了"山水天际线"和"城市天际线"的抽象形态，并以"loft"的方式生成了展览空间；最后在南向主立面上，推出3个进深不同的"Deep Facade"，形成"群山"的效果，以此向中国传统的抬梁式木构架屋顶致敬。

为了实现轻盈的屋面并满足大跨度的内部展览要求，中国馆创造性地采用了以胶合木结构、PVC防水层和竹编遮阳板组成的三明治开放性建构体系。作为中国传统建筑文化的一个当代表达，中国馆采用胶合木与钢的混合结构来实现大跨度的展览空间。屋面主体结构由近40根南北向的结构檩条和37根东西向的异型木梁结合组成，其形成的1400个不同的内嵌式胶合木节点是结构设计与施工工艺的完美结合。位于屋面最上层的是由竹条拼接的板材所组成的遮阳表皮系统。光线透过竹编表皮漫射进中国馆室内空间，在PVC表皮上布下了斑驳的投影。建筑师希望通过这个造化自然的"空"来表达属于中国的空间品质。

China Pavilion at the 2015 Milano Expo is the first building built individually abroad in international expositions. The concept of the Pavilion comes from understandings and reflections of the main theme of the Expo—"Feeding the Planet, Energy for Life" and embodies the project's theme—"The Land of Hope and Origins of Life." Through manipulations of two main facades of the main southern entrances and northern landscape waterfront, the profile of a city skyline is merged with the profile of a rolling landscape. Exhibition space is generated according to the loft pattern. Three "Deep Facades" of different depth are arranged on the main elevation facing south, in imitation of mountains, paying homage to traditional Chinese timber structure.

Designed as a freeform timber structure, the Pavilion roof uses advanced technologies such as a sandwich open system consisting of laminated panels, PVC waterproof layer and bamboo sunshade to create a large-span exhibition space. As a modern representation of traditional Chinese architectural culture, a mixed system of laminated wood and steel is used for the space. The main roof structure consists of about 40 north-south structural rafters and 37 west-east non-standard wooden beams, and the resultant 1400 joints bespeak the accomplishment of perfect combination of structural design and construction techniques. Shingle bamboo strips on the top of roof form a sunshade system. Light penetrates into the space through the sunshade system, casting mottled shadows on the PVC facade. The architects hope to convey the quality of Chinese space through the void in man-made nature.

2015米兰世博会中国馆
China Pavilion for Expo Milano, 2015

意大利 米兰 | 竣工时间 2015年

建筑设计
清华大学美术学院
Link-Arc建筑事务所
陆轶辰

摄影 吕恒中, Hufton Crow, Sergio GraZia, Roland Halbe

Architects
Academy of Art & Design; Tsinghua University & Studio Link-Arc
LU Yichen
Location Milano, Italy
Completion 2015
Photo LV Hengzhong, Hufton Crow, Sergio GraZia, Roland Halbe

公共建筑 观演／博览

屋脊内景

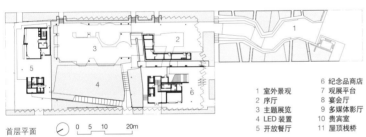

首层平面　0 5 10 20m

1 室外景观　　6 纪念品商店
2 序厅　　　　7 观展平台
3 主题展览　　8 宴会厅
4 LED装置　　 9 多媒体影厅
5 开放餐厅　　10 贵宾室
　　　　　　　11 屋顶栈桥

二层平面

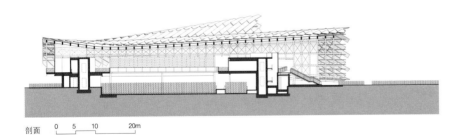

剖面　0 5 10 20m

景观桥下方屋面竹板

南侧幕墙外景　　景观桥望西侧屋面

从南侧艺术家村鸟瞰
西立面

　　银川当代美术馆坐落在黄河西岸河畔，以黄河河床冲刷千百年间地貌的变化为设计灵感，用流动的褶皱肌理还原了千年前塑造至今的沉积岩运动。美术馆的外观犹如一个"运动中"的地质断面，暗喻着这片土地千年间的变化，而其流畅、优雅的线条，又给人以轻盈的印象，仿佛是"降落"在大地上，自成大地的景观，与环绕周围的湿地自然风光融为一体。建筑的设计实际上是从内部开始的，美术馆顶部的玻璃天窗最大限度地将自然光线引入室内，通过空间的交替、材质的变换以及光影的变化，可以使参观者经历到一场有层次感的情绪上的旅途。

Museum of Contemporary Art Yinchuan is located at the border between lush wetlands and arid desert divided by the Yellow river, and the design concept is inspired by the local topography. The crafted museum's massing responds to geological forces visible in the sedimentary creases abundant on the facade, its smooth, elegant lines gives the impression of lightness, as if it was a part of nature. Internal programmatic arrangement dictated the envelopes appearance, and the skylight imparts light to the inside space which is considered of human needs and spatial perception. Through the alternation of space, the transformation of material and the change of light to dark, visitors can experience the traces of Time with greater awareness.

银川当代美术馆
Museum of Contemporary Art Yinchuan

建筑设计
waa 未觉建筑
宁夏建筑设计研究院有限公司
张迪 Jack Young

宁夏 银川　｜　竣工时间 2015 年　｜　摄 影 NAARO

Architects
waa (We Architech Anonymous);
Ningxia Architectural Design & Research Institute Co., Ltd.
ZHANG Di, Jack Young
Location Yinchuan, Ningxia
Completion 2015
Photo NAARO

中庭
出口——通向艺术家村

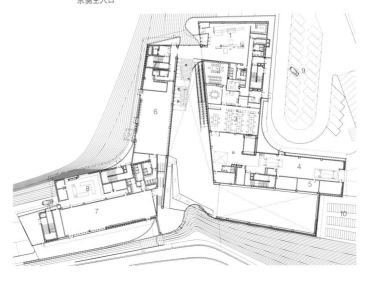

地下一层装置艺术展厅
东侧主入口

1 接待区
2 会议室
3 行政办公室
4 物流入口大堂
5 临时储藏
6 图书馆咖啡馆
7 餐厅
8 厨房/辅助
9 停车场
10 非机动车停车位

一层平面　　0　5　10　20m

北向全景

项目主体功能由博物馆、城市规划馆及文化局办公三部分组合而成。基地面向湿地河流，周边山峦叠嶂。设计通过序列化的排列组合生成成一种连续的差异性变化的建筑形体，诠释了张家界峰林地貌山体的形态特征。在处理场地和建筑的关系上，通过一系列地形操作的手法，使得大地与建筑进行局部形态整合，形成一种地景化的表达，在此构型肌理的基础上，通过景观水系与内部庭院的设置实现"溪流""峡谷"等自然地貌的抽象再现，形成一种人工的"自然"。选择石材作为建筑的表面材质，以强化建筑形态与地形地貌的关联性。经过加工后的石材具有一种竖向肌理，根据光线角度与强度的不同，其表面会产生微妙变化的阴影关系，从而使建筑在不同的阳光照射下拥有丰富的表情。

The project consists of three main parts: a museum, a city planning exhibition hall, and offices for the local cultural bureau. The site faces a river on wetland, surrounded by mountains. A continuously changing architectural form is generated by serialized arrangement and combination in design, explicating the characteristics of Zhangjiajie's peak forest. Through a series of manipulations of the terrain in treatment of the site and buildings, the environment and the architecture is partially integrated as an expression of the landscape. On this basis of formal texture, a sort of artificial nature is made as a result of the placement of water system for landscape purpose and internal courtyards that represent creeks and valleys in an abstract way. Stone is selected as the materials applied on facades to enhance the correlation between architectural form and the terrain. The vertical texture on the processed stone panels shadows on the surface with subtle changes according to different angles and intensity of sunlight, hence riches expressions of the building in the sun.

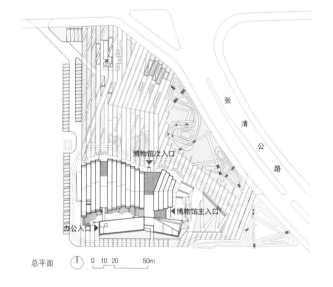

总平面

张家界博物馆
Zhangjiajie Museum

湖南 张家界　　竣工时间 2015 年

建筑设计
地方工作室
魏春雨　齐靖　刘海力

摄　影 姚力　胡骉

Architects
WCY Regional Studio
WEI Chunyu, QI Jing, LIU Haili
Location Zhangjiajie, Hunan
Completion 2015
Photo YAO Li, HU Biao

局部临水景观

入口门厅

地景肌理

峡谷内庭

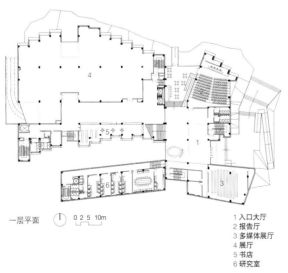

一层平面　0 2 5 10m

1 入口大厅
2 报告厅
3 多媒体展厅
4 展厅
5 书店
6 研究室

南侧堤台、下沉庭院和入口引桥

商丘博物馆位于商丘西南城市新区，收藏、陈列和展示商丘的历代文物、城市沿革和中国商文化历史。博物馆主体由三层上下叠加的展厅组成，周围环以景观水面和庭院，水面和庭院之外是层层叠落的景观台地和更外围高起的堤台（下面设室外展廊），文物、业务和办公用房组成L形体量，设置于西北角堤台之上，设南北东西四门。博物馆的整体布局和空间序列是对商丘归德古城为代表的黄泛区古城池典型形制和特征的呼应与再现，博物馆犹如一座微缩的古城。上下叠层的建筑主体喻示"城压城"的古城考古埋层结构，也体现自下而上、由古至今的陈列布局。

Shangqiu Museum is situated in the southwestern part of Shangqiu City, Henan Province. It collects, displays the cultural relics of Shangqiu City, the evolution of the city, and the culture and history of Shang dynasty. The main part of the museum consists of exhibition halls overlapping on three floors, surrounded by water landscape and a courtyard. Beyond that are cascading landscape terraces and a high bank at a even farther place beneath which is an outdoor exhibition corridor. The L-shaped volume consisting of cultural research, business and office is placed on the top of the bank at its northwest corner with four doors on each direction. Like a miniature ancient city, the museum has the overall layout and spatial sequence corresponding to the historic Gui'de City of Shangqiu, which represents the typical pattern and characteristics of city planning of ancient times in the flood region of the Yellow River. The main part of the building with overlapped floors has a metaphoric meaning of the archaeological structure buried in the historic city.

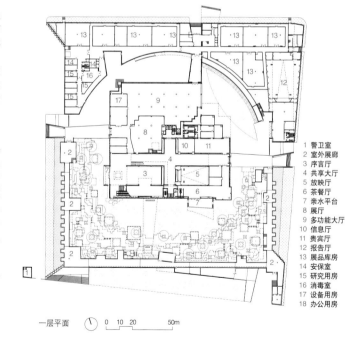

一层平面

1 警卫室
2 室外展廊
3 序言厅
4 共享大厅
5 放映厅
6 茶餐厅
7 亲水平台
8 展厅
9 多功能大厅
10 信息厅
11 贵宾厅
12 报告厅
13 展品库房
14 安保室
15 研究用房
16 消毒室
17 设备用房
18 办公用房

商丘博物馆
Shangqiu Museum

河南 商丘　　竣工时间 2015 年

建筑设计
中国建筑设计院有限公司
李兴钢 谭泽阳 付邦保

摄　影　夏至

Architects
China Architecture Design Group
LI Xinggang, TAN Zeyang, FU Bangbao
Location Shangqiu, Henan
Completion 2015
Photo XIA Zhi

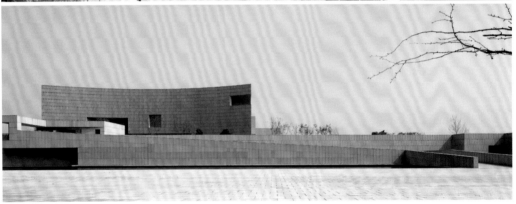

东南鸟瞰
南立面

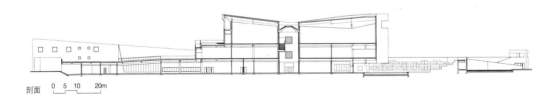

剖面　0　5　10　20m

十字中庭　　　　　序言厅

鸟瞰
胜利之墙

侵华日军南京大屠杀遇难同胞纪念馆三期扩容工程
The Memorial Hall of the Victims in Nanjing Massacre by Japanese Invaders, Phase III

江苏 南京　　竣工时间 2015 年

建筑设计　华南理工大学建筑设计研究院
何镜堂 倪阳 刘宇波

摄　影　夏至 战长恒 马明华

Architects Architectural Design & Research Institute of South China University of Technology
HE Jingtang, NI Yang, LIU Yubo
Location Nanjing, Jiangsu
Completion 2015
Photo XIA Zhi, ZHAN Changheng, MA Minghua

三期扩容工程建设纪念侵华日军1945年9月9日在南京投降事件，凸显抗战胜利为主题。构思立意为胜利、圆满的情感表达。三期扩容工程是侵华日军南京大屠杀遇难同胞纪念馆扩建工程的一个补充和延续，兼具开放性与公共性、日常性与纪念性。这里是一个容纳历史记忆与当前生活、胜利喜悦与死亡悲痛的场所，人们可以在这里纪念、休息、放松、漫步、玩耍。新馆是一个功能复合开放的综合体，除了胜利纪念广场、绿化公园外，还容纳了胜利纪念馆、车站、车库、商业配套、办公等功能设施。工程通过绿植屋面、光伏发电、中水回收、透水混凝土、下沉庭院及天井、热风压拔风效应等多个生态低碳措施，创造出一个既满足空间艺术气氛又能够体现可持续理念的生态绿色纪念性建筑。

The Memorial Hall of the Victims in Nanjing Massacre by Japanese Invaders, Phase III is built to commemorate the surrender of Japanese troops in Nanjing on September 9, 1945, highlighting the theme of the victory of Anti-Japanese War. The idea is to express the feelings of victory and fulfillment. The extension project complements and continues with the Memorial Hall of the Victims in Nanjing Massacre by Japanese Invaders, containing both openness and publicity, everydayness and commemoration. It is a place to accommodate historical memories and current lives, victory joy and death grieve, where people can memorize, rest, relax, walk and play. New museum is a multi-functional and open complex. Except for the victory memorial square, green park, it also contains a war memorial hall, a bus station, a community garage, commercial supporting facilities, office, etc. The projects adopts ecological and sustainable concepts in design, such as green roof, solar photovoltaic, gray water recycle, sunken garden and patio, chimney effects, and many other low-carbon measures, creating a ecological and sustainable architectural monument.

胜利之路
胜利火炬

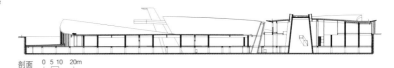

剖面 0 5 10 20m

东立面
西侧下沉广场

一层平面 0 5 10 20m

1 胜利广场　　6 庭院　　　11 桥
2 序厅　　　　7 报告厅　　12 大巴车站
3 展厅　　　　8 辅助用房　13 天井
4 胜利之路　　9 贵宾休息室 14 自行车库
5 商业配套　　10 下沉庭院　15 门厅
　　　　　　　　　　　　　　16 胜利之墙

大剧场外大堂

剧院坐落在松花江北岸江畔,以环绕周围的湿地自然风光和北国冰封的地貌特征为设计灵感——从湿地中破冰而出。消解的边界使庞大的建筑体量更多以局部的方式出现,融入自然的母体中。剧院顶部的玻璃天窗最大限度地将自然光线引入室内,室内大量采用当地的水曲柳天然木材墙面,自然的纹理和多变的有机形态,形成通透、温暖、柔和的氛围。观众厅空间消解了平面化的空间,呈现出一种流动的自然状态,单纯的材料和多变的空间组合为最佳的声效提供了条件。设计强调市民的参与与互动,人们可以通过建筑外部环绕的坡道从公园和广场一直走到屋顶平台。

Sitting on the northern side of the Sungari, the theater is inspired by the views of the wetland and topographic characteristics in North China in winter, hence the visualization as if extruding from the ice of wetland. The huge mass of the building is presented in parts due to dissolved boundaries, integrated into nature. The glass skylight on the top of the theater invites sunlight into the building as much as possible. Manchurian ash panels locally produced are used widely in the interior, and the natural texture and variant organic forms result in warmth and softness. A natural condition of flowing, the audience hall breaks the flattened layout, and the pure material and various spatial combinations achieve the best acoustic effects. The design emphasizes on public participation and interaction, and visitors may approach to the roof deck through an exterior ramp surrounding the building from the park and plaza.

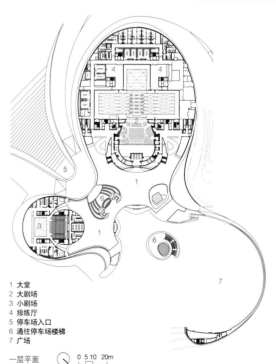

1 大堂
2 大剧场
3 小剧场
4 排练厅
5 停车场入口
6 通往停车场楼梯
7 广场

一层平面 0 5 10 20m

哈尔滨大剧院
Harbin Opera House

黑龙江 哈尔滨 | 竣工时间 2015 年 | 摄 影 Adam Mørk Hufton+Crow Iwan baan

建筑设计
MAD 建筑事务所
北京市建筑设计研究院有限公司
马岩松 党群 早野洋介

Architects
MAD Architect;
Beijing Institute of Architectural Design Co., Ltd.
MA Yansong, DANG Qun, Yosuke Hayano
Location Harbin, Heilongjiang
Completion 2015
Photo Adam Mørk Hufton+Crow Iwan baan

公共建筑 观演/博览

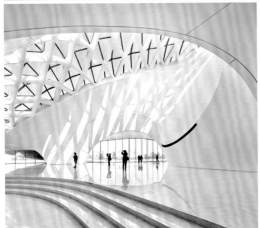

从水池处看大剧院夕阳景色
小剧场外大堂

冬天景致

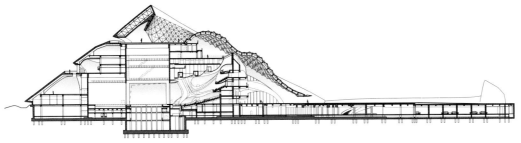

剖面　0 5 10 20m

大剧场大堂及室外景观

西北侧入口

　　博物馆在平面和形体上都采用了纯粹的方正几何形。建筑外墙交错折叠，表皮为镀铜金属穿孔板，以此呼应当地的地域特色。景观设计引入坡地元素，建筑坐落在绿坡之上。建筑空间设计以中庭为中心，中庭为一个 40m×31m，高 33m 的通高空间，围绕中庭逐层展开公共回廊和陈列展厅，中庭四周设置了通高的金属帷幕，在天窗光线的照射下形成丰富的动感和层次，与外墙的金属感形成呼应。

The museum is a pure square both in plan and by form. The external walls are folded and staggered, with pierced copper-plated panels as the material on surface, responding to the local building characteristics. Slopes are introduced in the design, and the building is sited on a green slope. Architectural spaces are arranged around an atrium of 40m×31m with a height of 33m. Public corridors and exhibition halls on each floor are laid out with the atrium as the center. Four sides of the atrium are decorated by metallic draperies of full height, conveying motions and layering to the audience in accordance with metallic on the exterior walls.

远景

俯视

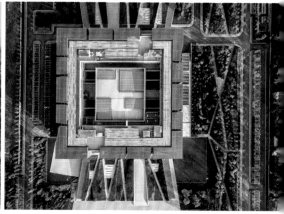

云南省博物馆新馆
New Yunnan Province Museum

云南 昆明　｜　竣工时间 2015 年　｜　摄　影　存在建筑

建筑设计
香港许李严建筑师事务所
深圳市建筑设计研究总院有限公司
严迅奇　陈邦贤

Architects
Rocco Design Architects Limited; Shenzhen General Institute of Architectural Design and Research Co., Ltd.
YAN Xunqi, CHEN Bangxian
Location Kunming, Yunnan
Completion 2015
Photo Arch-Exist Photography

二层平面

1 综合陈列区
2 专题陈列区
3 金玉满堂
4 展馆前厅
5 观景平台
6 中庭上空
7 庭院上空

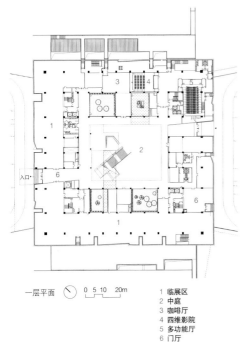

一层平面　0 5 10 20m

1 临展区
2 中庭
3 咖啡厅
4 四维影院
5 多功能厅
6 门厅

公共建筑　观演/博览

大厅休息区
大厅内景
展厅走廊

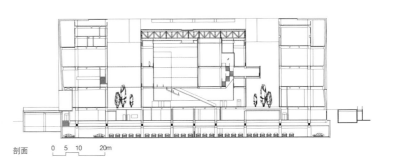

剖面　0 5 10 20m

主入口及商店外景

设计借鉴了木心先生的故乡乌镇的古老城市肌理,美术馆由水环抱,通过一座桥将人引入。建筑由大大小小的一组混凝土盒子错落堆叠而成,墙面采用了凹凸木纹清水混凝土,远观有浑然一体的粗犷感,近看又有细腻的木纹纹理。展厅整体色调以灰色为主,根据不同展厅采用不同的灰色以为区分。建筑师承担了建筑、室内、展陈、景观、标识、照明等全部设计工作,使得建筑的质量得到很好的把控。

Based on the historic urban context of Mr. Mu Xin's hometown, Wuzhen, this gallery is surrounded by water with a bridge connecting with the outside world. A group of concrete boxes of various sizes are put together. Bare concrete with recognizable wooden grains on surface is used for the wall, both rough viewed from afar and delicate if coming closer. The color of grey dominates exhibition halls, and the functional division is decided according to the tone of grey. The architect takes charge of all designs including architectural design, interior design, exhibition, landscape, signs, lighting, etc. to ensure better control of architectural quality.

主入口外景

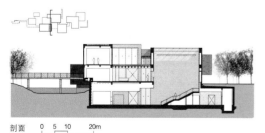

剖面　　0　5　10　　20m

木心美术馆
Muxin Art Museum

浙江 桐乡 ｜ 竣工时间 2015 年

建筑设计
OLI 建筑设计事务所
苏州华造建筑设计有限公司
冈本博 林兵 法比安

摄　影 沈忠海

Architects
OLI Architecture PLLC
Suzhou Huazao Architectural Design Co., Ltd.
Hiroshi Okamoto LIN Bing Fabian Servagnat
Location Tongxiang Zhejiang
Completion 2015
Photo SHEN Zhonghai

公共建筑 ｜ 观演/博览

入口大厅内景
入口大厅内外

图书馆内景

1 售票厅　　6 木心展厅1
2 贵宾室　　7 特展厅3
3 贵宾室码头　8 木心展厅5
4 入口大厅　　9 视听室
5 序厅　　　10 报告厅·休息室
　　　　　　11 假山庭院

一层平面　0 5 10 20m

文学厅内景

博物馆屋面平台　　　　　　　　　　　　　　　　　　大报恩寺塔基遗址及地宫
　　　　　　　　　　　　　　　　　　　　　　　　　报恩新塔回廊

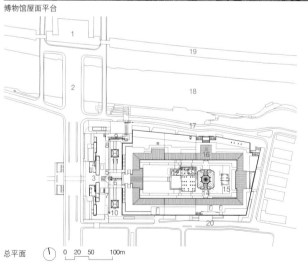

1 中华门
2 长干桥
3 主入口广场
4 西入口游客中心
5 香水河桥遗址
6 御道遗址
7 香水河河道复原
8 香水河水工遗址
9 香水河暗沟遗址覆土保护
10 永乐御碑亭
11 宣德御碑亭
12 月台遗址
13 大殿遗址
14 报恩新塔
15 观音殿遗址
16 遗址博物馆
17 扫帚巷
18 秦淮河
19 明城墙
20 规划道路

总平面

金陵大报恩寺遗址博物馆
Site Museum of Jinling Grand Bao'en Temple

建筑设计
东南大学建筑学院
东南大学建筑设计研究院有限公司
韩冬青 陈薇 王建国

江苏 南京　　竣工时间 2015 年　　摄　影 陈颢

Architects Southeast University School of Architecture; Architects & Engineers Co., Ltd. of Southeast University
HAN Dongqing, CHEN Wei, WANG Jianguo
Location Nanjing, Jiangsu
Completion 2015
Photo CHEN Hao

西入口

天王殿遗址

北画廊遗址

  金陵大报恩寺是明永乐年间在原宋朝寺庙基础上兴建的皇家寺庙。藏有佛祖舍利的琉璃塔曾被喻为中世纪七大奇观之一，享誉世界。该寺庙于19世纪中叶毁于战火。金陵大报恩寺遗址公园位于南京市城南古中华门外，规划设计历经众多学者长期的考古发掘、研究、竞赛、调整和论证，至2011年基本定案。

  金陵大报恩寺遗址博物馆是遗址公园的一期工程，其设计理念基于两个关键问题：其一，如何在严格的遗址保护要求下，使遗址本体的信息得到最恰当的呈现，并与现代博物馆的多元功能相得益彰？其二，如何在形式风貌上恰当地建立起历史与当下的关联？建筑创作通过置于城市格局中的遗址连缀、地层信息的叠合判断、围绕遗址展陈的空间经营和基于技术创新的意象再现等策略，实现了在地脉和时态的关联中传承和创新的初衷。

Jinling Grand Bao'en Temple was a royal temple rebuilt in Yongle, Ming Dynasty, based on the temple ruins that can backtrack to Song Dynasty. Inside the temple there was the world famous Glazed Pagoda enshrining Buddha relics, known as one of the seven wonders of the middle ages. However, the temple was destroyed in war in the mid-nineteenth century. In 2011, after years of archaeological excavations, research, competition, adjustment and demonstration by scholars from different fields, a design scheme of Jinling Grand Bao'en Temple Site Park, which is located outside of Zhonghua Gate in south Nanjing City, was finally decided.

Jinling Grand Bao'en Temple Site Museum is the first phase of the site park scheme, answering two crucial questions. First, how to appropriately present the site under strict protection regulations and coordinate with the multi-functional requirements of a modern museum; second, how to reflect the relationship between history and the present with architectural form and style? Inheriting and innovating the geographical context and temporal association become possible through the tactics of ruins clustering, strata information superposed judging, site presentation oriented space management and technology innovated image reproduction.

屋面平台与场地出入口

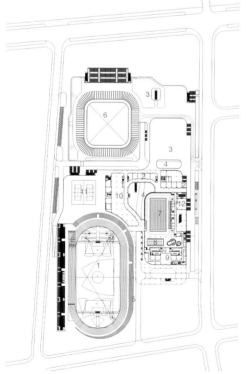

1 体育场
2 商业空间
3 屋顶平台
4 上空
5 露天看台
6 体育馆
7 游泳比赛池
8 游泳训练池
9 游泳馆运动员区
10 餐饮区
11 室外球场
12 竞赛用房

三层平面   0 10 20 50m

临安市体育文化会展中心位于临安锦南新城，项目定位为以体育赛事为主题的城市综合体，集体育健身、商业经营、市民休憩于一体。设计结合场地内低丘缓坡的地貌特征，并通过渐变穿孔铝板包裹建筑主体来营造半透明轻盈的视觉效果。其余建筑体量采用地景化的处理，形成逐层退台的绿化平台，各层平台均可与周边道路平接，极大丰富了场所的可达性和参与性，并融入整体的山水环境中。

Lin'an Sports and Culture center is located in Jinnan New Town of Lin'an. Oriented as an urban complex building with the main theme of sports events, the project contains functions including physical fitness, commercial management and public recreation. The design echoes to the geomorphic features of the gentle slope within the site, and creates a translucent and light visual effect by using gradient pierced aluminum plates that wrap the main building. The rest of the building is integrated with landscape through terraces of vegetation on different levels. Embedded in the surrounding environment, each terrace is connected with roads, maximizing participation of the public and accessibility to the site.

临安市体育文化会展中心
Lin'an Sports and Culture Center

建筑设计
浙江大学建筑设计研究院有限公司
董丹申 陈建 蔡弋

浙江 临安 | 竣工时间 2015 年 | 摄　影 赵强

Architects
The Architectural Design & Research Institute of Zhejiang University Co., Ltd.
DONG Danshen, CHEN Jian, CAI Yi
Location Lin'an, Zhejiang
Completion 2015
Photo ZHAO Qiang

街角空间与入口广场

体育馆周边台地高差衔接
东北侧鸟瞰

体育馆室内比赛大厅
体育馆通廊

山林中的美术馆

入口廊道开口

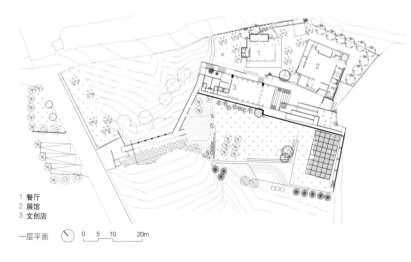

1 餐厅
2 展馆
3 文创店

一层平面　0　5　10　20m

毓绣美术馆
Yu-Hsiu Museum of Art

台湾 南投　|　竣工时间 2015 年

建筑设计
立联合建筑师建筑事务所
（立·建筑工作所）
廖伟立

摄　影　李国民

Architects
AMBi Studio
LIAO Weili
Location Nantou, Taiwan
Completion 2015
Photo LI Guomin

设计因地制宜地运用景观建筑的概念，以中国园林游线一阻、二引、三通的理念，将中介空间与时间同时拉长，使游园的理念融入"赏画"的过程中；试图将空间中的结构性、构造性、材料性耦合在一起，与自然、环境、人对话；将参观动线拉长，沉淀心情进入主馆，展示空间似滚动条般，经楼梯、走道缓缓展开，参观者在内/外、明/暗、松/紧之间游历于展品之间，最后来到顶层的特展间，后者对应于九九峰，体现人为与自然的辩证对话关系。

毓绣美术馆主题设定为当代"写实艺术"的私人美术馆，未来除与在地艺术家及九九峰艺术园区形成良好的关联外，还会肩负美术教育责任，让学童收获良多的"我的美术课在美术馆"体验，及邀请一辈子都没进过美术馆的阿公阿妈来参观，使毓绣美术馆成为一真正在地的山林中的美术馆。

The main concept of the design is inspired by the principles of blocking, introducing and revealing of a tour in a traditional Chinese garden. Intermediate space and time are elongated at the same time, and the concept of tour in garden becomes part of the process of appreciating paintings. The design combines material, structural and construction elements of the space in a dialogue with nature, the environment and human. Tourist circulation is also elongated so that visitors can calm down before entering the gallery. Exhibiting space is so arranged according to a scrollbar that visitors pass through exhibits between inside and outside, light and dark, relaxation and tension, etc. starting from staircases and hallways and ending with a special exhibition hall on the top floor which looks towards Jiujiu Peak, an indication of dialectic dialogue between human and nature.

This museum is registered as a contemporary "realistic art" museum. In addition to the expected good relation between the gallery and local communities, it also aims to promote art education to attract children to study in the gallery, and invite local senior people to the gallery. As such, Yuxiu Gallery will become a true local gallery in the woods.

美术馆入口
思空间与其他建筑形成聚落及巷弄

主馆通往顶楼的楼梯　　主馆室内空间

航拍实景

建筑设计灵感源于对贺兰山苍茫雄壮的感动,以及对当地居民因地制宜建造房屋方式的传承。利用山势高差,将建筑整体嵌入山体。规矩方正的主展厅与空间更丰富的互动展区有机结合,并在多元化空间中引入日光与山景。艺术馆外墙面毛石,均就地取材自贺兰山,该建筑是目前银川市最高的外装毛石砌筑建筑,同时也表现现代艺术与大自然的对话。

The design scheme is inspired by the vast and vigorous characteristics of the Helan Mountain, and by the inheritance of local construction techniques accommodating to local conditions. The design scheme utilizes the height difference of the mountain and embeds the entire building in it. The regular, square main hall is combined with the interactive area of more dynamic space in an organic way, inviting sunlight and the scenery of the mountain inside. As the tallest masonry wall structure in Yinchuan, the museum is built with local quarry stone, providing a conversation between modern arts and nature.

建筑与韩美林雕塑品形成互动场所

银川韩美林艺术馆
Yinchuan Han Meilin Art Museum

建筑设计
北京三磊建筑设计有限公司
张华 范黎 孙睿

宁夏 银川   竣工时间 2015 年   摄 影 存在建筑

Architects Sunlay Design Group
ZHANG Hua, FAN Li, SUN Rui
Location Yinchuan, Ningxia
Completion 2015
Photo Arch-Exist Photography

公共建筑 观演/博览

建筑营造的安全领域感

韩美林先生专属笔会厅
室内展览空间

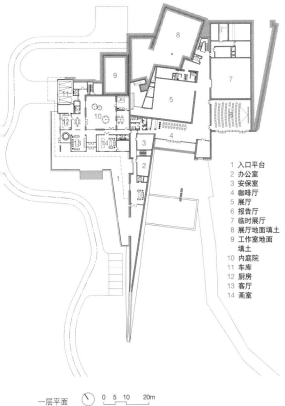

1 入口平台
2 办公室
3 安保室
4 咖啡厅
5 展厅
6 报告厅
7 临时展厅
8 展厅地面填土
9 工作室地面填土
10 内庭院
11 车库
12 厨房
13 客厅
14 画室

一层平面 0 5 10 20m

规划兼档案馆

规划兼档案馆剖透视

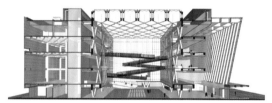

博物馆剖透视

项目由博物馆和规划兼档案馆两栋建筑组成。设计的策略首先是引入平民化的公共活动，将建筑布置在地块南侧，在北侧完整地退让出城市绿地，在两栋建筑之间置入城市广场，以满足当地大型民间艺术"花鼓灯"的表演以及市民日常的休闲娱乐活动；东西两栋建筑均采用正方形的建筑空间布局。博物馆立面效仿岩层断面的肌理，形成层叠的质感；规划档案馆利用银灰色穿孔铝板，形成与博物馆截然不同的建筑个性。博物馆采用回字型的空间布局，并向东侧广场敞开，主入口空间与中庭空间相连，斜撑支柱支撑起36m见方的井字桁架屋顶，五折的蛇形廊道通过10m左右跨度的拉杆从屋顶悬下，并通过斜拉杆形成超静定结构，避免晃动。在光的中庭中，吊桥既是一个空间雕塑，又是重要的公共空间。规划档案馆由公共性较强的规划馆和相对私密的档案馆组成。设计根据两馆的特性在空间组织上进行了区分。西侧紧邻广场部分设置规划馆，通高的公共空间连通广场空间，增强建筑的公共性。游客可通过缓缓上升的螺旋坡道从不同角度不同高度观看蚌埠市模型的全貌。建筑东侧临近城市道路部分为档案馆，为市民提供便捷的公共服务。

The project is composed of two buildings, i.e. a museum and a planning exhibition center in combination with planning archives. The design strategy is to introduce people-oriented public activity first through leaving urban green space in the north side, and building an urban plaza in the middle of the two buildings, which can satisfies the need of local performing local art "flower-drum lantern" and residents' daily entertainment. Both the buildings on west and east sit on the square plan. The facade of the museum resembles the sectional texture of rock stratum as stacked up. Grey perforated aluminum plates are used for the planning archives, in sharp contrast to the museum. A hollowed square is adopted for the layout of the museum, open to the plaza on west, and the main entrance is connected to the atrium. The inclined bracing pillars support the roof of ribbed beams with a span of 36m, while the serpentine corridor of five sections hangs with rods on the roof to form a super static structure to avoid vibration. In the sunny atrium, the suspension bridge is not only a statue, but also an important public space. The planning archives consists of a public planning exhibition hall and a more private archives, and a distinction is made between the nature of the two parts. The exhibition hall is placed close to the western plaza, to which the multi-storied public space is connected, enhancing the publicity of the building. Visitors can have a full view of the city from different angles when ascending to the top of the hall along the spiral ramp. The archives sits nearby urban roads on the east side, providing convenient service for the public.

蚌埠博物馆及规划档案馆
Bengbu Museum, Urban Planning Exhibition and Archive Centre

安徽 蚌埠 | 竣工时间 2015 年

建筑设计
深圳建筑设计研究总院有限公司
孟建民 邢立华 徐昀超

摄 影 张广源

Architects
Shenzhen General Institute of Architecture Design and Research Co., Ltd.
MENG Jianmin, XING Lihua, XU Yunchao
Location Bengbu, Anhui
Completion 2015
Photo ZHANG Guangyuan

公共建筑 观演／博览

蚌埠市博物馆、规划兼档案馆鸟瞰
城市规划展览馆内部

博物馆内部

1 博物馆
2 规划兼档案馆
3 城市广场

博物馆内部

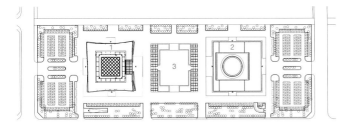

总平面　0 10 25 50m

南侧外景

主入口夜景

坐落于上海市中心的刘海粟美术馆，是以中国现代美术教育的奠基人刘海粟个人名字命名的重点国家美术馆。用地局促的条件下，底层仍尽可能多分配为绿地以实现社会效益最大化，将建筑往空间扩展。设计立意"云海山石"取意于刘海粟一生"为师为友"的黄山，也是国画永恒的主题。大手笔的体量切割表现出强烈的雕塑感，建筑体量外在表现为峻峭山石，意如璞之于外，室内则如玉石剖开所呈现的洁白玉质，整个美术馆呈现出浑然天成的整体感。简洁有力的折面勾勒出大气的建筑体量，亦为建筑赋予强烈的动感。

As a key national art gallery, Liu Haisu Art Gallery is located in downtown Shanghai, named after Mr. Liu Haisu who is the founder of Chinese modern art education. Confronting the constraints of the site, the ground floor is reserved for vegetation to maximize social benefits while architectural space can extend outside. The design idea, sea of clouds and mountain of rocks, comes from the imitation of sheer rocks of the Yellow Mountain, both the friend and mentor for Mr. Liu in his lifetime and the eternal theme of Chinese painting. Rough cuttings of the bulk of the building bring up a strong sculptural sense. The exterior of the building looks steep rocks and shows roughness on surface with delicate interior design as the jade, contributing to the natural sense of entirety. Simple but powerful folded surfaces make the profile of the building stand out, and impress the audience with the strong potential of movement.

刘海粟美术馆 / Liu Haisu Art Museum

上海 | 竣工时间 2015 年

建筑设计 同济大学建筑设计研究院（集团）有限公司
陈剑秋 吴靖杰 郭辛怡

摄影 章勇

Architects Tongji Architectural Design (Group) Co., Ltd.
CHEN Jianqiu, WU Jingjie, GUO Xinyi
Location Shanghai
Completion 2015
Photo ZHANG Yong

公共建筑　观演／博览

鸟瞰
中央大厅内景
主入口

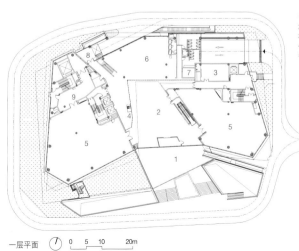

1 入口平台　6 咖啡／纪念品
2 门厅　　　7 衣帽间
3 贵宾室　　8 卸货平台
4 问询台　　9 设备平台
5 临时展厅

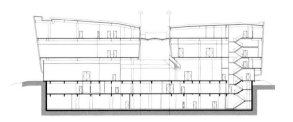

一层平面　　0　5　10　20m
剖面　　　　0　5　10　20m

龙美术馆（西岸馆）位于上海市徐汇区的黄浦江滨，基地以前是运煤的码头，现场有一列被保留的1950年代所建的煤料斗卸载桥和2009年已施工完成的两层地下停车库。

新的设计采用独立墙体的"伞拱"悬挑结构，呈自由状布局的剪力墙插入原有地下室与原有框架结构柱浇筑在一起，地下一层的原车库空间由于剪力墙体的介入转换为展览空间，地面以上的空间由于"伞拱"在不同方向的相对联接形成了多重的意义指向。机电系统都被整合在"伞拱"结构的空腔里，地面以上的"伞拱"覆盖空间，墙体和天花均为清水混凝土的表面，它们的几何分界位置也变得模糊。这样的结构性空间，在形态上不仅对人的身体形成庇护感，亦与保留的江边码头的煤料斗产生视觉呼应。

Long Museum West Bund is located at the bank of Huangpu River, Xuhui District, Shanghai. The site used to be the wharf for coal transportation, and a coal-hopper-unloading-bridge constructed in the 1950s is remained on the site with a two-story underground parking completed as early as two years ago.

The new design adopts the cantilevering structure featuring "vault-umbrella" with independent walls while the shear walls with free layout are embedded into the original basement so as to be concreted with the original framework structure. With the shear walls, the first underground floor of the original parking has been transformed to an exhibition space with the space above ground, highlighting multiple orientations because of relative connection of the "vault-umbrella" at different directions. In addition, the electrical and mechanical system has been placed inside the "vault-umbrella" structure. As to the space above ground covered by the "vault-umbrella", the walls and the ceiling feature as-cast-finish concrete surface with blurred boundaries for geometrical division. Such structure cannot only shield the human body in conformation but visually echoes with the Coal-Hopper-Unloading-Bridge at the wharf.

从一层小展厅望向大厅

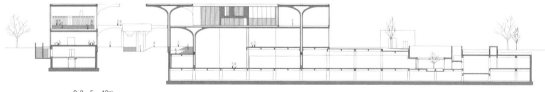

西南 – 东北剖面　　0 2 5　10m

龙美术馆（西岸馆）
Long Museum (West Bund)

建筑设计　大舍建筑设计事务所
柳亦春　陈屹峰　王龙海

上海　　竣工时间 2014 年　　摄影　夏至　苏圣亮

Architects Atelier Deshaus
LIU Yichun, CHEN Yifeng, WANG Longhai
Location Shanghai
Completion 2014
Photo XIA Zhi, SU Shengliang

二层展厅望向首层大厅入口处

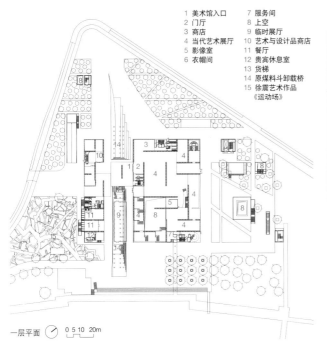

1 美术馆入口
2 门厅
3 商店
4 当代艺术展厅
5 影像室
6 衣帽间
7 服务间
8 上空
9 临时展厅
10 艺术与设计品商店
11 餐厅
12 贵宾休息室
13 货梯
14 原煤料斗卸载桥
15 徐震艺术作品《运动场》

一层平面 0 5 10 20m

煤料斗卸载桥与主体建筑间的通道
从阶梯式展厅看地下一层展厅

西南立面

建筑选址以"建筑修补山体"为理念，利用场地废弃的采石场山体缺口布置建筑主体，最大程度地减小建筑对现状自然环境的侵扰。竖向展开的博物馆布局在一个较高的地形上，突出了建筑与阖闾城遗址以及太湖景观的空间与视线联系。妥善利用现有的地形起伏，组织不同高程的竖向设计。精心组织穿越博物馆建筑屋顶的公众登山流线，有效整合了场地现有景观和历史文化遗存，将遗址本体、太湖景观整合在博物馆的参观体验之中，为城市公共活动提供了重要场所。尽最大可能保护基地内原有的山体、水面和绿化环境，并通过屋顶覆土绿化使建筑成为自然环境的延伸，在整体上构建出一个开放的博物馆形态，极大地丰富了太湖沿线景观。

Based on the concept of "building repairing the mountain", the museum chooses the location at the opening of the mountain, using the original abandoned quarry to site the main volume of museum to minimize impact on the environment. Vertically extending, the museum is set in a higher area, highlighting the visual connection among the architecture organization, Helvyu city historic site and the Lake Taihu's landscape. The buildings make good use of the undulating terrain and combine vertical designs with different elevations, carefully organizing the public climbing route through the roofs of the museum to integrate with the existing landscape and historical relics. As such, these elements are successfully integrated for tourist experience, also provide an important place for public activities. What's more, the design retains the existing mountain, water and vegetation as much as possible, and make the green roofs become the extension of the environment. On the whole, it has built an open museum in form, remarkably enriching the landscape along Lake Taihu.

从龙山看博物馆

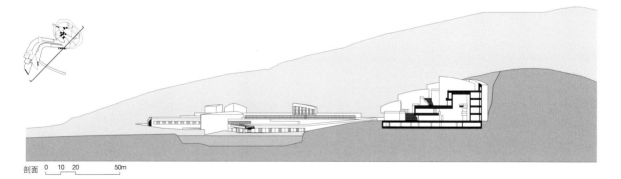

剖面 0 10 20 50m

无锡阖闾城遗址博物馆
Helv City Historic Site Museum

江苏 无锡 | 竣工时间 2014 年

建筑设计
同济大学建筑设计研究院（集团）有限公司
李立

摄　影 姚力

Architects
Tongji Architectural Design (Group)Co., Ltd.
LI Li
Location Wuxi, Jiangsu
Completion 2014
Photo YAO Li

公共建筑 观演/博览

从学术交流中心看博物馆
光线穿过形体之间

学术交流中心的曲折连廊

1 门厅
2 会议室
3 接待室
4 贵宾接待
5 报告厅
6 控制室
7 办公室
8 前台
9 连廊
10 上空
11 露台
12 地下车库

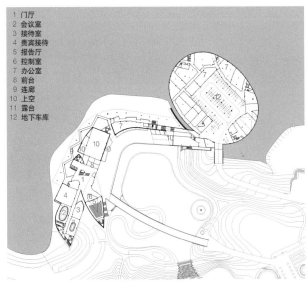

13.50m 标高平面   0 10 20 50m

中央大厅内景

西北向夜景
东南向外景

柯力博物馆
Keli Museum

建筑设计
佚人营造建筑师事务所
王灏

浙江 宁波　　竣工时间 2014 年　　摄影 刘晓光 高文仲

Architects
Architects Anonymeous Architects & Workshop
WANG Hao
Location Ningbo, Zhejiang
Completion 2014
Photo LIU Xiaoguang, GAO Wenzhong

本项目位于柯力电气厂区入口处，原用地是入口庭院广场，2010年业主决定在此兴建柯力博物馆，以展示柯力的产品系列以及企业对中国古代及近代衡器的收藏。

博物馆由地下一层的产品陈列馆、完全开放的一层花园及二三层的多功能空间构成。地下一层分5个展厅，出入口位于最北端的下沉广场，由两道直角楼梯缓慢引导而下。围绕5个展厅设置了4个下沉庭院，种植紫藤及竹子。一层的庭院采用了方格不规则路网，可以从厂区周围建筑自由穿梭出入，路面由露石混凝土浇注，在场地上满铺当地产的黑色小道渣石，形成松柔的辅助地面。整个地面建筑由30m×60m的巨型型钢混凝土框架举起。这个型钢混凝土框架形成整个建筑的特征和力量。地面建筑由巨框内的5个高低不同的玻璃砖几何体组成，根据力学和视觉平衡原理布置在巨框内，三层标高约10m处的楼板是整个结构的刚性平层。二层楼板转角柱从三层大梁处悬吊下来，承受了二层楼板的大部分重量。

The project is located at the entrance of Keli Sensing Technology Company in Jiangbei Industrial Zone, Ningbo. Originally the site was an entrance square, and the client decided to build a museum in 2010 to display their products as well as their collection of ancient and modern Chinese weighing apparatus.

The building consists of a exhibition hall of production on the underground, an open garden on the ground floor, and multi-purposeful space on the second and third floors. The exhibition hall of production is divided into 5 parts to be approached by two rectangular staircases of gently sloping gradient, with the entrance on a sunken square in at the tip of the northern side. 4 sunken courtyards are placed around the 5 exhibition halls, planted with vines and bamboos. An irregular road network paved with gravels and concrete is adopted for the courtyard on the ground with accesses from surrounding buildings. Small soil sediments are used to cover the roads on the site, forming a soft auxiliary surface. The whole volume is lifted by a huge reinforced concrete frame of 30m×60m, which contributing to the character and strength of the building. The volumes above ground are made up of 5 boxes in glass bricks with different heights, properly arranged in a huge rectangular frame. The slab of the third floor of about 10m in height is the rigid plane for the whole building. Corner pillars on the second floor hang from the main beam beneath the third floor, bearing most loads on the second floor.

二层平面

1 主入口
2 下沉广场
3 大厅上空
4 天井
5 巷弄
6 广场
7 大厅
8 展厅
9 办公

地下一层平面

一层广场入口巷弄局部
地下一层博物馆大厅
展厅

西向晨景
屋顶合院局部

位于水墨氤氲的人文之乡江苏南通的范曾艺术馆，为范曾书画艺术作品以及南通范氏诗文世家的展示、交流、研究、珍藏而建造。艺术馆以与传统文化有着紧密情感关联的"院"为切入点，依照3种不同院落的自发性生成秩序铺展略带松散的局部关系，进而构架起以井院、水院、石院、合院为主体的叠加的立体院落，从而突破性地将院落从物化关系中脱离，最终呈现游目与观想的合一，以期达到"得古意而写今心"的意境。

Situated in the cultural and historic town of Nantong, Jiangsu Province, Fan Zeng Art Gallery is built for exhibition, communication, research and the collection of calligraphy, paintings and poetries of the Fan Family of Nantong. The concept of the design is inspired by courtyard, an element of traditional space with close connections with traditional culture. According to the spontaneous generation order of three different types of courtyards, i.e. well courtyard, water courtyard and compound courtyard, the layout in a relatively loose way is decided. On this basis, the stereoscopic courtyard is constructed, which liberates from materiality of the courtyard to combine touring and thinking by means of "making modern creations under ancient rules".

范曾艺术馆
Fan Zeng Art Gallery

建筑设计 同济大学建筑设计研究院（集团）有限公司 原作设计工作室
章明 张姿

江苏 南通 ｜ 竣工时间 2014 年 ｜ 摄影 姚力 苏圣亮

Architects Original Design Studio, Tongji Architectural Design (Group) Co., Ltd.
ZHANG Ming, ZHANG Zi
Location Nantong, Jiangsu
Completion 2014
Photo YAO Li, SU Shengliang

二层水院

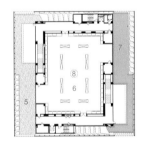

1 入口门厅
2 展厅
3 出口门厅
4 贵宾接待
5 石院
6 主展厅
7 水院
8 光井

二层平面

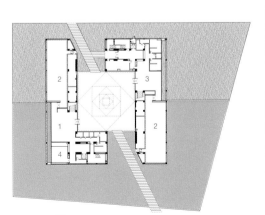

一层平面　0　5　10　20m

东南向外景
底层架空井院

广场夜景

乌海市黄河渔类增殖站及展示中心
Wuhai Yellow River Fishing Station and Exhibition Center

内蒙古 乌海 | 竣工时间 2014年

建筑设计 内蒙古工大建筑设计有限责任公司 张鹏举

摄影 张广源

Architects Inner Monglian Grand Architecture Design Co., Ltd. ZHANG Pengju
Location Wuhai, Inner Mongolia
Completion 2014
Photo ZHANG Guangyuan

入口水景

门厅休息区
内外通透交融的内院空间

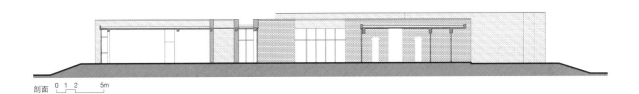

剖面 0 1 2 5m

项目位于内蒙古乌海市西邻黄河的对岸，是一座水利枢纽的配建工程，功能包括生产和展示空间。设计首先开放了交通动线，结合若干内院弹性地提供了展示空间，进而打碎体量，平铺于场地当中，与有限的树木相融布置，隐身于独特的河岸风景中。在此基础上，建筑用附近废弃砖窑的留存红砖直接建造。随后的设计就演变成了一场关于这种朴素材料的游戏：砖如何成墙，墙上如何开洞，墙与地又如何交接以及顶部如何保护等平实的建造活动。

Located on the opposite bank of the Yellow River on the west side of Wuhai, Inner Mogolia, this project is built to support a hydro-junction complex, used for fishing industry and exhibition. Combining several inner courtyards, the design opens up circulation routes to provide more flexible exhibition space. The bulk of the building is broken down to smaller parts spreading over the site, arranged with a limited number of trees to hide itself in the riverfront landscape. On this basis, it is built by bricks collected in a nearby discarded traditional brick kiln. In the latter stage of the design, the focus is put on the best method to build the walls, openings, joints between the walls and the surface of the ground, and protection of the roof.

1 孵化暂养车间
2 催产孵化车间
3 饲料药品车间
4 办公室
5 门厅
6 餐厅
7 会议室
8 休息区
9 厨房
10 实验室
11 宿舍

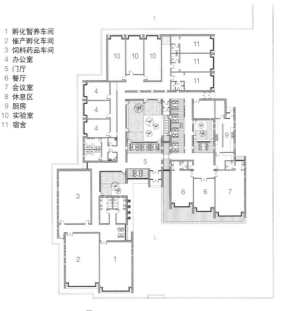

平面 0 2 5 10m

从山上回望清晨苏醒的梁希纪念馆

项目位于湖州梁希森林公园内，整体建筑展开布置与山体形态自然组合，采用防腐木、清水混凝土等与自然相融合的材料，空间相互渗透。建筑主入口与主水景有机结合，与室外坡道形成强烈导示感，水景平台的纪念雕像如同浮于水面，增强了纪念性。建筑周边景观设计与自然地形紧密结合，使建筑本体功能延展到更广阔的外部空间。主体建筑采用坡屋顶，建筑形体随山体层层跌落，使山体与建筑的视景连成一片。室内开窗充分利用天光及室外景观，游客在建筑内可以欣赏到自然的美景，将纪念的主体——梁希"美丽中国"的林学思想在建筑中得到体现。通过环境的衬托、景观的营造、建筑的融合、室内的渲染，整个建筑成为一个具有凝聚力且极具开放性的纪念性精神场所。

Situated in Liangxi Forest Park in Huzhou, the building integrates the layout with the profile of mountains using antiseptic wood, fair-faced concrete and other natural materials to allow spatial penetration. The main entrance of the building and the main water environment are combined organically-forming a strong sense of guidance with outdoor ramps. The statue of the waterscape platform seems like floating in the water, which enhanced the monumentality. The surrounding landscape is closely integrated with the natural terrain, extending the building functions to a wider outer space. Gabled roofs are used for the main building, and the shape responds to the cascading silhouette of mountains blending the building and the mountains into one organic piece visually. Indoor windows make full use of the daylight and outdoor landscape, and visitors can enjoy the beauty of natural inside the building. Liang Xi's forestry thought of "Beautiful China" as the main theme of the memorial, has been reflected in the building. Through the environment setting, landscape construction, building integration and interior rendering, the whole building has been forming a cohesive and highly open commemorative place.

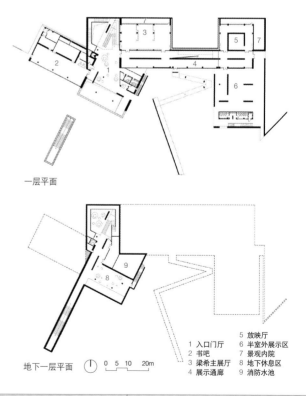

一层平面

地下一层平面

1 入口门厅
2 书吧
3 梁希主展厅
4 展示通廊
5 放映厅
6 半室外展示区
7 景观内院
8 地下休息区
9 消防水池

浙江湖州梁希纪念馆
Huzhou Liangxi Memorial Hall

建筑设计：苏州九城都市建筑设计有限公司　张应鹏　王凡

浙江　湖州　｜竣工时间 2014 年　｜摄影　姚力

Architects: Suzhou 9 Town Studio Co., Ltd.
ZHANG Yingpeng, WANG Fan
Location: Huzhou, Zhejiang
Completion: 2014
Photo: YAO Li

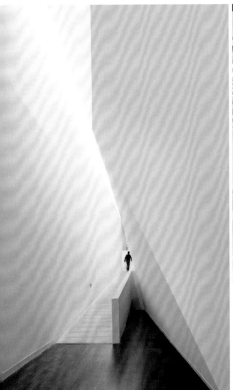

下沉半室外景观中庭坡道
室内坡道

西北侧回望裸露的鹿山山体
流动展厅入口

东南侧入口广场

总平面　0 20 50 100m

菜市口大街视角

项目包括220kV变电站主厂房及电力科技馆两部分内容。其中地下三至五层为变电站主厂房，是可参观的地下220kV智能变电站。地下二层以上为具有商业价值的附属设施。总图设计在北侧退让出30m范围的建控区，以园林景观设计营造出高品质城市空间。建筑形体分为高度不同的若干小体块，形成空间梯度，高度递减至建控区。此举消解了对城市历史街区的视觉压迫，同时形成丰富的建筑表情。建筑表皮设计着力于体现建筑与城市历史和文脉发展的关系，以超白玻璃反射天光云影和胡同院落，以传统纹样开洞的石材幕墙表皮暗合中国神韵。

The project consists of two parts, i.e. the main building as the 220kv transformer substation, and a Electric Power Science and Technology Museum. The main part of the transformer substation is arranged on the third to fifth floors beneath the ground, open to visitors as an intelligent 220kv transformer substation. Auxiliary facilities for commercial use are arranged from the second floor beneath the ground upwards. In the master plan, a 30m high building control area is reserved in north for better urban space by use of garden landscape. The buildings are grouped into smaller sections of different height, cascading towards the height-controlled area, reducing visual pressure on historic districts and enriching architectural representations. The design of architectural surface has highlighted the relationship between architecture and the city's history and culture, using super-white glass to reflect the sky, clouds, hutong compounds and embodying Chinese charm by the stone curtain walls with openings in traditional style.

立面细部

北京菜市口输变电站综合体（电力科技馆）
Beijing Caishikou Power Transformer Substation Complex (Electric Power Science and Technology Museum)

北京 | 竣工时间 2014年

建筑设计
清华大学建筑设计研究院有限公司
北京市电力设计院

庄惟敏　张维　杜爽

摄　影　姚力

Architects
Architectural Design and Research Institute of Tsinghua University; Beijing Power Design Institute
ZHUANG Weimin, ZHANG Wei, DU Shuang
Location Beijing
Completion 2014
Photo YAO Li

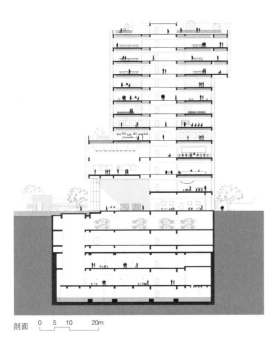

剖面　0　5　10　20m

1 展厅	5 服务区	9 卸货区
2 营业大厅	6 贵宾接待	10 空调机房
3 值班	7 门厅	11 弱电机房
4 消防控制室	8 走道	

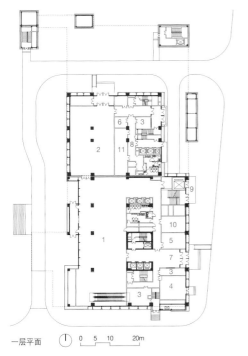

一层平面　0　5　10　20m

中山会馆视角
建筑局部
建控区外景

室外台阶形成休闲纳凉的公共场所

　　设计方案包含了两条重要的轴线：其一是由现有村落（一条客家古街和一组旧厂房）向内延伸，形成一条贯穿基地南北的"时光轴"，以示历史文脉的留存，旧建筑前的月牙形水塘正代表了客家文化的特征；其二是与基地上两座山丘制高点连线垂直的"景观轴"，遥望高尔夫球场。美术馆主体被抬高架设于两个山丘之间，美术馆形体折起，形成虚空的体量，让出时光轴，使之延续和山体相连。建筑主体的折起，为南方气候下的场所提供一个有阴影的覆盖，与"时光轴"垂直相交，成为一个汇集展览、工坊、咖啡厅等多种功能活动的、多元的公共开放空间。

The building has two important axises in design: one extends inwardly from the existing village (a traditional Hakka street and a group of old factory buildings), as a temporal axis running north and south of the site to show the preservation of the historical context. The crescent pond in front of the old buildings embodies the features of Hakka culture. The other one is the vertical "landscape axis" looking out to the golf course. The main building of the gallery is raised and built between two hills. The building is folded up to form an empty mass, allowing the temporal axis to stretch out and connect with mountains. The upward building provides a dynamic and multi-functional space for exhibition, workshop and cafe, which meets the temporal axis perpendicularly.

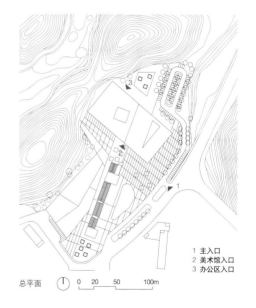

1 主入口
2 美术馆入口
3 办公区入口

总平面　　0　20　50　100m

中国版画艺术博物馆
China Scratchboard Art Museum

建筑设计　悉地国际　朱雄毅 凌鹏志

Architects CCDI Group
ZHU Xiongyi, LING Pengzhi
Location Shenzhen, Guangdong
Completion 2014
Photo FANG Jian

广东 深圳　　竣工时间 2014年　　摄影 方健

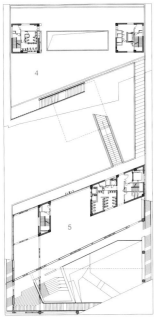

三层平面　　1 办公　4 平台
　　　　　　2 工坊　5 展厅
　　　　　　3 门厅

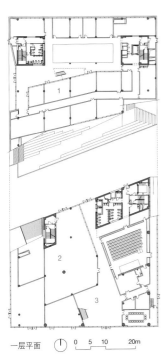

一层平面　　0　5　10　20m

新老建筑对话

结构的枝杈
从三层看入口与展厅
架空层空间

公共建筑　观演/博览

河边望博物馆

玉树州博物馆是"4·14"玉树地震灾后重建的十大重点公共建筑项目之一。建筑师力求创造一座融入藏区环境与市民生活,体现地域特色与具有场所氛围的现代博物馆建筑:在城市空间关系上,高低错落的台式体量、高耸的入口大厅与山上的结古寺遥相呼应,和谐地融入群山环绕的大自然背景与城市肌理中。在建筑空间关系上,将充满藏族氛围的庭院与光影弥漫的中庭结合,作为公共空间脊梁串起各个功能空间;空间序列始于入口圆台大厅,终于东侧眺望结古寺的室外庭院,形成层次丰富的空间序列。粗犷稳重的形体、高低向上的群落、极具特色的细部、质朴纯净的色彩与强烈对比的光影体现了神秘的地域特色和浓厚的文化气息。

State Museum is one of the ten public buildings of reconstruction in the aftermath of the earthquake on April 14, 2010. Yushu Museum aims to integrate the Tibetan environment with people's daily life, embodying regional features and the atmosphere in a modern way. Regarding the urban space relationship, the picturesque highland building style of uneven profile is adopted, and the high entrance hall responds to Jiugu Temple on the mountain at a distance, blending with the surrounding mountains and the unique urban context. Regarding the architectural spatial relationship, the mysterious atmosphere of the Tibetan courtyard and the radiant atrium are combined as a public space that connects various functional spaces. The dynamic sequence of space starts with a circular entrance hall and ends in an exterior courtyard on the east side overlooking Jiegu Temple. The dignified form, uneven profile of the building group, details with mysterious regional features, strong contrast of colors and the effects of light and shadow demonstrate mystery and cultural thickness of the region.

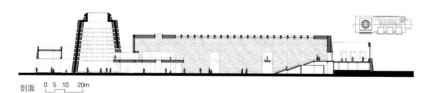

剖面 0 5 10 20m

玉树州博物馆
Yushu Museum

青海 玉树藏族自治州 | 竣工时间 2014 年

建筑设计
华南理工大学建筑设计研究院
何镜堂 郭卫宏 丘建发

摄影 姚力

Architects
Architectural Design & Research Institute of South China University of Technology
HE Jingtang, GUO Weihong, QIU Jianfa
Location Yushu, Tibetan Autonomous Prefecture
Completion 2014
Photo YAO Li

院子
围廊

1 玉树州博物馆
2 牦牛广场
3 治曲民族商城
4 民主路
5 团结路
6 结古大道

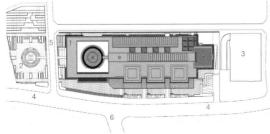

总平面 0 20 40 80m

中庭 | 外立面细部

西北鸟瞰

三层室外平台

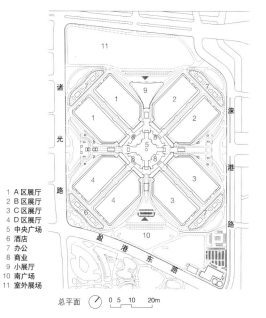

1 A区展厅
2 B区展厅
3 C区展厅
4 D区展厅
5 中央广场
6 酒店
7 办公
8 商业
9 小展厅
10 南广场
11 室外展场

总平面

国家会展中心（上海）
Shanghai National Exhibition and Convention Center

建筑设计
清华大学建筑设计研究院有限公司
华东建筑设计研究院有限公司

庄惟敏 张俊杰 单军

上海　　竣工时间 2014 年　　摄　影 姚力

Architects
Architectural Design and Research Institute of Tsinghua University; East China Architectural Design and Research Institute
ZHUANG Weimin, ZHANG Junjie, SHAN Jun
Location Shanghai
Completion 2014
photo YAO Li

上海国家会展中心目前是世界规模最大的会展综合体，提供了40万 m^2 的展览空间和10万 m^2 的室外展场，同时集聚与新型展会相关的现代服务业，打造城市全新的商业娱乐休闲中心。设计面临着要在有限的基地内将规模浩大的综合体集约在一个900m见方的体量内，又在机场临空区限高43m的制约下，而无法采取会展建筑典型的分散布局方式的困难。因而在建筑规划布局上采用了四叶草的造型，体现出向心汇聚和包容的特点。4组展览单元与四叶草的对称图形自然契合，达到了形式与功能的统一。该项目作为一个容纳数个超过2.8万 m^2 面积和30m高度的展厅的复杂综合体，在大跨度结构、超限防火、大屋面防水、人车交通流线、照明与采光方面都有着创新研究和设计。

Shanghai National Exhibition and Convention Center is so far the largest exhibition complex in the world, providing an indoor exhibition area of 400 000 m^2 and an outdoor exhibition area of 100 000 m^2. Meanwhile, it includes related contemporary service to be a commercial recreation center. Facing the height restriction of 43m from the airport, the huge complex has to house various functions based on a square of about 900mx900m of each side without adopt the typical decentralized layout. Therefore, the a clover-shaped plan is used, characterized by centripetal convergence and compatibility. The four groups of exhibition units and the symmetry of the clove fit naturally, perfectly combing form and function. With an area of more than 28 000m^2 below the height of 30m, the complex demonstrates innovative achievements in large span structures, extreme fireproof methods, large roof waterproofing, traffic flow design, lighting, etc.

南侧主入口鸟瞰
二层室内平台
中心广场鸟瞰

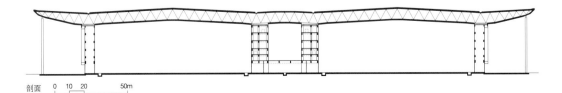

剖面　0　10　20　　50m

剧场夜景

首演盛况

剧场位于五台山景区南入口外、两座小山前的开阔场地上。由于大型情景演出的需要，剧场空间是一个长宽131m×75m、高21.5m的大空间。以730m长、徐徐展开的"经折"状的墙体置于剧场之前，由高至低排列形成渐开的序列，成为剧场表演的前奏。"经折"和剧场均采用了不同材质的表皮，包括石材、玻璃和不锈钢等材料，通过这些材料的反光和透射的特点，将体量化解为不同尺度的起伏的图案，不同程度地映照着周围的景象，蓝天、白云、山峦、树木，也包括身处其间的观众，一切尽在似有与似无之间，极大地消解着建筑物的轮廓线，破解建筑体量对周围环境的压力。

The theater is located at the open space of the south entrance of the Wutai Mountain Scenic Area, ahead of two hills. Due to the needs of large-scale scene performances, the theater has large space spanning of 131m in length, with 75m in width and 21.5m in height. To approach the building, visitors have to pass by a 730m-long folded wall set in front of the building with a sequential profile arranged from high to low, as the prelude to theater. Various materials are used in the folded wall and the theater, including stone, glass and stainless steel. The reflective attribute of these materials is used to dismantle the bulk into many patterns of various sizes reflecting surrounding scenes such as the blue sky, white cloud, mountains, trees, and visitors on site. All seems to be set in between nothing and something, merging the profile of the building into the environment with reduced impact of the bulk.

又见五台山剧场
Encore Wutai Mountain Theater

建筑设计
北京市建筑设计研究院有限公司；
北京建院约翰马丁国际建筑设计有限公司
朱小地 高博 朱颖

山西 忻州 | 竣工时间 2014年 | 摄　影 傅兴

Architects
Beijing Institute of Architectural Design; BIAD John Martin International Architecture Design
ZHU Xiaodi, GAO Bo, ZHU Ying
Location Xinzhou, Shanxi
Completion 2014
Photo FU Xing

剧场出口空间
剧场入口空间

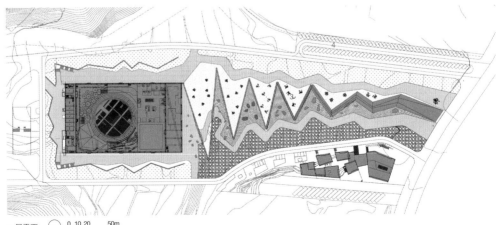

1 剧场
2 售票处
3 "经折"广场
4 停车场

一层平面 0 10 20 50m

文化区整体望海鸟瞰
南向主广场透视

项目位于新城文化核心区的中心位置，是城市功能区块与城市开放空间的衔接、转换枢纽，以展示辽东湾地方文化、城市建设历史、未来发展规划为主要功能。建筑以稳重的建筑形态塑造了一个具有仪式感的场所空间，幻化的建筑表皮再现了"红海滩"地域景观，内向的空间布局隐喻着"天圆地方"的传统内涵，将城市开放空间融入建筑主导空间，诠释了建筑对城市空间的尊重与回应。

The project is located in the cultural center of the new town. As a transitional hub of the city's open and functional parts showcasing the history of local culture and construction of Liaodong Bay and the prospective development plan is the main purpose. The solid architectural form informs a sense of rituality, and the illusive building skin represents the red beach landscape. Introversive spatial layout has a metaphoric meaning of "round sky and square earth" in traditional planning, bringing open urban space in the building-dominated space and explicating the respect and response of the building to the city.

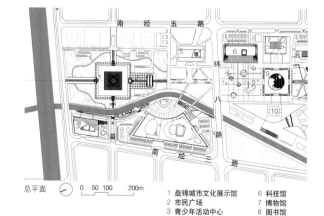

总平面   0 50 100   200m

1 盘锦城市文化展示馆　6 科技馆
2 市民广场　　　　　　7 博物馆
3 青少年活动中心　　　8 图书馆
4 大剧院　　　　　　　9 文化中心广场
5 创意馆　　　　　　　10 会展中心

盘锦城市文化展示馆
Panjin City Cultural Exhibition Center

辽宁 盘锦　　竣工时间 2014 年

建筑设计
沈阳建筑大学天作建筑研究院
沈阳建筑大学建筑设计研究院
中国中建设计集团有限公司
张伶伶 赵伟峰 王靖

摄　影 天作建筑

Architects
Shenyang Jianzhu University Tianzuo Architecture Research Institute;
Architectural Design and Research Institute of Shenyang Jianzhu Universit
China Construction Engineering Design Group Corporation Limited
ZHANG Lingling, ZHAO Weifeng, WANG Jing
Location Panjin, Liaoning **Completion** 2014
Photo Tianzuo Architecture

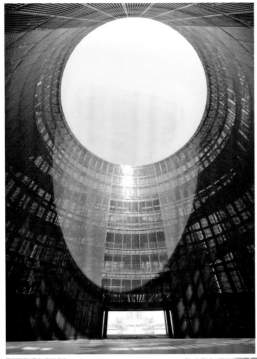

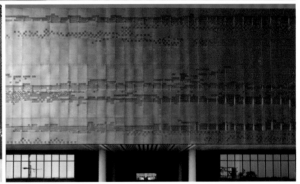

红筒室外展区
室外展区

室内展厅
模拟红海滩的穿孔板表皮

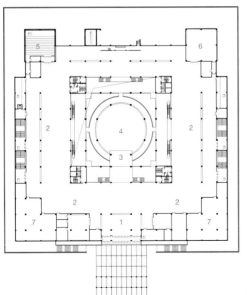

1 入口大厅
2 核心展厅
3 VIP 看台
4 城市展示沙盘
5 4D 虚拟展示中心
6 报告厅
7 精品展厅
8 红筒内庭
9 展厅
10 室外平台

一层平面　0 5 10 20m

二层平面

公共建筑 ｜ 观演/博览

东立面

阶梯广场
鸟瞰

设计的主要构思是将剧院设计成为"文化的万花筒"。如同将周围的光线导入，通过漫反射展现绚烂夺目光影效果的万花筒一般，剧场定位为自然与人、与文化碰撞的华丽场所。在形态操作上，建筑的基本形体以一个100m×100m×34m的立方体形式展开，在基地中构成了中心。所谓"万花筒"是通过5组直径18m的圆筒从不同的方向与立方体进行交错切割的挑空手法，将该空间自然分割，用以规划不同功能的公共空间。从5个圆筒中划分出入口大厅、休息厅、室外平台、地面和屋顶的露天剧院以及通向商业的天桥，并通过这些立体交错的空间构成整座剧院。通过不同的视线角度将周边的风景纳入其中，剧院内看到的风景也会随之出现各种各样的轮廓。大剧院的幕墙宛如一层轻纱包裹着青灰色的混凝土墙体，让建筑若隐若现于自然天地之间，同时通过泛光照明的衬托，凸显清水建筑高贵典雅的气质。

The design concept is "a cultural kaleidoscope". Like kaleidoscope that diffuses light rays to produce a series of dazzling light effects, the design of the opera house complex is oriented as a site for interactions of nature, people and culture. The basic building blocks, sized 100mx100mx34m, become the center of the site. The so-called kaleidoscope is realized through intersecting and cutting 5 cylinders of 18m in diameter that divides the space into various functional areas, hence the entrance hall, foyer, outdoor terrace, the open-air theater, and the overpass connecting the theater complex to the commercial tower originated from the 5 cylinders. Views of the outdoor scenes are introduced from various angles, diversifying the range of visual experiences within the complex. The theater wall evokes the image of grey concrete wrapped in a fine scarf, imbuing the architecture with a sense of natural beauty. The lighting of the theater is simple, highlighting the nobility and elegance of the fair-faced concrete venue.

上海嘉定保利大剧院
Jiading New Town Poly Grand Theater

建筑设计
同济大学建筑设计研究院（集团）有限公司
安藤忠雄建筑研究所

安藤忠雄 陈剑秋 戚鑫

上海 | 竣工时间 2014 年 | 摄 影 章勇

Architects
Tongji Architectural Design (Group) Co., Ltd;
Tadao Ando Architect & Associates
Tadao Ando, CHEN Jianqiu, QI Xin
Location Shanghai
Completion 2014
Photo ZHANG Yong

室外剧场 | 入口大厅连廊
剧场大厅

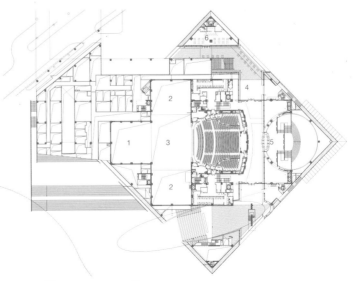

1 后台上空
2 侧台上空
3 主台上空
4 室外平台
5 主门厅
6 咖啡厅

二层平面　0　10　20　50m

入口前广场
自文成公主庙方向看纪念馆

玉树文成公主纪念馆
Memorial of Princess Wencheng, Yushu

建筑设计
中国建筑西南设计研究院有限公司
钱方 黄怀海

Architects
China Southwest Architectural Design & Research Institute Co., Ltd.
QIAN Fang, HUANG Huaihai
Location Yushu, Tibetan Autonomous Prefecture
Completion 2014
Photo Arch-Exist Photography

青海 玉树藏族自治州　竣工时间 2014 年　摄　影 存在建筑

纪念馆位于玉树藏族自治州文成公主庙东北的崖壁之侧。建筑依山就势，以藏式蹬道主体隐喻"天路"，以唐风门楼点明"大唐"，通过两者的结合阐释汉藏融合的理念。唐风门楼在西安建造，采用重走进藏之路的方式运抵现场，与藏式主体相结合，这一建造行为，则进一步回应纪念馆的主题——千百年来汉藏两族延绵不断的动态交流。

建筑主体采用当地毛石砌筑，墙体收分显著，厚重自然；土黄色主调的门楼古朴雄浑。屋盖的单向密肋梁为不规则平面赋予了内部感受的整体性，并形成了屋面的"天路"台阶。人们走过蓝灰色砾石铺就的广场，推开木构大门，可以在转折抬升的台地内部漫步观展，并在顶部倾泄而下的阳光中，感受路径的曲折与空间的变化。

The memorial is located on the cliff side, northeast of Princess Wencheng Temple in Yushu. The building conforms to the terrain, adopting Tibetan-style steps with the metaphoric meaning of "heaven", and using Tang-style gatehouse to clarify the connections with Tang dynasty. The combination explains the concept of a fusion of ethnic Han and Zang cultures. The gatehouse was built in Xi'an and was transported to Yushu, echoing to the theme of constant and dynamic cultural exchange between ethnic Han and Zang communities for thousands of years.

The main building built by local quarry stones with a distinct tapering effect is thick and solid, while the gatehouse built in yellow rammed earth demonstrates tolerance and generosity. Single-direction rib beams densely arranged below the roof unifies the irregular interior plan, which are also used as steps on the roof. When visitors walk through the square paved by blue grey gravels, they can push the heavy door open, enter the memorial and enjoy the exhibition, experiencing change of the interior space.

自纪念馆主入口内看入口广场
展厅内景

1 文成公主庙
2 文成公主纪念馆
3 唐风门楼
4 消防水池及泵房
5 入口前广场
6 公共卫生间及水处理机房
7 原有白塔
8 原有转经楼
9 玛尼石堆
10 庭院

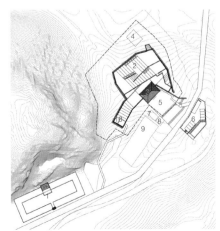

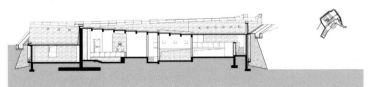

总平面　0 5 10 20m

剖面　0 2 5 10m

建筑与周边民居及远山

　　绩溪博物馆是一座包括展示空间、4D影院、观众服务、商铺、行政管理和库藏等功能的中小型地方历史文化综合博物馆。

　　整个建筑覆盖在一个连续的屋面下，起伏的屋面轮廓和肌理仿佛绩溪周边山形水系，又与整个城市形态自然地融为一体。为尽可能保留用地内的现状树木，建筑设置多个庭院、天井和街巷——营造出舒适的室内外空间环境；内部沿街巷设置东西两条水圳，汇聚于主入口大庭院内的水面；南侧设内向型的前庭——"明堂"，符合徽派民居的布局特征；主入口正对方位设置一组被抽象化的"假山"。围绕"明堂"、大门、水面设有对市民开放的立体"观赏流线"，将游客缓缓引至建筑东南角的"观景台"，俯瞰建筑的屋面、庭院和远山。规律性组合布置的三角屋架单元，适应连续起伏的屋面形态；在适当采用当地传统建造技术的同时，灵活地使用砖、瓦等当地常见的建筑材料，并尝试使之呈现出当代感。

Jixi Museum is a comprehensive local museum of medium size, consisting of exhibition halls, a 4D theater, visitor service, shops, administration and warehouse. The entire building is covered under a continuous roof with undulating profile and texture that mimics the mountains and waters surrounding the county, and fits in naturally with the entire town.

In order to preserve as many of the existing trees on the site as possible, multiple courtyards, patios and lanes are introduced into the overall layout of the building, and creating comfortable outdoor and indoor spaces. Two streams run along the lanes, finally converging into the pool in the large courtyard at the main entrance. In the south part of the building is "Ming Tang", an interior courtyard. Directly opposite the main entrance is a group of abstract "rocks". Surrounding "Ming Tang" is a "sightseeing route" that guides tourists to the "sightseeing platform" at the southeast corner of the building, where they can have a bird's eye view of the roofscape, courtyards and distant mountains. Triangular steel structural trusses adapt well to the undulating roof. Local building materials such as bricks, clay roof tiles are used in modern and innovative ways to pay respect to history yet in response to our own times.

西侧鸟瞰

绩溪博物馆
Jixi Museum

安徽 绩溪　　竣工时间 2013 年

建筑设计
中国建筑设计研究院
李兴钢 张音玄 张哲

摄　影　夏至 李哲 邱涧冰

Architects
China Architecture Design Group
LI Xinggang, ZHANG Yinxuan, ZHANG Zhe
Location Jixi, Anhui
Completion 2013
Photo XIA Zhi, LI Zhe, QIU Jianbing

公共建筑 观演/博览

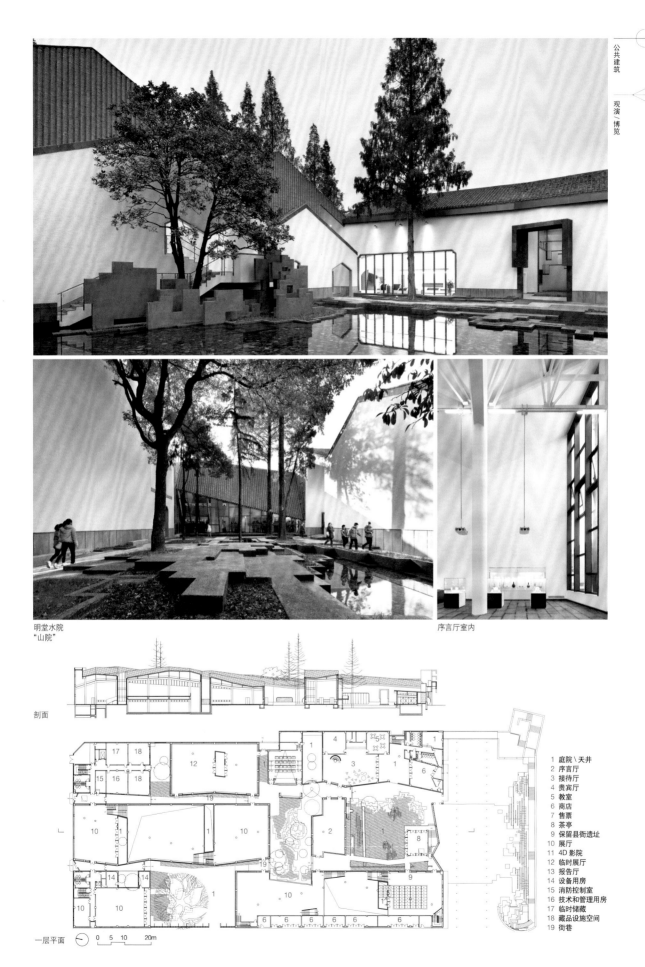

明堂水院 "山院"　　　　　　　　　　　序言厅室内

剖面

一层平面　0 5 10 20m

1 庭院\天井
2 序言厅
3 接待厅
4 贵宾厅
5 教室
6 商店
7 售票
8 茶亭
9 保留县衙遗址
10 展厅
11 4D影院
12 临时展厅
13 报告厅
14 设备用房
15 消防控制室
16 技术和管理用房
17 临时储藏
18 藏品设施空间
19 街巷

南沿街立面

方案构思以贾平凹朴实、内敛的性格为切入点，用简洁大气的外部造型与富于变化的内部空间，使观者充分感受到他作品中传统与现代的碰撞、融合。形体上利用建筑实体之间的旋转、交叉、重叠营造出建筑的空间张力。建筑主体的凹字形平面与单坡的屋顶形式采撷自传统关中民居建筑，同时也巧妙地与他名字中的"凹"字形成呼应。展馆内部空间围绕中心庭院展开，建筑外立面采用充满质朴感的混凝土为主材料，以传统关中民居灰黄色为底色进行打磨，使整个建筑充满历史的厚重感，同时透出时代气息。

The concept of design is inspired by Mr.Jia's simple and introverted personality. The grand facades and interior space are combined for visitors to experience the impact from his traditional and modern work. This unique spatial tension is presented by our design technique including "rotation" "intersection" and "overlapping". This unique plan and shed roof are inspired by his name in Chinese character—"AO". It is the reason why this building has echoes of him. By the way, the indoor space of the exhibition hall is centering around the central courtyard. Material of exterior facades is grey-yellow concrete. It marks this building full of history and modern atmosphere.

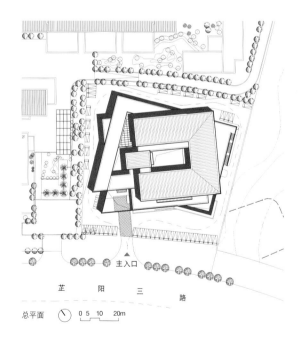

总平面

贾平凹文化艺术馆
Jia Pingwa Culture & Art Gallery

陕西 西安　　竣工时间 2013 年

建筑设计
中国建筑西北设计研究院有限公司
屈培青　阎飞　李大为

摄　影　张广源　屈培青

Architects
China NorthWest Architecture Design and Research Institute Co., Ltd.
QU Peiqing, YAN Fei, LI Dawei
Location Xi'an, Shaanxi
Completion 2013
Photo ZHANG Guangyuan, QU Peiqing

公共建筑 | 观演/博览

主入口庭院
室外楼梯　　　　鸟瞰

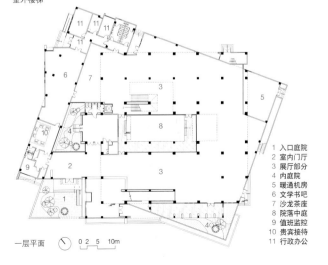

展厅

1 入口庭院
2 室内门厅
3 展厅部分
4 内庭院
5 暖通机房
6 文学书吧
7 沙龙茶座
8 院落中庭
9 值班监控
10 贵宾接待
11 行政办公

一层平面　　0 2 5 10m

南京博物院中轴线景观

南京博物院位于南京中山门内西北侧，于1933年兴建。改扩建方案的设计理念是：补白、整合、新构。"补白"是对不同时期的建筑和场地环境进行分析梳理，将新扩建的建筑恰如其分地布置在合适的位置，从而使得新老建筑以及建筑与场地环境相协调。"整合"，分别通过对新老建筑功能布局、交通流线体系、新老馆内外部空间、新老建筑形式与材料以及展览与休闲功能5个方面进行整合梳理，使其达到一体化。在"补白"与"整合"的基础上，设计重点通过对中轴空间、建筑形式、环境景观的整体塑造达到"新构"的目的。"老大殿"原地抬升3m，既改善了原来建筑低于城市道路3m的不利现状，也减少了地下空间大面积的填挖土方，也为地面上下空间的流线组织创造了有利条件。

Nanjing Museum is located to the northwest side of Zhongshan Gate, first built in 1933. The concept of design is "filling vacancies, integrating all parts and constructing the new". "Filling vacancies" means to seek harmoniousness between old and new buildings by analyzing and classifying the building and site environments of different periods, and by appropriate arrangement of the new and expanded buildings in good places. "Integrating all parts" aims to integrate functional layout between the old and new buildings, the traffic flow system, the internal and external spaces of new and old halls, forms and materials, exhibition and leisure and features of old and new buildings. On this basis, the design intends to realize "new construction" through an overall integration of spaces, architectural form and landscape along the central axis. The base of "Old Main Hall" is uplifted by 3m, which not only solve the problem that the original building was 3m lower than the level of urban roads, but also reduces enormous earthwork and excavation for underground space. Besides, it offers favorable conditions for above-ground and underground circulation organization.

1 老大殿
2 历史馆
3 科研办公及武警用房
4 特展馆
5 艺术馆
6 非遗馆
7 入口广场
8 中央广场

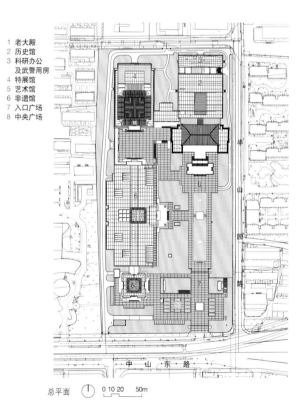

总平面

南京博物院改扩建工程
Nanjing Museum

江苏 南京　　竣工时间 2013年

建筑设计
杭州中联筑境建筑设计有限公司
江苏省建筑设计研究院有限公司
东南大学建筑设计与理论研究中心
程泰宁 王幼芬 王大鹏
摄影 张广源 赵伟伟 陈畅

Architects
CCTN Design; Jiangsu Architectural Design Research Institute Co., Ltd.; Architectural Design and Theoretical Research Center of Southeast University
CHENG Taining, WANG Youfen, WANG Dapeng
Location Nanjing Jiangsu Completion 2013
Photo ZHANG Guangyuan ZHAO Weiwei CHEN Chang

历史馆与老大殿的过渡空间

休息廊看老大殿
连接各馆的地下长廊

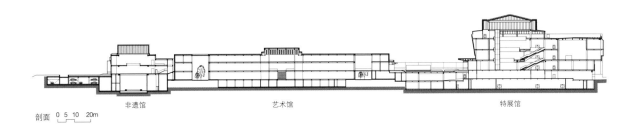

剖面　0　5　10　20m

非遗馆　　　　　艺术馆　　　　　特展馆

艺术馆正立面

全景
从西栅景区入口看乌镇大剧院

本案位于江南水乡梦境似的古镇——乌镇。业主将乌镇设定为国际重要戏剧节的活动据点。剧院设两个剧场：1 200席的主剧院及600座的多功能剧场，它们背对背，满足现代剧场机能却又不显突兀地融入这片古典、精巧的水乡。设计应用代表吉兆的"并蒂莲"的隐喻，将这个寓意祥瑞蓬勃的形象，转化为一实一虚的两个椭圆体量，分别配置两座剧场，重迭并蒂的部分则为舞台区，舞台可依需求合并或单独利用，以创造多样的表演形式。

由于兼具戏剧节表演与观光的双重机能，剧院将满足不同形式的使用需求。访客搭乘乌篷船或经由栈桥步行到达剧院。多功能剧场位于建筑右侧，一片砌上京砖的斜墙，宛如花瓣层迭，包围出剧场的前厅空间；西侧的大剧院则以清透光亮的体量展现对比，折屏式的玻璃帷幕，外侧披覆一圈传统样式窗花，在夜晚泛出的幽幽光影反射在水面上，为如梦似幻的水乡增添另一番风情。

The project is sited in Wuzhen, a historic, romantic and surreal water village. The owner conceived that Wuzhen would be an important place in the global atlas of theaters. The building consists of two theaters of 1 200 and 600 seats, respectively, placed back to back, both of which meet the functional requirements of modern theaters and become part of the traditional and exquisite water village. Using the culturally auspicious "twin lotus" as its metaphor as a vital part of the two theaters sharing one stage area, the design is composed of two oval shapes interlocking one another. The stage can be combined or separated flexibly for various forms of performance.

Given the dual purposes for theater festivals and tourism, the functions of the theaters are multiple. Visitors arrive at the theaters by wooden boats or on foot from an island across a bridge. The multi-purposeful theater on right is shaped by pedal-like segments of thick reclining walls that enclose the foyer, cladded in ancient super-sized bricks. The grand theater on left is enclosed in a zigzag fan-shaped glass front with traditional Chinese window motifs, glowing in the evenings with reflections in water. This adds charm to the already misty and surreal atmosphere of this fairy water village.

乌镇剧院
Wuzhen Theater

浙江 桐乡 竣工时间 2013年

建筑设计
姚仁喜 | 大元建筑工场（会元设计咨询）（上海）有限公司
姚仁喜 袁建平 朱文弘
摄 影 龚娅 陈伯熔 郑锦铭 等

Architects
Kris Yao | Artech
Kris YAO, YUAN Jianping, ZHU Wenhong
Location Tongxiang, Zhejiang
Completion 2013
Photo GONG Ya, CHEN Borong, Jeffrey Cheng, etc.

公共建筑 — 观演/博览

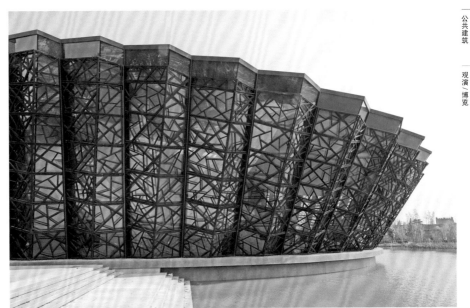

立面细部
大剧场

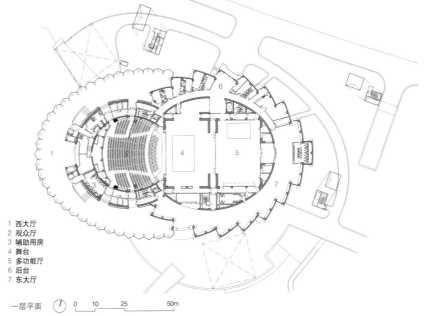

1 西大厅
2 观众厅
3 辅助用房
4 舞台
5 多功能厅
6 后台
7 东大厅

一层平面　0　10　25　50m

三条营街巷视角局部

金陵美术馆的前身是一个建于 20 世纪六七十年代的工业厂房。建筑师通过设计，将旧有厂房转变为一个富于吸引力的中国艺术博物馆，促使工业遗产与传统文化友好相处，并成为历史街区中最有活力的公共设施。

空间组织中，将传统街巷延伸至老工业建筑内部，打通原先封闭的历史街区，形成一个迷人的、开放的城市艺术广场，增进了传统街区的活力。通过空间叠加，将一个寻常的工业厂房，转变为一个迷人的立体花园，创造了一个不同于国际化的独特的艺术空间，将传统生活与中国艺术巧妙地结合在一起。通过精心设计的与传统砖瓦肌理一致的金属打孔板的应用，在工业建筑与传统建筑之间植入一层半透明表皮，巧妙地调和了两类不同遗产的相互关系，修补了传统街区的历史肌理。通过新材料、新技术以及太阳能等新能源的应用，减少能源消耗，减少碳排放，使金陵美术馆成为与环境友好的一个绿色艺术博物馆。

The formal buildings of Jinling Art Museum were built in the 1960s and 1970s. The project aims to renew the old factory buildings as an attractive museum of Chinese art, promoting the coherence of industrial heritage and tradition culture as the most energetic public facility in this historic area.

In spatial organization, traditional alleyways are extended into the buildings to break down closeness of the historic area, forming an intriguing public art plaza to invigorate local communities. Through "overlaying" traditional urban forms, a stereoscopic garden is realized from a normal factory with unique art space combining traditional daily life and Chinese art together. The use of carefully designed pierced metal plates in accordance with texture of traditional bricks creates a translucent skin in between industrial architecture and traditional dwelling houses, cleverly reconciling the relationship between the two types of heritage and adding historical texture to the traditional area. Through the application of new materials, new technology and new energy sources such as solar energy to reduce energy consumption and carbon emissions, it shapes a green and environmentally friendly art museum.

金陵美术馆
Jinling Art Museum

建筑设计
西安建筑科技大学建筑学院刘克成建筑工作室
建学建筑与设计所有公司

刘克成 肖莉 吴超

Architects
Liu Kecheng studio, Xi'an University of Architecture and Technology;
Jianxue Architecture and Engineering Design Institute Co., Ltd.
LIU Kecheng, XIAO Li, WU Chao
Location Nanjing, Jiangsu
Completion 2013
Photo WEN Zongbo

江苏 南京　　竣工时间 2013 年　　摄　影 文宗博

城墙上视角东南透视

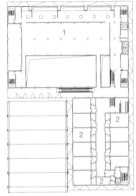

1 南京书画院藏品展
2 美术工作室
3 展厅
4 城南记忆馆

7.4米标高层平面

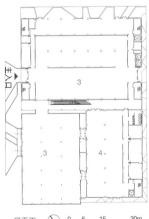

一层平面　　0　6　15　30m

总平面　　0　12　30　60m

东南角局部透视
室外斗形公共活动空间
锯齿形厂房内部

公共建筑　观演·博览

中区院落

核心教学区位于津南新校区中心，包括公共教学楼及综合实验楼东西两个建筑单体，在建筑设计上以"新书院风格"为核心理念，形成融合东方意蕴与时代精神的"现代书院"特征。以院落嵌套的方式演绎"书院"风格，以3组不同高度的院落空间组合，在外部形成清晰组群特征的同时在内部产生了丰富的空间效果。围绕"书院"的组团间道路、文化谷与校园公共道路系统的梳理使区域内实现"人车分流"，方便快捷；结合"书院"内部空间布局形成的立体交通网络，通过坡道、连廊、台地与文化谷不同形制的组合形成趣味性的交通、交流空间，在区域内形成的立体交通体系使内外交通方便快捷的同时强化了内部空间的活跃度。在"书院"内部将不同形式的教学空间、公共空间、休憩交流空间功能充分融合，形成综合体，丰富内部空间层次。

Core teaching area is located in the centre of Jinnan New Campus, including two isolated buildings on both sides of the main north-south landscape axis. The east one is for public teaching and another for complex lab. The core idea for this design is "A new style of traditional Chinese Academy", which in consequence leads to the formation of clear characteristics from outside and various spatial effects in interior by organizing three combinations of courtyard space in different height. The inter-group road surrounding academy groups, the cultural valley and the public road system of the college contribute to the realization of a clear circulation for both people and traffic under the reorganization in this design. A three-dimensional transportation system has been created considering of the interior space, consisting of different connecting forms such as ramp, corridor, platform, valley...Several points are created to activate the interior space and the connection between inside and outside has been enhanced at the same time. All kinds of functions like teaching, activities and communication are well mixed inside the "Academy Courtyard", making the buildings a complex with abundant spatial levels.

南开大学新校区核心教学区
Core Teaching Quarter of the New Campus of Nankai University

建筑设计 同济大学建筑设计研究院（集团）有限公司 原作设计工作室
章明 张姿

Architects Original Design Studio,Tongji Architectural Design (Group) Co., Ltd.
ZHANG Ming, ZHANG Zi
Location Tianjin
Completion 2016
Photo ZHANG Yong

天津　竣工时间 2016 年　摄影 章勇

公共建筑 教育/科研

西立面
中区底层架空层

共享中庭

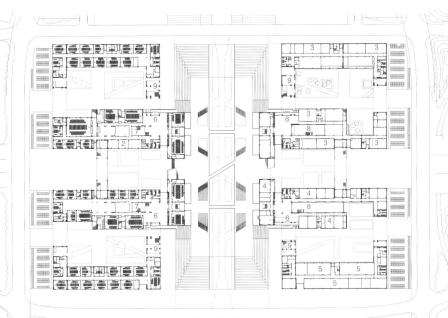

1 普通教室
2 教务办公
3 标准实验室
4 信息办公
5 电信机房
6 入口门厅
7 共享中庭
8 内庭院
9 教学区入口

一层平面　0 10 20 50m

从主入口楼梯看风雨操场

德富中学是位于上海嘉定新城德富路上的一所24班初中。校内建筑物共两栋,分别为主体教学楼、风雨操场及食堂。西面的主体教学楼呈田字形布局,可容纳24班教室、教师办公及附属设施。南北向为主要教室,东西向为特殊教室。建筑从一层到三层向太阳错落,从而形成丰富的屋顶平台区域。主教学楼与风雨操场及食堂采用四条斜向无障碍坡道相连。四个庭院通过建筑底层相互连通,为学校的老师、学生提供一个自由行走的场所。主教学楼采取内外双廊设计,除去基本的垂直交通外,建筑师还设计了丰富的漫游式交通系统,自由舒展的廊道与错落的屋面紧密结合,它们使建筑的内外界限变得模糊起来,行走变得有趣。

De Fu Junior High School is located on Defu Road of Jiading New Town, Shanghai, consisting of 24 classrooms. There are two buildings on campus, i.e. the main teaching building and the gymnasium and canteen. The main teaching building which sits on the west part of the site houses 24 classrooms, teachers' and ancillary facilities. Normal classrooms are oriented north and south while special rooms oriented east and west. The building descend towards the sun from the third floors to the ground floor and generates a vigorous roof platform. Four wheelchair accessible ramps connect the main teaching building with gymnasium and canteen. Four courtyards are connected on the ground floor, providing a space for free walk. A double corridor layout for both vertical circulation and enriched walking system is adopted for the teaching building, and the freely spread corridors are interwoven with terraced roofs closely, contributing to more interesting walking experience that blurs the boundaries between the interior and exterior of the building.

连廊

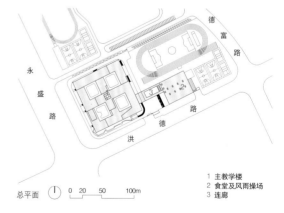
总平面

1 主教学楼
2 食堂及风雨操场
3 连廊

上海德富路初中
De Fu Junior High School

建筑设计 上海高目建筑设计咨询有限公司 / 江苏省第一工业设计院上海分院
张佳晶

Architects Atelier GOM; First Industry Design Institute of Jiangsu Province, Shanghai Branch
ZHANG Jiajing
Location Shanghai
Completion 2016
Photo SU Shengliang

上海 | 竣工时间 2016年 | 摄影 苏圣亮

鸟瞰

1 普通教室　　8 实验室
2 教室更衣间　9 计算机教室
3 生物实验室　10 美术教室
4 化学实验室　11 选修教室
5 多功能教室　12 史地教室
6 劳技教室　　13 教师办公室
7 合班教室　　14 辅助用房

主教学楼一层平面

主教学楼西南庭院
体育馆

从中心庭院室外楼梯观教学楼
宿舍楼外观

设计基于两个主要思想,一是在传统教育制度及现行建筑规范的限制中寻找突破,创造能释放更多自由精神的校园建筑;二是在建筑的当代性和地域性之间寻求一种对话。规划中,两条长廊南北向贯穿校区,校园被长廊分为3个功能区,长廊提供了丰富多变的空间,以激发学生更多行为发生,长廊的屋顶以连续坡折线出现。建筑另一特征是提供了大量户外空间,教室之间的一系列院落提供了游戏空间,教室和公共用房之间的带状室外空间通过架空等多种手段形成了多元立体的交往空间。公共空间部位的分隔和围护墙采用了镂空处理,增添了生动的趣味性。

The design is based on two main concepts: the way to break through current educational system and building codes to release more free spirit on campus, and a dialogue between modernity and locality. In the planning scheme, two long corridors oriented north-south winds through the whole campus, dividing it into three functional sections. The corridor of rich variations encourages students for more activities, with continuous folded roof on top. The other architectural characteristics is the ample playground as a result of a series of courtyards. Partially stilted, the strip outdoor space, in between classrooms and public rooms, provides a stereoscopic space for communication. The partitions and enclosing walls of public spaces are hollowed out in an interesting way.

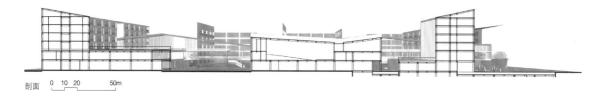

剖面　0　10　20　　50m

苏州实验中学原址重建项目
Reconstruction Project of Suzhou Experimental Middle School

江苏 苏州　　竣工时间 2016 年

建筑设计　同济大学建筑设计研究院(集团)有限公司
曾群 文小琴 汪颖

摄　影　章勇

Architects Tongji Architectural Design (Group) Co., Ltd.
ZENG Qun, WEN Xiaoqin, WANG Ying
Location Suzhou, Jiangsu
Completion 2016
Photo ZHANG Yong

教学楼内院
整体鸟瞰

庭院局部
教学楼二层连廊

中轴连廊内部

一层平面

1 实验室
2 录像室
3 教室
4 校园超市
5 学生餐厅
6 图书阅览室
7 校史室
8 乒乓球室
9 舞蹈教室
10 多功能馆

Public Architecture   Educational Architecture / Scientific Research Architecture

入口广场

项目位于山西兴县新区。建筑物从地面缓缓升起，通过屋顶台阶及斜坡处理手法，建筑屋顶与地面融为景观的一部分。前后错动的形体有利于每个房间及活动场地均能享受到阳光。课堂教育之外的活动空间，在现代教育中显得尤为重要。建筑物从当地窑洞智慧获得启发，上一层用户充分利用下一层用户的屋顶作为生活院子，教学楼每层公共空间均有通向屋顶的开口，在有限的课间活动时间使得学生非常便捷地使用屋顶平台空间，有效拓展学生课外活动场地。墙体构造采用复合墙体形式，表层为山西传统材料——青砖，中间空气隔层有效避免常年温差较大的当地气候对室内舒适度的干扰。

The project is located in the new district of Xing County, Shanxi Province. Buildings are raised from the ground, and the roof becomes part of the landscape through terraces slopes on it. The staggered form allows sufficient sunlight on each room and the playground. Activity space outside the classroom is particularly significant in modern education. Inspired by local cave buildings, users of the upper level can make full use of the roof of the level below as livelihood courtyard. Public space of classrooms on each level has exists to the roof for students to easily access as an efficient extension of after-class activities. Composite walls are used in construction, adopting the grey brick, a traditional material in Shanxi on facades. The air interlayer effectively reduces negative impacts of temperature variance on interior comfort of the building.

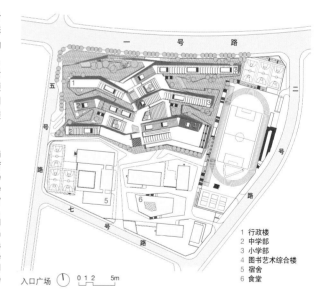

入口广场

1 行政楼
2 中学部
3 小学部
4 图书艺术综合楼
5 宿舍
6 食堂

山西兴县 120 师学校
Instruction Building of the 120th Division School in Xing County, Shanxi Province

山西 吕梁　　竣工时间 2016 年

建筑设计
WAU 建筑事务所
深圳市清华苑建筑与规划设计研究有限公司
吴林寿 赵向莹

摄　影　马明华 战长恒

Architects
WAU Design; Shenzhen Tsching-Hua Yuan Ltd.
WU Linshou, ZHAO Xiangying
Location Lvliang, Shanxi
Completion 2016
Photo MA Minghua, ZHAN Changheng

公共建筑 | 教育/科研

风雨操场活动屋面 | 图书馆
小学部活动屋顶

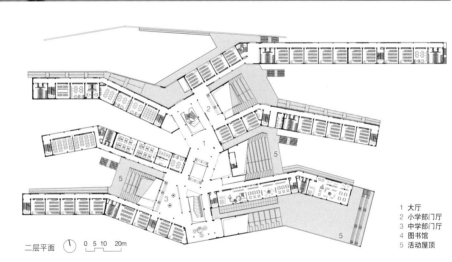

二层平面　0 5 10 20m

1 大厅
2 小学部门厅
3 中学部门厅
4 图书馆
5 活动屋顶

中学部二层门厅　　　　　　　　　　　　　小学部走廊

东南侧夜景

屋顶鸟瞰 | 次入口

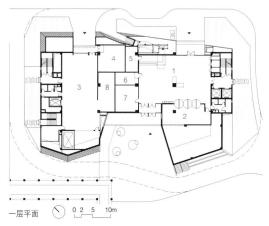

一层平面 0 2 5 10m

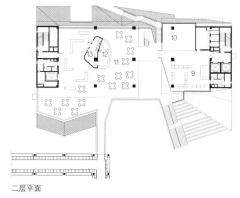

二层平面

1 门厅
2 展厅
3 卸货点
4 消防控制室
5 办公室
6 网络机房
7 标本室
8 仓库
9 便利店
10 多媒体空间
11 公共空间
12 咖啡吧收银台
13 服务室

清华大学海洋中心
Tsinghua Ocean Center

建筑设计
OPEN 建筑事务所
深圳市建筑科学研究院
李虎 黄文菁

广东 深圳   竣工时间 2016 年   摄　影 张超

Architects
OPEN Architecture;
Shenzhen Institute of Building Research Co., Ltd.
LI Hu, HUANG Wenjing
Location Shenzhen, Guangdong
Completion 2016
photo ZHANG Chao

清华大学海洋中心是为清华大学新成立的深海研究创新基地设计的一栋实验及办公楼,其场地位于深圳西丽大学城清华研究生院。设计从科学、人文、自然三个总体概念入手,在保证了所有办公、实验功能的同时,于其间注入现在校园所缺乏的公共机能,以形成充满活力的"垂直校园"。

该建筑由多个研究中心组成,在每两个中心之间插入了一个水平的园林式的共享空间,包括岛屿状的会议室、头脑风暴室、展厅、科普中心、交流中心、咖啡厅等。另外每个中心里的实验室部分和办公服务区又被水平地拉开,形成垂直贯通的缝隙,穿梭其间的室外楼梯将这些水平及垂直的共享空间蜿蜒地联系起来。随着时间的推移,这些共享空间里的绿植也将日渐繁茂,将地面的花园向上一直延伸到60m高的屋顶。

Sited right next to the main campus entrance, Tsinghua Ocean Center, a laboratory and office building for the newly established Deep Ocean Research Base of Tsinghua University, is located at the eastern end of the Tsinghua's graduate school campus in Shenzhen Xili University Town. The design uses science, humanity and nature as the three main themes, guaranteeing all functional requirements for the lab and office with previously lacking social spaces injected into this energetic "vertical campus".

The building is composed of different research centers, with a horizontal garden-like public space in between each two research centers, including island-style rooms for conference, brainstorm, exhibition, scientific popularization, cafe, etc. Labs and offices in each center are separated horizontally, forming a vertical gap to place stairs to connect different horizontal and vertical public spaces together. As time goes by, plants in the shared spaces will flourish and extend from the ground all the way up to the 60m high roof garden.

主入口

望向野生动物园的孔洞
户外楼梯

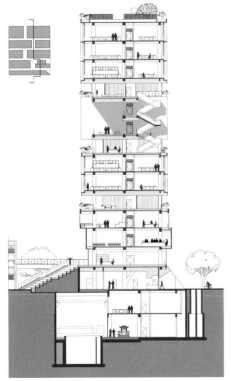

剖面 0 2 5 10m

北侧鸟瞰

水纹立面夜景

外景

天颐湖儿童体验馆
Tianyi Lake Children's Edutainment Mall

山东 泰安 | 竣工时间 2016 年

建筑设计
iDEA 建筑设计事务所
上海同建强华建筑设计有限公司
高岩 郭馨

摄 影 吴清山

Architects
iDEA Architects; Tongjian-Qianghua Architectual Design Co., Ltd.
GAO Yan, GUO Xin
Location Tai'an, Shandong
Completion 2016
Photo WU Qingshan

东北向立面
沿街建筑实景

东西两侧的山包和南侧湖岸把建筑的初始体量挤压成三条弧线。前期研究的城市空间模式在基地上的拓扑变形进一步将项目体量细化拆分，形成多条弧形的组合体量，降低了建筑大体量造成的压迫感，和湖光山色相映成趣。建筑的外立面设计为彩色的抽象水纹图案，水纹取自环境，色彩基于儿童，希望能够为儿童带来富有想象、反复琢磨玩味的立面效果。在立面深化中，充分利用数字化技术进行优化，最终将原本1000多种单板样式减少到17种单板样式，大大缩减了成本及工期。

Two small hills and the lake front compress the original mass into an outline of three arcs. The previous study on urban pattern dismantles and breaks down the mass through the topological transformation of the site, resulting in a combination of various arcs that reduces the pressure of bulk and fits in the surrounding environment. The facade is designed as the colorful ripples pattern. Ripples derive from the lake, while color is based on children. The facade design involved computational generation and optimization, using only 17 different types of panels as a result of the original one thousand samples to effectively shorten the construction period with reduced cost.

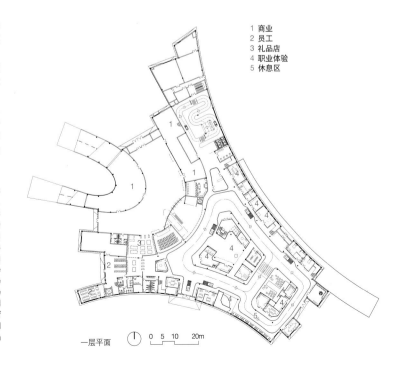

1 商业
2 员工
3 礼品店
4 职业体验
5 休息区

一层平面

东立面

东侧风雨球场下方连通东西的公共活动空间

由于本项目用地呈南北进深小、东西向面宽大的特点，设计将各功能区进行重构、整合，通过风雨廊、中庭、多功能通道等连接，共同形成学校综合体。既节约用地，保证了内部空间的使用效率，又提供了最大的户外活动空间。学校的公共空间位于校园的入口等核心位置，并紧临主要的交通空间，以空间优先的方式强调素质教育的地位与特点。主入口处设计了可供接送家长的休息等待和交流互动的空间，地块中央多功能通道的设置，为解决上下学家长接送车辆的拥堵以及全校性活动的开展提供场所。

The site of the project is characterized by broad width of the east-west orientation, and narrow depth from south to north. All functional sections are reconstructed and integrated in the overall plan, connected by covered corridors, atriums and pathways of multiple purposes. As such, land is saved but the efficiency of use of internal space is guaranteed with maximized space for outdoor activities. Public spaces are located at the entrance of the campus and other crucial positions, close to the main circulation space, highlighting the spatial priority of quality education. At the main entrance, the space for waiting and communication of parents is reserved. The central multi-purpose passage solves the problem of traffic jam caused by vehicles, and provides additional room for school activities.

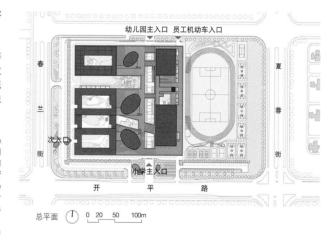

总平面

苏州湾实验小学
Suzhou Bay Experimental Primary School

建筑设计
苏州九城都市建筑设计有限公司
张应鹏 黄志强

江苏 苏州　　竣工时间 2016 年　　摄影 姚力

Architects
Suzhou 9 Town Studio Co., Ltd.
ZHANG Yingpeng, HUANG Zhiqiang
Location Suzhou Jiangsu
Completion 2016
Photo YAO Li

东侧大报告厅与风雨球场之间的院落

空中步行系统　　　　　　　　北部幼儿园入口共享大厅

办公楼外景
园区鸟瞰

三栋研发办公楼在一片开放景观公园内呈组团布局，围合基地中央的餐厅建筑。建筑群风格统一、体量均衡。四层高的办公楼采用"8"字形平面——两个彼此连接的正方形，中间掏空形成内庭院。每个建筑组团的两座内庭院面向园区景观打开，成为贯穿建筑和景观的过渡空间。建筑首层幕墙采用浅色的天然石材百叶板，外形上构成一个坚实的基座，三、四层深色的建筑外立面为金属镶板和封闭玻璃幕墙的组合搭配，玻璃幕墙之间留有细长的通风窗扇。基座与上层建筑之间的缝隙为一道退进式玻璃窗带，为建筑二层带来自然采光。餐厅建筑采用由自然向人工过渡的空间布局，浅色纪念碑式的单体有着独特的个性，表达了作为员工相遇交际空间所承载的开放意义。

Three buildings for research and development are grouped around the canteen at the center of the site, unified by style and balanced in scale. Built on a 8-shaped plan, the four-storied research and development buildings are made up of two intersecting square building volumes, and enclose an entrance courtyard and a garden area. Both the courtyard and the garden area open towards the landscape, creating a transitional zone connecting the buildings and the park. Light-colored natural stone louvers are used for curtain walls on the first floor, designed as a solid plinth, while the third floor and the fourth floor's facades are covered by metal panels in darker color, leaving room for slender ventilation casements inside the curtain wall. The upper stories are visually separated from the plinth by the recessed glazed horizontal band that invites natural light into the second floor. The canteen building features a similar sequence of spaces, from natural landscape to manmade construction. With its monolithic, light-colored appearance, the building presents a different character in accordance with its more public function as a meeting place for the employees.

北京华为环保园 J 地块数据通信研发中心
Beijing Huawei R&D Center

北京　竣工时间 2016 年

建筑设计　中国建筑设计院有限公司　德国 GMP 国际建筑设计有限公司

陆静　林蕾　郑飞

摄影　张广源

Architects China Architecture Design Group; Architecten von Gerkan, Marg and Partner
LU Jing, LIN Lei, ZHENG Fei
Location Beijing
Completion 2016
Photo ZHANG Guangyuan

公共建筑 | 教育、科研

员工餐厅
办公内庭院

餐厅室内

办公中厅

1 研发办公
2 员工餐厅

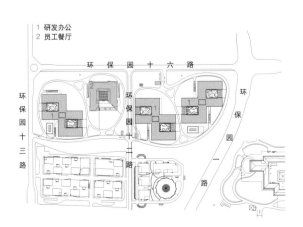

总平面　0 20 50 100m

主入口

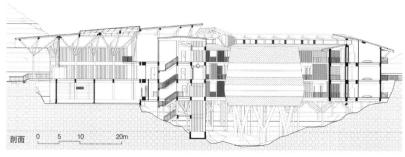

剖面 0 5 10 20m

俯视

  项目基地选址地处谷地，视野开阔，四周的山坡均为大片浓密的松林，地面由裸露的喀斯特地貌岩石和丰富的绿草植被组成。基地内自然高差较大，四周高差在12m左右，场地中央有一处10m深的小窝凼。设计方案充分利用自然地形，减少土方量，节约投资。设计对大片成型的林带及树木予以保护，采用集中式布局，最大限度地保护自然环境。办公、客房、宿舍三位一体的综合楼——一个在深山密林中的科研基站，远离城镇，没有手机信号，没有WiFi，甚至没有电视信号，是一种与现代信息化社会的"与世隔绝"状态。不管是科学家还是普通员工，大家的生活处于一种"同质"状态。因此，建筑师提出了"家"的设计理念，最终的建筑造型汲取了当地民居的设计语言，为不同身份的人设计一个共同的"家"。建筑以木构架为主要特点，采用吊脚楼的形式使整个建筑隐于环境中。

The site is located in the valley of wide view. The hills surrounded are forested by pine trees, and the ground is composed by Karst landform bare rock and rich grass vegetation. Natural height difference is enormous on site, about 12m on perimeters, with a depressed pit of 10m in depth at the center. The design makes full use of the natural terrain to reduce earthwork to save investment. The forest and trees are preserved in the design, and a centralized program is adopted to preserve the natural environment. A complex building contains offices, guest rooms and dormitories is built as a scientific research center far away from urban settings with no WiFi, cell phone or television signals. Insulated from modern information society, both scientists and the common employees live in a homogeneous world. As such, the architects proposes the concept of a mutual home for all users from different walks of lives, deriving architectural language from vernacular houses. The building is characterized by wooden frames, adopting stilted structure to merge itself in the environment.

FAST 工程观测基地综合楼
Comprehensive Building of FAST Observation

建筑设计
中国中元国际工程有限公司
北京构成维森建筑装饰设计有限公司
于一平 李凯 马婕

贵州 黔南　　竣工时间 2016 年　　摄　影 一界建筑摄影

Architects
China Ippr International Engineering Co., Ltd.;
Beijing Gouchengweisen Building Decoration Design Co., Ltd.
YU Yiping, LI Kai, MA Jie
Location Qiannan, Guizhou
Completion 2016
Photo UNO-J Studio

内院
内部环廊

露天平台

二层平面 0 5 10 20m

1 入口大厅
2 贵宾接待室
3 办公室
4 多功能厅
5 会议室
6 宿舍

西立面

小学入口广场
幼儿园东南角外景

这是一所建在坡地上的小学与幼儿园，位于南京岱山保障住宅区内，东侧为城市道路。基地内高差较大，用地面积紧张。

为使小学显得低矮、亲切，原本三层的建筑看上去只有两层高，从而减少了体量的压迫感，照顾儿童视觉感受。设计关注公共活动场所空间的营造，一条8m宽的南北向长廊贯穿建筑，串联成校园的"内部街道"，街道内部形成了两层通高的儿童公共活动交流空间。西侧保持25m的间距的普通教室和东侧采光、隔声要求不高的专用教室通过"内部街道"串联成一个整体，这样布置，既有利于分区，功能相互不干扰，又能做到流线最短，每部分使用者可以很方便到达经常活动的区域。因东侧北侧靠近公路，为了塑造静谧的环境，东侧北侧尽量减少开窗。为增加采光将各种大大小小的庭院穿插在房间之间，共同构成老师、学生互动交往的"小社会"。

为了更高效率地利用土地，幼儿园设计利用退台，将二层以上部分退让出大平台，从而使建筑提供更多分班室外活动场地，并保证每班得到充足日照。此外，建筑中间的内院使一层北侧房间也可得到足够的采光。在立体交通组织方面，人流通过大平台进入幼儿园后，可直接上一层或下一层，方便到达各个班级。在景观方面，每个班级都有对应的绿地供幼儿活动。

West to city roads, the primary school and kindergarten sited on a slope, within the affordable housing district in Daishan, Nanjing. The site has relatively enormous height differences with limited area for construction.

In order to lower the profile of the primary school, the original three-storied buildings look as if one story shorter to reduce the pressure of bulk from a child's view. This design pays full attention to the creation of public space. A north-south corridor of 8m in width runs through the building as an internal street-like pathway, which provides a two-story-high public communication space for children. The internal street connects normal teaching rooms of 25m intervals in the west wing and special teaching rooms with lower acoustic and lighting requirements in the east, so as to facilitate functional division and minimize circulation routes of easy accessibility. Due to the highway on the east and north of the building, the design minimized windows on these two elevations in order to create a quiet environment. To increase lighting, a variety of courtyards of different sizes are put in these rooms as interactive small communities for teachers and students together.

To make full use of the land, set-back terraces are adopted above the ground floor to offer more outdoor playground with sufficient sunshine for each class at the same time. The internal courtyards invites sufficient sunlight into rooms on the northern side on the first floor. After entering the kindergarten from the large platform, people can go upstairs or downstairs with access to each class. A certain area with vegetation is assigned for each class.

岱山小学 岱山幼儿园
Daishan Primary School and Kindergarten

建筑设计
南京大学建筑与城市规划学院周凌工作室
南京长江都市建筑设计股份有限公司
周凌

江苏 南京　　竣工时间 2015 年　　摄　影 侯博文

Architects Atelier Zhouling School of Architecture and Urban Planning, Nanjing University ; Nanjing Changjiang Metropolitan Design Co., Ltd.
ZHOU Ling
Location Nanjing ; Jiangsu
Completion 2015
Photo HOU Bowen

小学内庭院

小学西侧外景

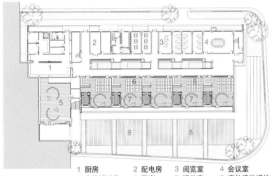

1 厨房　2 配电房　3 阅览室　4 会议室
5 音体活动室　6 卧室　7 活动室　8 室外活动场地

幼儿园一层平面　　0 2 5 10m

1 食堂　2 科学教室　3 普通教室
4 计算机教室　5 多媒体教室

小学二层平面　　0 5 10 20m

1 岱山幼儿园
2 岱山小学

总平面　　0 20 50 100m

幼儿园内庭院

项目位于苏州工业园区独墅湖科教创新区。建筑空间布局采用"城市空间－园林空间－灰空间－室内空间"的形式，完成由动而静、由内而外的过渡。塔楼以共享花园串联，打破垂直方向的疏离。裙房体块虚实错动，园林景观错位叠加，形成移步异景的动态园林空间体系；中庭空间多元导向，是交流空间，是艺术展览空间，亦是多功能活动空间。庭院延伸至屋顶花园，构建水平与垂直双向立体的园林空间。传统园林以现代手法融入建筑室内外空间，在建筑经度与纬度上得到延续，打破建筑的所谓"边界"，延续地域文化的同时，探索城市、园林与建筑的关系，实现有界到无界的升华。

Arts Group Co., Ltd. is located in Science and Education Innovation Park of SIP. The spatial arrangement according to "urban space, garden space, gray space, and interior space", following the transition from movement to standstill and from outside to inside. Space in the tower is connected by the shared gardens, breaking down the alienation in the vertical direction. The podiums are alternately arranged and garden elements are overlaid, froming vivid scenes along the route. The atrium is multiply oriented used as a communicative space, art exhibition space, and multi-functional space. The courtyard extends to the roof garden, constructing a three-dimensional garden space both horizontally and vertically. Traditional garden design is blended in the building by modern methods with architectural continuation both in longitude and latitude, which breaks the so-called "border" of the building. The whole design concerns about the concept of "continuation" of regional culture, and explores the relationship of the city, garden and architecture, sublimating itself from "bounded" to "unbounded".

北侧立面
南侧立面

中衡设计集团研发中心
The Design and Research Building of ARTS Group Co., Ltd.

江苏 苏州 竣工时间 2015 年

建筑设计
中衡设计集团股份有限公司
冯正功 高霖 平家华

摄 影 侯博文 姚力 赵江

Architects Arts Group Co., Ltd.
FENG Zhenggong, GAO Lin, PING Jiahua
Location Suzhou, Jiangsu
Completion 2015
Photo HOU Bowen, YAO Li, ZHAO Jiang

空间层次与空间渗透
图书馆

设计院门厅

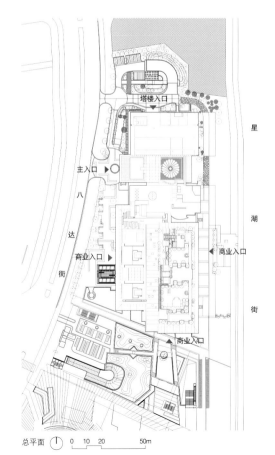

总平面　0　10　20　50m

中庭

南立面
鸟瞰

轴测分析

华东师范大学附属双语幼儿园
East China Normal University Affiliated Bilingual Kindergarten

建筑设计
山水秀建筑事务所
祝晓峰

上海 | 竣工时间 2015 年 | 摄影 苏圣亮

Architects
Scenic Architecture Office
ZHU Xiaofeng
Location Shanghai
Completion 2015
Photo SU Shengliang

一层廊道

教室单元内景

活动空间

走廊

这是一次在用地十分紧张的条件下建设庭院式幼儿园的实践。结合场地情况，建筑师采用了以六边形单元体组合建筑的方式，单元体和庭院是不规则的六边形，其中三个边等长，可以灵活组合。班级单元将教室和卧室两个空间合并，使室内空间成为一个灵活的整体，集中活动围绕中心的圆柱展开，分区活动则可以和六边形的墙面结合。每一间教室都与室外场地相连，二三层通过退台设置庭院。通过精心组织，将各种尺度的室内空间和庭院空间串联在路径上，孩子们的每一次外出都能够通过庭院获得更多与自然接触的机会。

To practice a courtyard-style kindergarten under strict constraints of land shortage, the architect uses hexagonal units for architectural spatial organization. All units and the courtyard are design in irregular six-sided shapes, of which three sides are equal in length for flexible grouping. Classroom and bedroom are combined in each classroom as a flexible large space, and group activities are organized around the central pillar, while activities of smaller groups can accommodate to the hexagonal walls. Each classroom is connected with outdoor playground, and a courtyard is arranged as a result of set-back terraces. Carefully designed routes string up all indoor and outdoor rooms of various sizes so that children have maximized opportunities to approach nature by the way of walking through the courtyard.

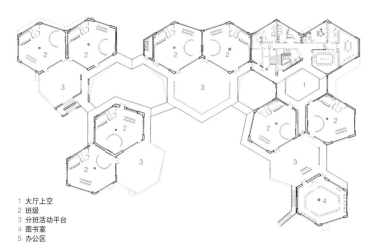

1 大厅上空
2 班级
3 分班活动平台
4 图书室
5 办公区

二层平面

主入口

本案由20个日托班的幼儿园和1个学前师资培训中心组成。总体布局上建筑靠北、东布置，以留出南侧大片的户外活动场地。一层将整3个单体建筑贯通，放置公共活动设施和管理用房。所有日托班都在二、三层的南侧，每个班配置了可延展活动的放大的走廊空间。由于幼儿园的规模较大，如何控制尺度成为设计重点，在形态上将3栋主体建筑以小体量错落叠置，并用双坡屋顶单元的重复拼接来消解建筑的体量，使建筑更接近小房子的抽象聚集。建筑的室内空间、构造、材料及色彩设计都力图为幼儿营造一个温馨的童话世界。

The kindergarten consists of twenty day-care classrooms and one pre-school childhood education & teacher training center. Buildings are placed on north and east, leaving a large area on south for outdoor activities. The ground floor connects all three individual buildings for public service facilities and managerial use. All day-care classrooms are arranged along on the southern side on the second and third floors, with enlarged corridors to extend outdoor activities for each class. Given a relative large scale of the kindergarten, size control becomes the key to the design. The three main buildings are stacked and overlapped with one another, and double-eaved units are used repetitively to reduce the bulk of the buildings as an abstract aggregation of small houses. Architectural internal space, construction techniques, materials and colors contribute to a fairy world for children altogether.

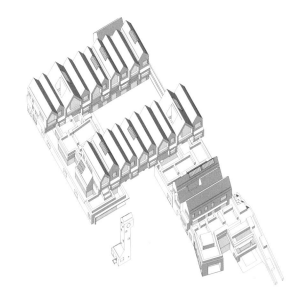

轴测

中福会浦江幼儿园
Pujiang China Welfare Institute Kindergarten

建筑设计　同济大学建筑设计研究院（集团）有限公司　周蔚　张斌

上海　｜　竣工时间 2015 年　｜　摄影　苏圣亮

Architects Architectural Design and Research Institute of Tongji University (Group) Co., Ltd.
ZHOU Wei, ZHANG Bin
Location Shanghai
Completion 2015
Photo SU Shengliang

培训教室
日托活动室

一层平面

0 5 10 20m

1 主入口门厅　　6 更衣沐浴区
2 晨检　　　　　7 音乐活动室
3 图书室　　　　8 活动室
4 室内活动空间　9 室内游泳池
5 多功能厅　　　10 消防控制室

西南外观

夜景

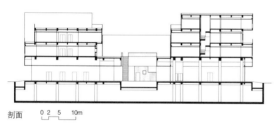

剖面　0 2 5 10m

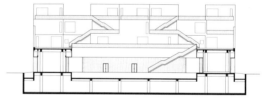

　　基地处于上海西北郊区的安亭新城，整体环境意象不明确，弥漫着一种挥之不去的疏离感。建筑师希望为研发人员营造一个"居所"，给他们带来切实的归属感。

　　设计把单体建筑定义成包含多个环境层次的聚落。每个聚落由研发空间和试制车间上下两个部分叠加而成。研发空间位于2~4层，中央是带形的内广场，构成整个聚落的"中心"，广场两侧集聚了众多通用研发单元，朝内广场这侧逐层后退，形成了一系列露台，被室外楼梯连成整体，成为内广场的有机延伸，并在深度和宽度之外为内广场带来一个向上的维度。建筑一层被分解为大小不等的体量，以内院为核心，游离在地面与研发空间底面所持持的具有强烈水平维度的架空空间内。相对于聚落上部的明亮，这里秩序微妙而暧昧。聚落由此也有了一种垂直性，一种对天空与大地的明确回应。

The project is situated in Anting New Town in the northwestern outskirts of Shanghai. The site is ambiguous by nature, haunted by alienation. The design intends to bring a sense of belonging to the "home" for the researchers and engineers, apart from fulfilling the functional needs and flexibility for various modes of inhabitation.

Each settlement unit consists of two overlapped parts. The upper part ranges from the 2nd to 4th floors, accommodating office units for research and development with a strip plaza in the middle as the core of the whole research units on both sides of the square. The units was intentionally set back on various levels to form terraces, connected by outdoor staircases as an organic extension of internal square, resulting in an upward dimension in addition to depth and width. The first floor is divided into several parts centering around the internal courtyard, arranged in between the space of strong horizontality defined by the ground and research units. The upper part is bright and well organized, while the lower is ambiguous and subtle, hence a sense of verticality in explicit response to earth and sky is formed.

上海国际汽车城研发港 D 地块
Plot D, The R&D and Innovative Port of Anting International Automobile City

上海	竣工时间 2015 年	摄　影　苏圣亮

建筑设计　大舍建筑设计事务所
陈屹峰　柳亦春　宋崇芳

Architects Atelier Deshaus
CHEN Yifeng, LIU Yichun, SONG Chongfang
Location Shanghai
Completion 2015
Photo SU Shengliang

公共建筑 教育/科研

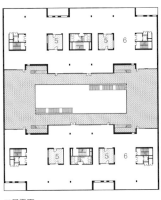
二层平面

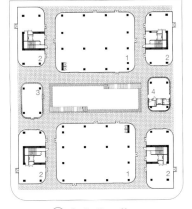
一层平面 0 5 10 20m

1 试制车间　4 公共服务
2 辅助　　　5 庭院
3 设备服务　6 研发

二层内广场沿长边方向
研发区与试制区上下并置

试制区夜景　　　　　　　一层内院沿长边方向

凹凸的围护成为儿童游戏的乐园
童趣园内景

角落中的乐趣

芭莎·阳光童趣园
BAZAAR · Sunshine Playhouse

甘肃 白银 | 竣工时间 2015 年

建筑设计
香港中文大学建筑学院
深圳元远建筑科技发展有限公司
朱竞翔 韩国日 吴程辉

摄 影 刑涛 赵妍 金宇轩

Architects
School of Architecture, The Chinese University of Hong Kong;
Shenzhen Yuanyuan Architecture Science & Tecnhology Co., Ltd.
ZHU Jingxiang, HAN Guori, WU Chenghui
Location Baiyin, Gansu
Completion 2015
Photo XING Tao, ZHAO Yan, JIN Yuxun

这座建筑物运用了轻型预制模块系统，融合了气候、结构、家具、用具、制造、建造、运输的多重考虑。它能够于七天内完成组装，亦能被拆除再易地重组。趣味十足的设计深受孩子喜爱，孩子们可在墙上、地面上的"格子"中攀爬探索。简便的建构过程让没有受过专业训练的村民，甚至学生也可参与建造，大大提升了社区的凝聚力。

This building uses the light prefabricated module system that integrates the climate, structure, furniture, fabrication, construction, transportation and other considerations in design. It can be erected within 7 days, and be dismantled and reassembled in a different location. The playful design makes it a playhouse which children like a lot as they can climb and sit for explorations in the boxes on walls and on the ground. The easy assembling process allows participation of local villagers and even students, which enhances the identity of local communities.

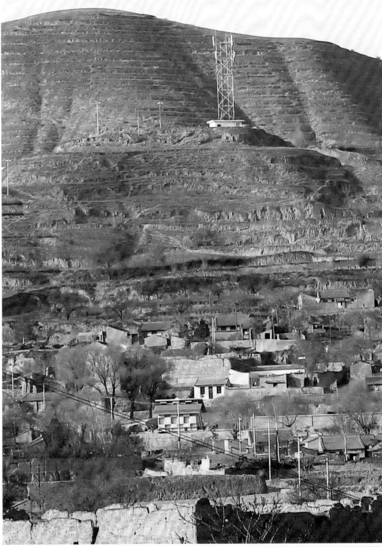

立面

剖面

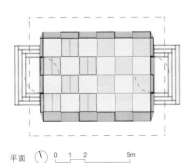

平面　0　1　2　　5m

村落中的童趣园
新建筑与旧式的夯土建筑

东出入口

本案设计方向是"以学生为本",建筑师希望做一个平和、自然、安静、利于交流和使用的建筑。建筑仅有4层,尽量做到压低高度,放大平面,并在尺度较大的平面里挖一个72m见方的院子作为阅读广场。图书馆东、西两侧中部的底层用异形曲面架空,形成通往内院的出入口,人们可以在其中穿行。庭院不仅仅是图书馆的庭院,更像是整个校园的公园。图书馆的空间模式由此变得特殊,不再是传统的那种高大、庄重的形象,而是通过中心"庭院"弱化建筑体量,使其更加平实近人。

The design is student-oriented, in hopes of creating a peaceful, natural and quiet building that facilitates communication and use. The building has only four floors, and the architects try to lower the height of the building while enlarging its plane. A courtyard of 72 m² was carved out in the relatively larger plane as a reading square. On the east and west sides, the building is stilted by curvilinear surfaces on the middle of the ground floor, forming entrances to the internal courtyard for people to walk through freely. The courtyard is not only the courtyard of library, but also the park of whole campus. The spatial pattern of the library hence becomes special—it is no longer tall and solemn, but rather plain and friendly by reducing architectural volume through the central "courtyard".

天津大学新校区图书馆
Library on the New Campus of Tianjin University

天津 | 竣工时间 2015 年

建筑设计
天津华汇工程建筑设计有限公司
周恺

摄 影 姚力

Architects
Huahui Architectural Design & Engineering Co., Ltd.
ZHOU Kai
Location Tianjin
Completion 2015
Photo YAO Li

入口左侧大台阶

临窗阅览空间

藏阅一体空间

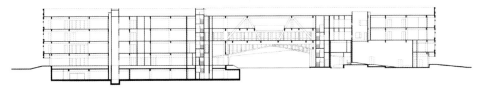
剖面 0 5 10 20m

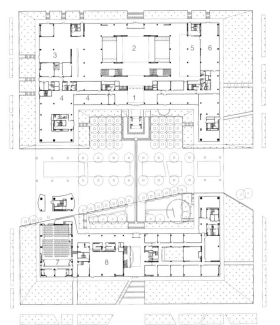

一层平面 0 5 10 20m

1 入口门厅　4 采编中心　7 报告厅
2 公共大厅　5 展厅　　　8 多功能厅
3 新书推荐区　6 电子阅览

内院

校园北区中心院落
校区鸟瞰

浙江音乐学院
Zhejiang Conservatory of Music

建筑设计
浙江绿城建筑设计有限公司
浙江绿城六和建筑设计有限公司
王宇虹 张微 朱培栋

浙江 杭州　　竣工时间 2015 年　　摄　影 姚力 陈兵 赵强

Architects
GAD; GLA
WANG Yuhong, ZHANG Wei, ZHU Peidong
Location Hangzhou, Zhejiang
Completion 2015
Photo YAO Li, CHEN Bing, ZHAO Qiang

图书馆前灰空间
舞蹈学院内院 | 大剧院公共门厅

浙江音乐学院项目沿杭州转塘象山而筑，用地线性展开，城市和自然分别从不同角度对场地进行限定。面对城市边缘这一特殊的场所，设计试图跳出传统大学校园的围城式布局，面向城市和自然，分别采用两种截然不同的边界策略对自然和场地予以回应。远山一侧，设计通过连续的建筑形体限定出校园场地范围。面向校园内部则以富有节奏感的形式语汇结合曲线流动的建筑体型，与沿山一侧的自然山势，以及与自然山势融为一体的地景建筑相互呼应。人工构筑的边界与自然生长的边界在此交融对话，形成了生机勃勃且充满戏剧化场景活力的音乐艺术院校氛围。

Zhejiang Conservatory of Music is built along Xiangshan, Zhuantang in Hangzhou, and the site spreads linearly with both urban and natural restrictions. In the face of the special site on the edge of the city, it attempts to get out of the traditional siege-style layout of campus design. Facing the city and nature, respectively, it uses two distinct border strategies to respond to nature and the site. On the far side of the mountain, the design specifies the area of the campus through a continuous architectural form. Facing the interior of the campus, the design chooses an architectural form combining a rhythmic modal vocabulary with the curvilinear shape, echoing to the mountains and the landscape integrated with it. The natural and artificial boundaries blend with each other in this place and form the vibrant conservatory full of dramatic scenes.

1 大剧院
2 音乐厅
3 继续教育中心
4 综艺楼
5 舞蹈学院
6 戏剧学院
7 图书馆及行政楼
8 音乐学院
9 体育馆
10 食堂
11 人文学院
12 文化管理学院
13 艺术工程学院
14 学生宿舍

总平面 0 50 100 200m

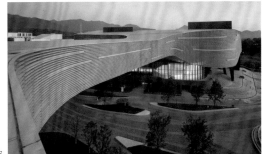

非线性的剧院表皮

建筑西侧学堂路全景

清华大学南区食堂项目，是对清华中心校园环境的一次修补更新，是用传统材料延续校园百年文脉，用现代的建筑语言创造校园公共空间的一次尝试。开放式底层流线的设置，实现了建筑各个界面的平层入口，更通过交通组织让建筑底层成为有围墙的开放空间，为来自校园各方向的师生提供公共活动的场所。精心延续的场地环境文脉，不仅延续了原始的地形和绿化，更保留了东南角的梧桐树，尊重场所历史的同时营造了丰富生动的现代校园生活。与建筑空间结合的可持续设计策略，不仅让建筑室内空间舒适宜人，更降低了建筑的造价与运营能耗。技艺合一的手工砌筑工法，借助传统材料的现代构造工艺，形成了整体而变化丰富的界面，配合校园丰富的生活场景。

The refractory on Tsinghua's southern campus repairs and updates the central built environment in the university, and continues with its context of a history of 100 years with traditional materials as an attempt of modern architectural language in creating public space on campus. The design of open circulations on the ground floor realizes exits of all interfaces on the same level, inviting teachers and students to use this public space. The continuation of the context not only extends original topography and vegetation, but preserves the plane trees on the southeast corner, paying homage to history and creating vivid modern college life. The sustainable design strategies combines design of architectural space, making better comfortable interior space and reducing construction cost and operation energy. Brickworks that combines modern techniques and traditional craftsmanship creates holistic and rich interfaces to supplement vigorous scenes on campus.

室内中庭

清华大学南区学生食堂
Central Canteen of Tsinghua University

建筑设计
SUP 素朴建筑工作室
清华大学建筑设计研究院有限公司
北京中元工程设计顾问有限公司
宋晔皓

北京　　竣工时间 2015 年　　摄　影　夏至

Architects
SUP Atelier; Architectural Design and Research Institute of Tsinghua University; Beijing Zhongyuan Engineering Design & Consultants Co., Ltd.
SONG Yehao
Location Beijing
Completion 2015
Photo XIA Zhi

公共建筑 | 教育/科研

西北场地及楼梯
东侧入口门厅及室内立体交通空间

走廊端头花砖光影

1 入口门厅
2 就餐厅
3 后厨
4 电梯厅
5 多功能厅
6 东侧广场

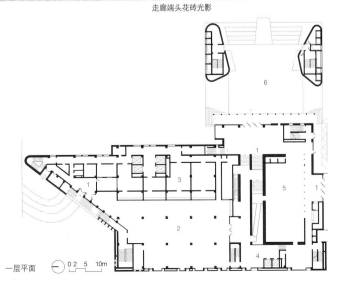

一层平面

中央广场

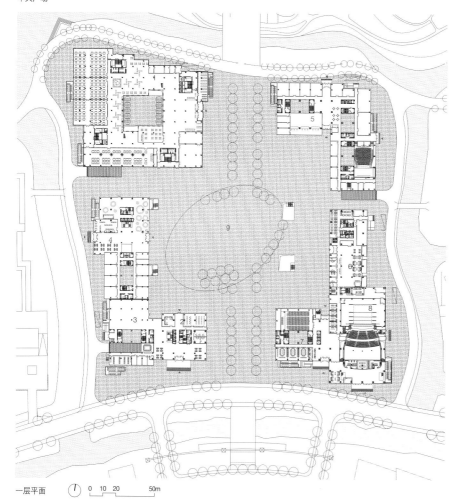
一层平面

1 中心图书馆
2 校总部行政楼
3 档案馆
4 杭州国际城市学研究中心
5 师生活动中心
6 接待中心
7 会议中心
8 大剧院
9 中央广场

杭州师范大学仓前校区
Hangzhou Normal University Cangqian Campus

浙江 杭州　　竣工时间 2015 年

建筑设计
维思平建筑设计
浙江大学建筑设计研究院有限公司
吴钢　陈凌　曲克明

摄影　锐景摄影 张辉

Architects
WSP Architects; The Architectural Design & Research Institute of Zhejiang University Co., Ltd.
WU Gang; CHEN Ling; QU Keming
Location　Hangzhou; Zhejiang
Completion　2015
Photo　Ruijing Photography; ZHANG Hui

西南向鸟瞰

图书馆阅览室北望湿地和山峦
会议中心与宴会厅连廊

百年历史的杭州师范大学在西溪湿地西侧五公里处的大学城建设新校区。新校区规划由十几个独立而又完全开放的学院组成，在这些学院的中心，安排了一个服务全校所有学院的核心区综合体，包括行政中心、档案馆、中心图书馆、师生活动中心、酒店和接待中心、国际会议中心、杭州城市学研究院、成人再教育研究中心等9大功能。

构思从城市设计开始，首先把巨大的场地分割成几个较小尺度的街区，每个街区以一个功能为主，同时各街区相互之间也有联系，满足方便灵活的使用需求。建筑设计基于一个严格的模数体系，用于控制这个复杂功能综合体的所有平面、立面和建造细节，确保用经济合理的成本，实现一个高完成度高质量的建筑，并为日后的运营维护减轻负担。在标准化模数体系和最直接的框架结构之外，建筑呈现给师生的是一个柔和多变的形像。联桥、退台、内庭、挑檐等形式既是功能互联、室外活动、通风、遮阳等使用功能的直接反映，同时也减小了建筑的体量感。

The one-century-old Hangzhou Normal University builds a new campus 5km west of the Xixi (West Creek) Wetland. The new campus planning consists of more than a dozen independent but fully open colleges. In the center of these colleges, a complex is arranged serving the whole campus with functions including an administrative center, the archives, a central library, an activity center, a hotel and reception center, and an international conference center.

The concept of the design starts with urban design. First of all, the huge site is divided into several blocks of smaller scales, each with a specific function. In the meantime, all blocks are connected for flexible use. The design is based on a strict modular system to control over budget and quality of all plans, elevations and other details in such a complex, and to reduce maintenance fees in the future. In addition to the standard modular system and most direct frame structure, the building image as presented to the users is mild and variegated. Connecting bridges, set-back terraces, internal courtyards, and overhanging eaves not only directly demonstrate functions such as connection, outdoor activities, ventilation and sun shading, but reduces the visual effect of volume.

从操场看东北外观

学校的场地一侧是舒缓的河道自然景观,另一侧则是高密度的城市公寓,用地十分紧张。总体设计上将标准化教室组合成3个独立的教学院落,布置在基地西侧靠近城市界面一侧;同时,把图书馆、风雨操场、食堂等管理和公共用房以分组退台跌落方式组织起来,布置在基地靠近自然界面一侧,面向东侧山体层层退台,为室外活动提供了丰富的空间;运动场置于东侧衔接起建筑和自然;通过一个脊椎状的公共空间系统将东西两侧不同的空间串联起来,最南侧形成一个四层通高的开放的入口中庭。这样高密度的水平和垂直相结合的空间模式,构建了一个亲近自然的多层次校园空间。

Peaceful landscape of a river spreads on one side of the site, while urban apartments of high density stand on the other, hence a shortage of land usable. Standardized classrooms are organized as three individual courtyard for teaching in the overall plan, placed on the west of the site close to the city. Meanwhile, the library, an indoor sports field, dining halls and other managerial rooms are organized with cascaded terraces on the other side close to nature, which face the eastern mountains and provide abundant space for outdoor activities. The athletic field is placed on the east side to connect buildings and nature. A spine-like public spatial system strings up various spaces on the east and west. An open 4-story-high atrium at the entrance appears on the southern fringe. Through the spatial combination of high density both in horizontal and vertical dimensions, the college space of multiple layers is hence created in close relation to nature.

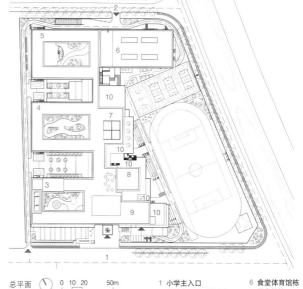

总平面 0 10 20 50m

1 小学主入口　　6 食堂体育馆栋
2 基地车行及后勤主入口　7 图书馆栋
3 专业教学楼栋　　8 报告厅栋
4 普通教学楼A栋　　9 行政楼栋
5 普通教学楼B栋　　10 活动平台

苏州科技城实验小学
Experimental Primary School of Suzhou Science and Technology Town

江苏 苏州　　竣工时间 2015 年

建筑设计
致正建筑工作室
大正建筑工作室
中铁工程设计院有限公司
张斌 李硕

摄　影　陈颢 夏至

Architects Atelier Z+ ; D-plus Studio ; China Railway Engineering Design Institute Co., Ltd.
ZHANG Bin , LI Shuo
Location Suzhou , Jiangsu
Completion 2015
Photo CHEN Hao, XIA Zhi

主入口

活动平台 | 半室外剧场
主入口及平台

南侧立面细部

本项目是清控人居建设集团与英国BRE机构合作的示范项目,目标为建成符合BREEAM标准的近零能耗示范实验建筑。建筑功能集展陈和游客接待中心于一体。为满足建筑作为实验平台的要求,同时减弱现场施工对生态环境的影响,本项目采用多系统并行建造方式,既便于更新和实测,又缩短了现场施工时间。作为可持续建筑的示范,本项目采用被动式的设计原则,优先通过建筑布局、空间、形态、材料的设计来回应当地的气候,以争取更多的自然通风、自然采光,有效地控制太阳辐射等;以简单的方式获取最大的舒适度。鼓励采用当地特有的乡土材料与工艺,例如将传统藤编工艺用于表皮系统。

The THE-Studio is a demonstration project by joint efforts of Tsinghua Holdings Human Settlement Group and BRE of the UK, aiming to build an experimental building of nearly zero energy consumption according to BREEAM standards. The building combines both functions of reception and exhibition. In order to meet the requirements as an experimental platform and reduce ecological impacts of construction, the strategy of a multi-system for simultaneous construction is adopted to facilitate updating and real time testing and shorten construction period. As a demonstration project of sustainable buildings, passive design principles are observed. The design of architectural layout, space, form and materials responds to the local climate to make full use of natural ventilation and sunlight with efficient control of solar radiation. The best degree of comfort is obtained in a simple way. The design encourages the application of local materials and craftsmanship, such as traditional rattan weaving is used on architectural surface.

1 主入口
2 展厅
3 贵宾室
4 办公室
5 观景平台
6 镜面水池

一层平面   0 1 2   5m

清控人居科技示范楼
THE-Studio

建筑设计
SUP 素朴建筑工作室
北京清华同衡规划设计研究院有限公司
宋晔皓

贵州 贵安　　竣工时间 2015 年　　摄　影 夏至

Architects SUP Atelier; Beijing Tsinghua Tongheng Urban Planning and Design Institute
SONG Yehao
Location Gui'an, Guizhou
Completion 2015
Photo XIA Zhi

二层建筑室内局部

室内主展陈空间

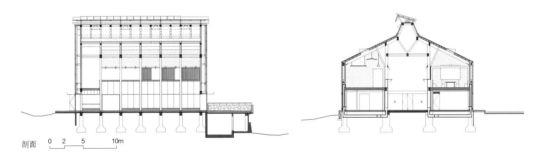

剖面　0　2　5　　10m

建筑西南角外景

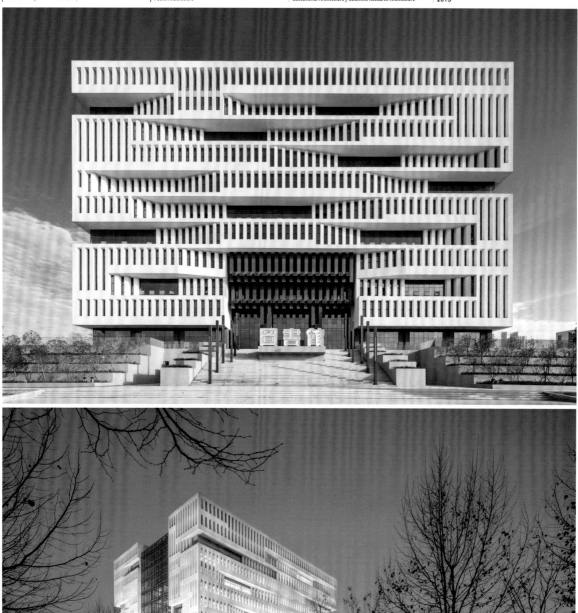

南立面
东北向夜景

武汉理工大学南湖校区图书馆
Wuhan University of Technology Nanhu Campus Library

建筑设计 华南理工大学建筑设计研究院
武汉理工大设计研究院有限公司
陶郅 郭钦恩 陈健生

湖北 武汉 | 竣工时间 2015 年 | 摄影 邵峰

Architects
Architectural Design and Research Institute of SCUT; Wuhan University of Science and Technology Design Research Institute Co., Ltd.
TAO Zhi, GUO Qin'en, CHEN Jiansheng
Location Wuhan, Hubei
Completion 2015
photo SHAO Feng

中庭　　　　　　　　　　　　　　　　　　　主入口

图书馆位于平坦宽阔的校园轴线中心，是整个校园的标志性建筑。建筑必须有足够大的体量创造具有控制力的广场界面。因此建筑师将首层设计为基座——五级的跌台绿化，图书馆主体立于高台之上。图书馆主入口通过对传统斗拱、排架的拆分和重组，以钢结构构架演绎传统木构的形态。东西立面覆以木色金属板，并将校园历史文化通过抽象篆书文字镂刻于其上。以立体庭院的方式营造绿谷的意象，玻璃中庭悬挂绿色垂幔，形成对直射阳光的有效遮挡，也营造出绿意盎然的中庭效果。同时在首层基座跌级绿化设置喷雾，改善图书馆建筑的微气候。

1 综合大厅
2 北门厅
3 跌级阅览区
4 阅览区
5 查新办公服务区
6 读者服务室
7 自动还书
8 书店
9 接待室
10 办公室

The library is located at the center of the flat and broad axis as the landmark of the whole campus. So the building has to be enormous enough to dominate the square. Therefore, the ground floor is designed as the base in form of cascading vegetation with five levels, on top of which sits the main part of the library. By dismantling and reassembling components of traditional bracket units, the main entrance of the library uses steel structure to represent the form of traditional timber architecture. The east and the west facades are covered with wood-like metallic plates, engraved with abstract characters which represent the history and culture of this university. A stereoscopic atrium is placed to display the imagery of a green valley. The glass atrium is decorated with green mantles to effectively block direct sunlight, enriching the atmosphere full of green and vigor. At the same time, water sprayers are set on the base of the building to improve the microclimate.

二层平面　0 5 10 20m

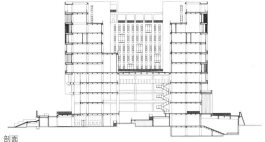

剖面

入口

南侧外观

As a continuation of the planning features of campus culture, the indoor sports field on the Hunnan new campus of Northeastern University is located in the planning area of livelihood sector. The building is raised from edges to the center, showing the characteristics of the planned marginal space and transitional urban area in a balanced and dynamic form. With the introduction of slopes, a large buffering area is placed between the building and the teaching facility to its south. The roof is designed to be covered with vegetation, playing a role of waterproof, thermal insulation, environmental protection to improve the campus landscape environment.

The building creates an entire form of tension in ascendance by expansion and deformation. The contrast of the solid upper part and the translucent lower glass curtain adds a sense of drifting to the building. The playground is an important functional space on campus. Major athletic activities such as basketball, volleyball and badminton are placed in the large space on the east side of the building. Fitness, martial arts, yoga and teachers' offices are arranged on the west, connected by the main entrance hall.

风雨操场位于校园规划分区的生活圈内，整体布局延续了校园文化的规划特征。建筑向内升高，四周收敛降低，用均衡而富有动势的形式体现出规划结构边缘空间与城市过渡空间的特征，并用斜坡的处理方式，使之与南侧的教学楼之间形成了一个较大的缓冲空间。坡顶设计为种植屋面，起到防水、保温、隔热和生态环保的作用，改善校园环境面貌。

建筑造型以整体的伸展和变形，创造出具有张力和升腾感的完整形态。上部实体和下部轻盈通透玻璃幕的对比为建筑增加了漂浮感。风雨操场是校园中重要的使用功能空间，将篮球、排球、羽毛球等主要的学生运动空间设置在建筑东侧的大空间内，健身、武术、瑜伽及教师办公空间布置在西侧的空间内，并将两部用主要入口大厅联系起来。

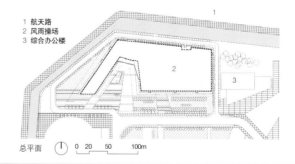

1 航天路
2 风雨操场
3 综合办公楼

总平面

东北大学浑南新校园风雨操场
Hunnan Campus of Northeastern University Gymnasium

建筑设计　哈尔滨工业大学建筑设计研究院
梅洪元　苑雪飞

Architects The Architectural Design and Research Institute of Harbin Institute of Technology
MEI Hongyuan, YUAN Xuefei
Location Shenyang, Liaoning
Completion 2015
Photo WEI Shuxiang

辽宁 沈阳　　竣工时间 2015 年　　摄　影 韦树祥

远景
鸟瞰

室内楼梯　　　　　　　　　　　　　室内球场

操场东面全景

总平面 0 10 20 40m
1 教学楼 3 管理楼
2 实验楼 4 宿舍楼

食堂

北京四中房山校区
Beijing No.4 High School Fangshan Campus

建筑设计
OPEN 建筑事务所
北京市建筑设计研究院有限公司

李虎 黄文菁

北京 | 竣工时间 2014 年 | 摄　影 夏至 苏圣亮

Architects
OPEN Architecture; Beijing Institute of Architectural Design
LI Hu, HUANG Wenjing
Location Beijing
Completion 2014
Photo XIA Zhi, SU Shengliang

宿舍楼
竹园上空

位于北京郊区的一个新城的中心，这个占地 4.5 公顷、拥有开明的管理团队和先进教育理念的新建公立中学，对新近城市化的周边地区起着至关重要的作用。

创造更多充满自然的开放空间的设计出发点，加上场地的空间限制，激发了在垂直方向上创建多层地面的设计策略。学校的功能空间被组织成上下两部分，并在其间插入了 6 个花园。下部空间包含一些大体量、非重复性的校园公共功能，它们从半地下推动地面隆起成不同形态的绿色山丘，支撑起上部的建筑。上部建筑是连通一体的根茎状的板楼，包含了教室、实验室等相对重复性的功能。教学楼的屋顶被设计成一个有机农场。这个项目是中国第一个获得绿色建筑三星级认证的中学。主动和被动节能策略同时被运用于减少碳足迹。

Situated in the center of a new town on the outskirts of Beijing, this new public school occupies an area of 4.5 hectares with an enlightened administration team advocating an advanced educational philosophy as vital elements to the newly urbanized surrounding area.

The design intends to create more open space immersed in nature, and restrictions of the site result in a vertical strategy of separating the programs on various levels. The functional space of the school is organized into the upper and tower parts, inserted by six gardens. The lower part of the building contains large and non-repetitive public functions which push the ground up from semi-underground to form green mounds of various shapes, supporting building above; the upper part is a super slab-like building that contains more repetitive functions including classrooms and labs. The roof of the upper building is designed as an organic farm. The project is the first high school in China that is certified by three-star Green Building Standards. Both passive and active strategies of energy saving are adopted to reduce the carbon emission.

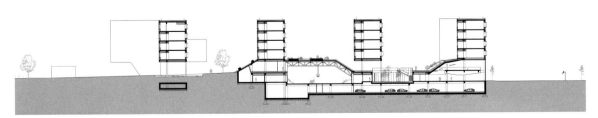

剖面 0 5 10 20m

建筑单体侧立面

组团 A 外景

组团 A 室外环境

松江名企艺术产业园区
Songjiang Art Campus

上海 | 竣工时间 2014 年 | 摄　影 苏圣亮

建筑设计
上海创盟国际建筑设计有限公司
袁烽　韩力　孟浩

Architects Archi-Union Architects
Philip F. YUAN, Alex HAN, MENG Hao
Location Shanghai
Completion 2014
Photo SU Shengliang

名企艺术产业园区的构思是通过创造步行街区与均质绿化体系的系统融合，整理空间布局关系。紧凑式的道路布局可以实现亲近的空间布局与邻里关系；同时，组团绿化编织了场地的基底，设计尤其注重发掘场地上现有河流与场地的对望关系，塑造水景、绿化与建筑相互交融的新场所精神。

建筑单体的设计尝试创造多层高密度的混合型产业园区。在单体建造方面，重点考虑的是降低成本材料及如何与数字化设计相结合。因此在设计中采用了直白的理性几何逻辑，通过简洁的构造方式，实现了传统材料的当代价值。通过对梁、柱、板等基本建筑元素的细节化处理，让结构直接呈现出建筑的本真之美。红色页岩砖、玻璃和混凝土等材料彼此真实反映了建造关系，营造出一种朴素和简洁的整体性氛围。

The design of this project attempts to address the local culture and context by creating new pedestrian districts integrated with the homogeneous green system to optimize urban space. A compact street network realizes close spatial layout and good relationships with the neighborhood. Meanwhile, green areas are weaved into the site in accordance with the existing river, resulting in an integrated space of water, green and architecture.

The design of individual buildings aims at a mixed industrial park consisting of multiple storied buildings of high density, and the focus has been placed on the application of digital design within a tight budget. A rational geometric logic with a simple construction method is adopted to embody the contemporary value of traditional building materials. Through careful treatment of architectural elements like beams, columns and panels on facades, the structure is exposed showing true beauty of the building. The use of red shale brick, glass and concrete amongst other materials reflects the tectonic relationship of different materials, indicating a harmonious atmosphere of the building.

服务中心侧面

总平面

1 服务中心
2 组团A
3 组团B
4 组团C
5 组团D
6 组团E

园区入口服务中心

教学楼西南侧外景
教学楼东南侧外景

  中新生态城滨海小外中学部位于天津中新生态城，是一座拥有36个标准班并达到国家绿色三星标准的低能耗中学。由于学校的用地非常紧张，为节省空间，建筑师尽可能地将各种使用功能紧凑结合布置，基地北侧为综合教学楼，包括教室、实验室、办公室、图书馆、报告厅、食堂、展厅、自行车库等功能。建筑设计中关于生态的考虑一直贯穿始终。作为绿色三星达标建筑，对各种节能设备和节能材料的应用是必不可少的，其中包括高性能的外围护结构，高效的空调机组、风机、水泵、电梯，智能的照明方式，能量回收的新风系统，可再生能源的利用，雨水收集净化再利用，非传统水源的利用，以及各种新型可回收材料的使用等。

As a low energy consumption high school with 36 standardized classes and rated as National Green Building Level-3, Binhai Xiaowai High School is located in Sino-Singapore Eco-City in Tianjin. Since the site area is limited, the architect tries to arrange various functionalities compactly. On the north side of the site is the comprehensive teaching building for multiple uses, including classrooms, laboratories, offices, libraries, presentation rooms, dining halls, exhibition halls, parking space for bikes, and etc. Sustainability is consistent throughout the design. As a National Green Building Level-3 building, the application of energy saving materials and equipments is indispensible, including high efficiency external structure, high performance air conditioners, smoke extractors, water pumps, elevators, intelligent lighting system, energy recycling ventilation system, use of renewable energy, rainwater harvesting and purification system, use of untraditional water sources, and various types of recyclable materials.

中新生态城滨海小外中学部
Binhai Xiaowai High School, Sino-Singapore Tianjin Eco-City

建筑设计　华汇设计（北京）　天津华汇工程建筑设计有限公司
王振飞　王鹿鸣　李宏宇

天津　　竣工时间 2014 年　　摄　影　王振飞

Architects HHDFUN; HHDesign
WANG Zhenfei, WANG Luming, LI Hongyu
Location Tianjin
Completion 2014
Photo WANG Zhenfei

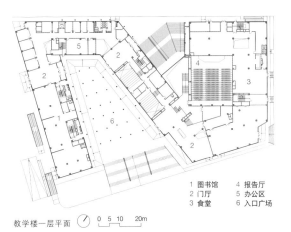

教学楼一层平面　0 5 10 20m

1 图书馆　4 报告厅
2 门厅　　5 办公区
3 食堂　　6 入口广场

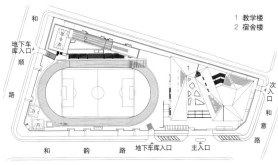

1 教学楼
2 宿舍楼

总平面　0 10 20 50m

教学楼南侧外景
宿舍楼二层平台

教学楼走廊空间　　　　　　　　　教学楼地下一层大厅空间

东向立面外景

西向外景

寒地建筑研究中心是我国东北地区首个寒地建筑科学重点实验室。设计以"平凡表达"的思想传递出建筑形体对环境的尊重、建筑形式对功能的反映以及建构方式与材料之间的逻辑。在保证建筑形体简洁、规整的基础上营造出环境亲和的室外庭院，创造出适宜寒地户外活动的积极空间。屋顶绿化采用适于寒地气候生长的树种，确保屋顶花园冬夏常绿。在建筑材料表达上，采用色彩温暖的碳化木来体现极寒地区的地域特征。经过碳化处理的材料不仅保持了木材原有的质感肌理，也增强了材料的耐久性与耐冻性。局部采用双层错缝构造，提升了建筑的保温性能。

Cold Area Architecture Experiment Center is the first critical laboratory for researching cold area architectural science in the Northeast. The design attempts to express the sufficient respect for environment by the shape of architecture, actual reflection of functions by the form of architecture and construction logic between construction method and materials by "ordinary expression", A friendly courtyard which is an active space suitable for outdoor activities in cold area, guaranteeing a simple and regular architectural figure at the same time is adopted. Vegetation on the roof adopts the kinds of plants suitable in cold area to ensure the roof garden is full of green plants throughout the year. Carbide wood of warm color is used to express the regional characteristic of cold area. The materials treated by carbonization not only maintains the original quality and texture of wood material, but also increases the durability and anti-frost property of materials. Double-layered structure is placed jaggedly on part of the building, improving the thermal insulation properties.

寒地建筑研究中心
Cold Region Architecture Research Center

建筑设计
哈尔滨工业大学建筑设计研究院
梅洪元 王飞

黑龙江 哈尔滨　竣工时间 2014 年　摄　影 韦树祥

Architects
Architecture Design and Research Institute of Harbin Institute of Technology
MEI Hongyuan, WANG Fei
Location Harbin, Heilongjiang
Completion 2014
Photo WEI Shuxiang

公共建筑 教育/科研

室外连廊
建筑入口

门厅

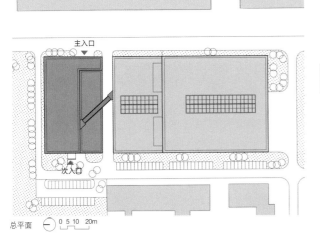

总平面　0 5 10 20m

一层平面　0 2 5 10m

1 门厅
2 全消声室
3 半消声室
4 热环境与节能实验室
5 混响室
6 发声室
7 陶艺造型室
8 展厅
9 物料间

内部街道

为解决复杂功能空间、高容积率与紧张用地之间的矛盾，考虑到现代大学的开放性和综合性，打破艺术院系与工科院系的壁垒，设计将功能按照公共性和使用需求，梳理为公共功能层、平台活动层、标准使用层，并垂直分布，成为既可面向校园开放、又能保证院系使用的新式教育综合体。艺术主楼朝向南侧花园和体育场，局部扭转形成教学楼群的"灯笼"。二层平台的艺术家工作室采用合院与聚落的概念，外墙与屋面以黑色压花钢板及彩色水泥加压板为饰面，形成色彩、质感丰富的新合院。结合艺术馆室外展场、图书阅览及地下实验室需求，设置4个特色下沉庭院。由此形成功能复合、流线分级、尺度亲人的师生学习交流的多维度立体校园空间。

Aiming at eliminating the isolation between different faculties, meeting the complicated functional requirements and, solving the contradiction problems between the high plot ratio and the limited site scale, 3 layers including public level, activity level and functional level are created and arranged vertically. Thus it brings an integrated modern complex open to the campus. The Art College is facing the garden and gymnasium. Part of the building is twisted and shaped as a "lantern", making itself a landmark in the area. The artists' studios, at the second floor, adopted the concept of courtyards and settlements. Its exterior walls and roofs are decorated with black embossed steel plates and colored cement pressboards presenting a friendly and lively posture. 4 sunken courtyards with unique characteristics are designed to meet the spatial need of outdoor exhibition and reading, which offers a multi-dimensional campus experiences to the teachers and students.

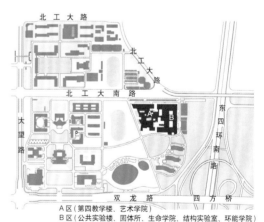

A区（第四教学楼、艺术学院）
B区（公共实验楼、固体所、生命学院、结构实验室、环能学院）

总平面 0 50 100 200m

轴测

北京工业大学第四教学楼组团
A Complex of the Teaching Facilities at Beijing University of Technology

建筑设计 中国建筑设计院有限公司
崔愷 柴培根 于海为

Architects China Architecture Design Group
CUI Kai CHAI Peigen YU Haiwei
Location Beijing
Completion 2014
Photo ZHANG Guangyuan

北京 | 竣工时间 2014年 | 摄　影 张广源

艺术学院教师工作室
艺术家工作室入口平台 | 平台上看教学楼
艺术学院南立面

中庭

东北侧外观

地下一层多功能厅

阅览室

同济大学浙江学院图书馆
Zhejiang Campus Library, Tongji University

浙江 嘉兴 | 竣工时间 2014 年

建筑设计
同济大学建筑设计研究院（集团）有限公司
张斌 周蔚 袁怡

摄　影 苏圣亮

Architects
Tongji Architectural Design (Group) Co., Ltd.
ZHANG Bin, ZHOU Wei, YUAN Yi
Location　Jiaxing, Zhejiang
Completion　2014
Photo　SU Shengliang

东侧外观

二层公共空间连廊

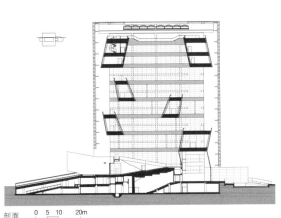

剖面 0 5 10 20m

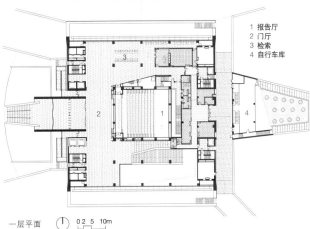

一层平面 0 2 5 10m

1 报告厅
2 门厅
3 检索
4 自行车库

  图书馆在区位上位于浙江学院的核心位置，建筑师将图书馆的体型定义为一个完整的立方体，以"独石"的姿态嵌固在圆形的微微隆起的场地中，只有西侧的主入口前厅以及东侧的室外草坡及小报告厅从独石中伸展出来。方形体量是由南北两侧相对独立但互有沟通的两栋板式主楼和它们之间的半室外开放中庭组成的。中庭的底部从地下层至三层横亘着一条由一系列大台阶和绿化坡地组成的往复抬升的地形化的景观平台，沿东西方向伸展，并将主要公共部分组织在一起。

The project is located in the axial road of the planned campus, defined as a pure cube, like a " monolithic " rock solidly embedded in round yet slightly elevated grounds, with only the main entrance of the west side and the green slope covering the logistics entrance and small lecture hall to the east stretching out from the monolith. The cube volume of the library is composed of two relatively independent buildings on the north and south sides and the semi-outdoor open atrium between them. The bottom of the atrium from the basement to the third floor is organized along a big landscaped platform that follows a series of terraces and vegetated slopes. This platform stretches from east to west, organizing the major public spaces together.

图书馆、餐厅水面景观

中德工程师楼与校史纪念馆

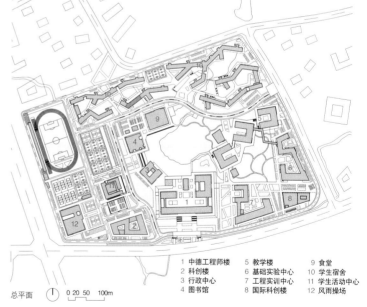

1 中德工程师楼	5 教学楼	9 食堂
2 科创楼	6 基础实验中心	10 学生宿舍
3 行政中心	7 工程实训中心	11 学生活动中心
4 图书馆	8 国际科创楼	12 风雨操场

总平面 0 20 50 100m

浙江科技学院安吉新校区
New Campus of Zhejiang University of Science and Technology in Anji City

浙江 湖州　　竣工时间 2014 年

建筑设计
浙江大学建筑系
浙江大学建筑设计研究院有限公司
秦洛峰　魏薇　沈晓鸣

摄影　俞淳流　韩敬然

Architects
Zhejiang University, Department of Architecture; The Architecture Design and Research Institute of Zhejiang University Co., Ltd.
QIN Luofeng, WEI Wei, SHEN Xiaoming
Location Huzhou, Zhejiang
Completion 2014
Photo YUN Chunliu, HAN Jingran

教学楼组团景观

校区规划为3大功能区块，基地东南部为教学行政区，北部为生活服务配套区，西部为文化运动区。在保留场地内山水地貌景观基础上，依山就势组织建筑、道路，最大限度减少土方开挖。保留了基地内自然水系并进行改造疏导，营造出小范围的湿地校园景观。对校园内建筑风格确定了简洁理性的基调，中德工程师楼、行政楼和文化组团临水布置，形成校区标志性建筑群。教学组团结合坡地灵活布置，半开放的院落建筑组团朝中心绿岛打开，形成安静的教学空间。设计从概念开始就对建筑细部节点、材料及建造效果进行了研究，以保证较高的完成度。

The campus is divided into three parts of different functions: instruction and administration in southeast, livelihood facilities in north, and cultural and sports facilities in west. Natural water system and landscape are preserved, and buildings are organized accommodating the rolling terrain, minimizing earthwork and excavation. Based on a scientific and simplistic rationale, Sino-German Engineering School, the administrative building and the group of cultural buildings are arranged on the waterfront as the landmarks of the campus. The cluster of instruction buildings are placed conforming to the contour of slopes, opening the semi-public courtyards to the central green island as the quiet reading area. The design has studied architectural details materials, and construction effects from the very beginginning to guarantee high quality of the completed works.

餐厅立面
中德工程师楼内院景观

结合了操场看台和景观植被的水平活动界面
学校东侧外观

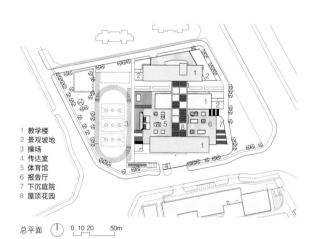

1 教学楼
2 景观坡地
3 操场
4 传达室
5 体育馆
6 报告厅
7 下沉庭院
8 屋顶花园

总平面　0 10 20 50m

1 多功能交通廊
2 会议接待
3 开放式阅览区
4 行政办公

二层平面　0 5 10 20m

北京育翔小学回龙观学校
Beijing Yuxiang Primary School Huilongguan School

建筑设计　北京市建筑设计研究院有限公司
石华　周娅妮　王小工

Architects Beijing Institute of Architectural Design Co., Ltd.
SHI Hua, ZHOU Yani, WANG Xiaogong
Location Beijing
Completion 2014
Photo YANG Chaoying, SHI Hua

北京　｜　竣工时间 2014 年　｜　摄影　杨超英 石华

东侧廊道空间

结合展示功能的坡道空间

共享空间"脊轴"
校园屋顶花园

　　北京育翔小学回龙观学校是一个以48班规模配置的完全小学。这个校园建筑设计的出发点源于对场地、校园学习生活模式的思考，设计希望营造一个绿色的、充满开放空间的校园，为孩子们在这里的快乐成长带来助力。设计将公共性的校园功能空间和单元化的教学空间进行分类，将公共性的校园空间安排在地下一层与首层，通过一个与操场看台和景观结合在一起的二层屋面将这些空间整合在一起，形成一个水平向的校园活动界面；同时，校园半开放的公共空间与教学单元通过一个共享的中庭联系在一起，形成一个垂直向的校园共享空间系统。垂直向共享空间系统与水平向活动界面交织在一起，共同构成了学校的开放空间网络。

Beijing Yuxiang Primary school Huilongguan school is an complete primary school with 48 classes. The design is inspired by the thinking on the site and the modes of education. The architects look forward create a green campus full of open spaces so that children could can be happy growing up here. The architects put the main functions into two categories: one is the public campus space, the other is unit-style campus space. The architects put the public campus space on the ground floor and the first floor beneath the ground. This part is connected with the roof of the second floor through an observation deck of the playground and landscape, as a horizontal interface open to school activities. At the same time, semi-public space and instruction units are connected by a shared atrium, forming a vertical shared space. The horizontal "interface" and the vertical "ridge" are combined, forming an open space system for the school.

教学楼沿南边河流外观

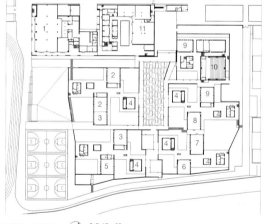

1 学生餐厅	4 教师办公室	7 微格教室	10 报告厅 (359座)	
2 美术教室	5 多媒体教室	8 公开课教室	11 学生阅览室	
3 计算机教室	6 科技教室	9 音乐教室		

中学部一层平面　0 5 10 20m

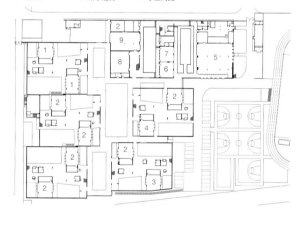

1 公开课教室	4 乒乓球室	7 形体教室
2 教室	5 学生餐厅	8 多功能教室
3 电脑房	6 跆拳道房	9 学生阅览

小学部一层平面　0 5 10 20m

中学部剖面　0 2 5 10m

上海嘉定桃李园实验学校
Shanghai Jiading Tao Li Yuan Experimental School

建筑设计
大舍建筑设计事务所
柳亦春 陈屹峰 高林

Architects Atelier Deshaus
LIU Yichun, CHEN Yifeng, GAO Lin
Location Shanghai
Completion 2015
Photo SU Shengliang

上海　　竣工时间 2015 年　　摄　影 苏圣亮

桃李园实验学校位于嘉定城区以北的开发区内,周边空旷,南侧有一条小河,隔河尚有部分未拆迁的村庄和农田,仍能感受到江南水乡的地理特征。为了与之呼应,设计尝试再现江南传统书院的空间形态,为中小学生营造一处受教与自由天性互动、且具地方气质的校园空间。

学校的每个院子就是一个年级,建筑的上下层采用不同功能和空间相叠加的方式,底层为专业教室及教师办公,上层为普通教室。平台之上是安静和常规的教学场所,平台之下通过部分架空形成可以全天候活动的公共空间,它既和灵活机动的课外教学相结合,又是楼前楼后院落相互渗透的地方,这些架空层让整个校园的地面层成为一个庭院空间整体。

Tao Li Yuan Experimental School is located in an open field in Juyuan New Development Zone, northwest of the old town of Jiading District. The south border is defined by a small river. The geographic features of traditional Jiangnan river villages can still be felt. With an attempt to resonate with the geography of the water village and to reshape the spatial configuration of conventional academies in the lower Yangtze region, the project creates an endemic campus which stimulates the interaction between education and the pupils' instinctive pursuit of freedom.

Each courtyard organizes classes from the same grade. Different functional and spatial layers overlap, with specialized classrooms on the ground level and ordinary ones on the first floor. Above the platform are the quiet classrooms for regular classes while underneath is the public space, leaving the ground floor a free plan for public activities, suitable for all weather conditions. These public space flexibly runs in between classrooms and connects the courtyards within that project, integrating the ground-level space with different courtyards.

中学部主庭院
中学部教学单元内院

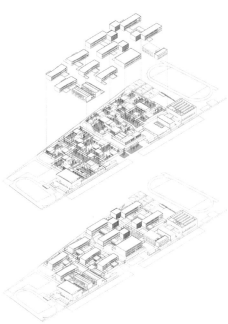

轴测

小学部主庭院
小学部二层平台俯瞰视角

内庭院

北京大学光华管理学院西安分院是除北京本部之外，继深圳、上海之后的第三所分院，选址于西安临潼国家级旅游休闲度假区骊山脚下，景色优美。设计尊重自然环境与地域文化，注重院落的塑造。院落中植入一间北京老宅，作为会客场所，唤起人们对北京大学老建筑的情感共鸣，将北大的人文精神融入建筑，强调高等教育建筑的地域特征与时代特征。学院的几组建筑伴随着地势高低错落，富有节奏变化的坡屋顶彼此交织、相互掩映，自然地融入这片土地，与不远处的骊山组成一幅唯美的画卷。

The Xi'an Branch of Guanghua school of Management of Peking University is the third branch in China in addition to that in Shenzhen and Shanghai, located in a tourist resort at the Mount Li in Lintong District, Xi'an city. The design pays homage to the natural environment and local culture, with an emphasis on the making of the courtyard, in which a compound house in the Beijing style is placed as a meeting venue to arouse emotional resonance towards the old buildings at Peking University. The integration of the humanity spirit of Peking University into the new buildings emphasizes the characteristics of the region and the times. The several clusters of buildings are arranged accommodating to the terrain with the roof profile rich of rhythms, blended with the environment in an organic way, responding to the nearby Mt. Li as a beautiful painting.

1 教学区
2 报告厅
3 学员培训酒店
4 贾平凹文化艺术馆
5 前庭－礼仪广场
6 内院－中心庭院

总平面

北京大学光华管理学院西安分院
Peking University Xi'an Branch of Guanghua School of Management

陕西 西安　　竣工时间 2014 年

建筑设计
中国建筑西北设计研究院有限公司
屈培青工作室

屈培青　高伟　张超文

摄影　张广源　屈培青

Architects
Qu Peiqing Studio, China Northwest Architecture Design and Research Institute Co., Ltd.
QU Peiqing, GAO Wei, ZHANG Chaowen
Location Xi'an, Shaanxi
Completion 2014
Photo ZHANG Guangyuan, QU Peiqing

入口广场
燕园堂主立面

内庭院蔡元培雕像

酒店内庭

园区全貌夜景

软件园位于旅顺南路北侧连绵山丘半山腰的台地上。远远望去建筑表皮整体透明轻盈，走近建筑并仰视，直至天空，建筑是消隐的。抽象构成的建筑手法，应对当今办公空间规模的巨大化，表现出可操作的灵活性。室内平面布局的开放性，垂直空间的流动性，宽走廊相连着会议室、咖啡厅都适宜于北方寒冷地区一日之内室内高效舒适的工作模式。采光与通风分离化的单元式幕墙，确保了建筑外观整体的统一和自然通风调节的便利。把山地环境引入室内，阳光泻下，山风流动，人通过建筑与自然共生。

The software park lies on a terrace halfway up a hill to the north of Lvshun South Road. With a transparent surface, the building looks light, graceful, fresh and clean from a distance. If approached and viewed upward, the building is somewhat invisible. The architectural technique characterized by abstractness has shown operable flexibility in tackling the enormity of office spaces at present. The openness of the internal floor plans, the fluidity of vertical spaces and the connection between wide corridors and conference rooms and coffee houses are suitable for cold regions in north China, creating an environment where staff can stay indoors all day long without going out and work efficiently and comfortably. The curtain wall assembled in units separating lighting and ventilation ensures the overall unity of appearance and the convenience of natural ventilation. The mountainous environment integrates with the indoor environment. With sunshine and wind from the mountains, people stay in the building harmonious with nature.

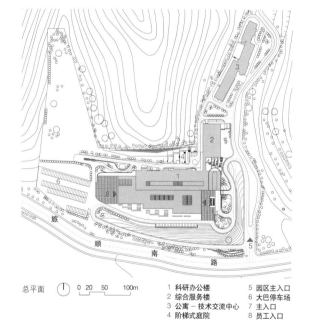

总平面 0 20 50 100m

1 科研办公楼　　5 园区主入口
2 综合服务楼　　6 大巴停车场
3 公寓－技术交流中心　7 主入口
4 阶梯式庭院　　8 员工入口

大连华信（国际）软件园
Dalian Hi-Think (International) Software Park

辽宁 大连　　竣工时间 2014 年

建筑设计　中国电子工程设计院　黄星元

摄影　陈鹤

Architects China Electronics Engineering Design Institute HUANG Xingyuan
Location Liaoning, Dalian
Completion 2014
Photo CHEN He

技术交流中心外景

西侧外景
从北侧看通道过街楼

生态中庭

下沉式庭院

本项目位于同济嘉定校区的教学区和宿舍区相邻的地带。布局上采取了"对偶"的方式。两组学生公寓采用两个5层的h形体量展开，4组不同的中介空间分别形成半围合的内院、下沉广场和中心场地。1栋8层专家公寓尽量收紧，成为局部的制高点。学生公寓按1人/2人间标准布置，但增加了一种合住式6人间"WG"，成为一种变形。内部采光顶空间和直跑楼梯相结合，推动了空间和形式的透明性。建筑外立面运用低成本的面砖和穿孔板阳台栏板，虚与实、红与白对比较为规律。局部运用错峰波浪形的挑阳台形成戏谑气氛，南北形成呼应。

This project is located in the border area between the teaching zone and dormitory zone on the Jiading campus of Tongji University. The method of "antithesis" is adopted in layout and the transition with the neighboring texture is shaped with effort. Two 5-floor h-shaped volumes are adopted for the two groups of student dormitories, with the central symmetry forming echo and 4 groups of different intermediary space forming half-closed garth, sunken plaza and central site respectively. An 8-floor expert apartment building shrinks as much as possible and becomes the commanding height of the area. The student dormitory adopts the standard layout of 1 person/2 person room. But a sharing room "WG" accommodating 6 people is added and becomes a variant. The combination of interior daylighting roof space and straight running stairway is adopted to promote the transparency of the space and form. The building adopts low-cost facing brick and perforated plate balcony sideboard. The contrast of virtual and actual, red and white is relatively regular. The whimsy atmosphere is fostered by alternating peak wave balcony partially, and an echo is formed by south and north part.

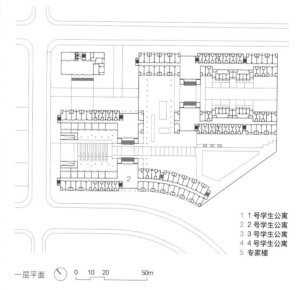

1 1号学生公寓
2 2号学生公寓
3 3号学生公寓
4 4号学生公寓
5 专家楼

一层平面

同济大学嘉定校区留学生宿舍及专家公寓
The Foreign Students' Dormitory and Experts' Apartment in Jiading Campus of Tongji University

上海　竣工时间 2014年

建筑设计
同济大学建筑设计研究院（集团）有限公司
李振宇　刘红　卢斌

摄影　吕恒中

Architects
Tongji Architectural Design (Group) Co., Ltd.
LI Zhenyu LIU Hong LU Bin
Location Shanghai
Completion 2014
Photo LYU Hengzhong

专家公寓北立面
留学生宿舍西入口

留学生宿舍室内

楼间庭院

留学生宿舍室内

从象山向南俯瞰瓦山

项目位于中国美术学院象山校区,是学校的专家楼,建筑内设茶室、餐厅、会议室及客房,临水依山而建。在这样一个实验性的建筑中,融汇了建筑师长期以来对于中国传统文化、建筑、园林的研究和思考,并依靠当代技术使其落地。建筑总体沿河一字排开,以聚落式铺陈,东西、南北方向均设有多条穿越的通道,产生了丰富的空间和视觉效果。建筑探索了生土材料的应用,将夯土建筑的技术应用于墙体。同时,以一个130米长的巨大的木屋架覆盖整个建筑,屋顶铺满青黑小瓦,犹如一瓦山呈现在面前。

Close to the hills and water, this project is sited on the Xiangshan Campus of China Academy of Art, used as a hotel to accommodate invited experts. It contains tea rooms, cafes, conference rooms and guest rooms. In such an experimental building it embodies the architect's research and thinking of Chinese culture, tradition, architecture and gardens in the past decades. Local techniques are used for construction. The building spreads out along the river as a cluster of settlements, and various pathways are arranged in all directions to walk through the building with enriched spatial and visual effects. The application of rammed earth is explored in the design, as seen in the wall. Meanwhile, an enormous wooden truss of span of 130 meters covers the whole building, roofed with grey and black tiles of small size as if a hill of tiles.

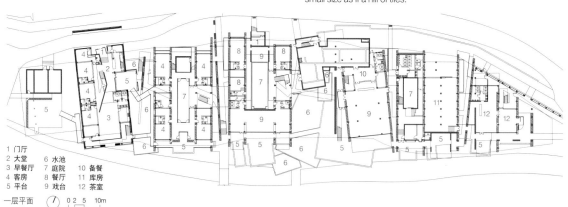

1 门厅　2 大堂　3 早餐厅　4 客房　5 平台　6 水池　7 庭院　8 餐厅　9 戏台　10 备餐　11 库房　12 茶室

一层平面　0 2 5 10m

瓦山——中国美术学院象山校区专家接待中心
Tiles Hill: New Reception Center for the Xiangshan Campus, China Academy of Art

浙江 杭州　竣工时间 2013 年

建筑设计　业余建筑工作室　王澍 陆文宇

摄影　张广源

Architects Amateur Architecture Studio　WANG Shu, LU Wenyu
Location Hangzhou, Zhejiang
Completion 2013
Photo ZHANG Guangyuan

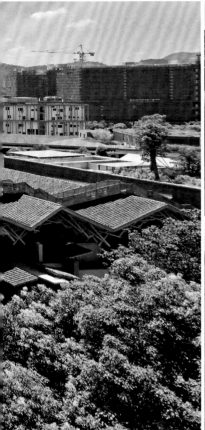

隔岸望瓦山中段
从瓦山顶上山道远望

隔岸望瓦山西尾

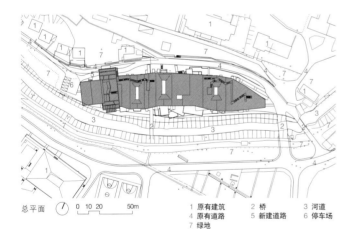

	1 原有建筑	2 桥	3 河道
总平面 0 10 20 50m	4 原有道路	5 新建道路	6 停车场
	7 绿地		

隔岸望瓦山腹内

西立面外景
主入口万物互联广场

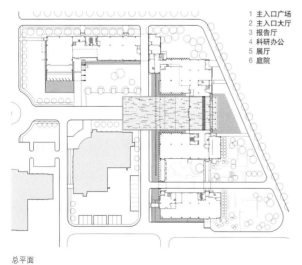

1 主入口广场
2 主入口大厅
3 报告厅
4 科研办公
5 展厅
6 庭院

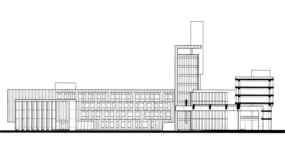

总平面　　　　　　　　　剖面　　0 5 10 20m

南京三宝科技集团物联网工程中心
Networking Engineering Center, Nanjing Sample Sci-Tech Park

江苏 南京　　竣工时间 2013 年

建筑设计
东南大学建筑学院
东南大学建筑设计研究院
张彤　殷伟韬　耿涛

摄　影　姚力　耿涛

Architects
School of Architecture, Southeast University;
Architects & Engineers Co., Ltd. of Southeast University
ZHANG Tong, YIN Weitao, GENG Tao
Location Nanjing, Jiangsu
Completion 2013
Photo YAO Li, GENG Tao

南京三宝科技集团物联网工程中心是该集团自用办公、研发综合楼，场地位于南京市紫金山以东，沪宁高速马群互通东北象限。园区内原有5栋建筑，与周边城乡结合带的碎片化肌理类似，空间环境简单、粗糙，缺乏秩序感和识别性。

项目设计实践的是一种"空间织补"策略，试图在片段性和差异化的空间环境中，重新建构城市肌理的连续性和识别感。新建建筑的L型体形，与原有建筑一起，闭合了园区的中心空间。通过将主楼与入口广场置于东西向道路的终端，揭示了园区中潜在的东西向轴线，建筑的主入口广场成为300米长空间序列的礼仪性终端。由此，园区外部环境的边界与中心、轴线的转接得以明确，空间的秩序建立起来。在建筑的立面设计中，4种特别设计的材质——定制陶板（复合窗）墙面、框栅玻璃幕墙、金属网板遮阳表层与锯齿板遮阳立面，成为"空间织补"的组织性元素。它们在视觉和触觉的多个层级上叠合，构成另一种经纬，参与室内外空间的组织，赋予空间织补以可感知的丰富肌理。

Located to the eastern part of Purple Mountain and in the northeastern quadrant of Maqun on Shanghai-Nanjing Highway, Networking Engineering Center, Nanjing Sample Sci-Tech Park consists of offices, a complex of research and development of this company. There used to be 5 buildings in the park, fragmented as the surrounding urban texture on the fringe of the city, and the space used to be simple, rough, lacking order and identity.

This project practices the design strategy of rear-striking "space re-fabrication", endeavoring to reconstruct the continuity and identity of the urban texture. The new L-shaped building connects with the existing buildings, enclosing the central part of the park. By locating the main building and entrance plaza at the end of the road oriented west-east, the underlying axis of the area is revealed. In the design of façades, 4 special materials are used—customized ceramic panels (compound windows), glass curtain walls in framed grids, shading surfaces of metal mesh and shading façades of serrated panels, together making the organizational elements of space re-fabrication. Unlike individual enclosed volumes, they overlap with one another on multiple levels both visually and tactually, and form another interweaving structure as part of the organization of indoor and outdoor space with enriched texture on space re-fabrication.

北翼的陶板复合窗立面
阅读空间
办公空间

校园群体建筑实景

本项目位于辽东湾新区起步区，南临辽东湾体育中心，与营口市隔辽河入海口相望。校园规划形成了"一轴""三核""八区"的总体结构：长达870米的彩虹长廊沿着校园水系纵贯南北，形成空间秩序的控制主轴，标志性建筑教学主楼、图书馆、国际交流中心分别构成校园中、北、南部的功能与景观核心，统领文理科教学、生活、科技交流、图书信息、行政、后勤保障等8大功能分区。作为新区引导性建设项目，大学承担整合区域秩序、激活城市未来的引领作用，在设计构思中强调衔接城市功能与空间结构，塑造以水系与院落为主要特征的校园空间秩序，以建筑与景观手段传达滨海湿地特色的地域文化，营造展现当代高校本体文化导向的场所精神与人文氛围。

The project is located in the front area of Liaodong Bay New Town, on the south of Bay Sports Center and opposite to Yingkou City across the bay. The campus planning is based on a "axis", "three cores", and "eight sectors". A 870-meter-long rainbow-like corridor winds throughout the campus river as the dominating axis. Landmark buildings such as instruction buildings, the library, the international exchange center constitute the functional and landscape core of the parts in center, north and south, respectively, overseeing the eight sectors for teaching, livelihood, scientific exchange, library information, administration, and logistics. As a new guiding project of the new town, the construction of the university is responsible to integrate regional order and invigorate the new urban area for the future. Continuation of functional and spatial structure has been emphasized, and the water system and courtyards are constructed as the main spatial characteristics of the university. The message of regional culture is conveyed by architecture and landscape, contributing to culture-oriented genius loci and the cultural atmosphere.

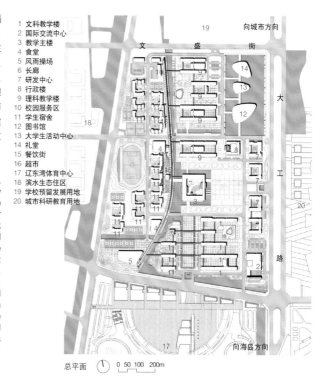

1 文科教学楼
2 国际交流中心
3 教学主楼
4 食堂
5 风雨操场
6 长廊
7 研发中心
8 行政楼
9 理科教学楼
10 校园服务中心
11 学生宿舍
12 图书馆
13 大学生活动中心
14 礼堂
15 餐饮街
16 超市
17 辽东湾体育中心
18 滨水生态住区
19 学校预留发展用地
20 城市科研教育用地

总平面　0 50 100 200m

大连理工大学辽东湾校区
The Liaodong Bay Campus of Dalian University of Technology

建筑设计 沈阳建筑大学天作建筑研究院；沈阳建筑大学建筑设计研究院；黑龙江省建筑设计研究院；辽宁省建筑设计研究院

张伶伶　赵伟峰　刘万里

辽宁 盘锦　｜　竣工时间 2014 年　｜　摄影 天作建筑

Architects Shenyang Jianzhu University Tianzuo Architecture Research Institute; Architectural Design and Research Institute of Shenyang Jianzhu University; Architectural Design and Research Institute of Heilongjiang Province; Architectural Design and Research Institute of Liaoning Prov.

ZHANG Lingling, ZHAO Weifeng, LIU Wanli

Location Panjin, Liaoning　Completion 2014
Photo Tianzuo Architecture

校区整体鸟瞰
漂浮在水面上的图书信息中心

教学区跨越水系的连廊系统
图书馆内庭院

教学主楼西南向外景

内院透视图
局部透视图

太阳能板局部透视

院落式的室外环境为科研人员创造了丰富的交流空间。生态、环保理念贯彻于园区设计的始终，并着力发挥企业优势。光伏发电是业主非常有特色的技术领域，设计用 3 万 m^2 太阳能光伏电池板覆盖所有研发试验单元的屋盖，并将其引入到整体建筑造型设计当中，太阳能技术总安装容量 2.54MWp，实现风电、光伏并网运行，利用风能、太阳能节约运行能耗，可提供 18% 的建筑用电，使洁净能源的使用真正成为园区的主题并发挥实际作用，同时也突出了企业的自身特色和技术优势。

The outdoor environment with courtyards supplies more space for communication amongst scientific researchers. The concepts of ecological, environmental protection sustain all stages of the design, maximizing the advantages of the enterprise. The owner's technology of photovoltaic power generation is distinctive, and the design uses 30,000m^2 solar photovoltaic panels to cover the roof for all research and development units, essential to the architectural form. The total capacity for the solar equipment is 2.54MWp, realizing the paralleling operation of solar and wind energy to cover 18% of electric consumption of the building. As such, clean energy has been a true theme of the park, highlighting the technological advantage of the client.

国电新能源技术研究院
Guodian New Energy Technology Research Institute

建筑设计
北京市建筑设计研究院有限公司
叶依谦 刘卫纲 薛军

Architects Beijing Institute of Architectural Design
YE Yiqian, LIU Weigang, XUE Jun
Location Beijing
Completion 2013
Photo YANG Chaoying

北京 | 竣工时间 2013 年 | 摄影 杨超英

东南方向外景
交流中心

科研楼夜景

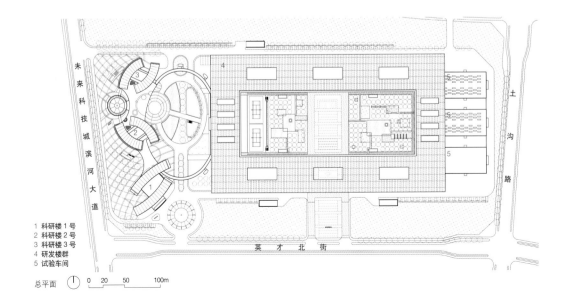

1 科研楼1号
2 科研楼2号
3 科研楼3号
4 研发楼群
5 试验车间

总平面

大厅与北侧古树与远山的对景

西北角街景

西南凉廊区与中心体量南侧外景

大乐之野庾村民宿
Lostvilla Boutique Hotel in Yucun

建筑设计　直造建筑事务所
水雁飞　苏亦奇　马圆融

浙江 湖州　　竣工时间 2017 年　　摄影 陈颢

Architects Naturalbuild
SHUI Yanfei, SU Yichi, MA Yuanrong
Location Huzhou, Zhejiang
Completion 2017
Photo CHEN Hao

项目位于莫干山镇庾村国营时期蚕种厂的西南角。场地由于各种历史地权的遗留问题，呈现一种极不规则、犬牙交错的边界状况。设计的考量是在调和当下周边多变的限制，以及对未来动态的预判下同时来展开的。内化的策略不仅是建立对于不利外部的防御性，反之也让被渗透的内部成为景观中的一部分。同时希望为小镇提供一些可共享的公共空间，例如咖啡厅、小型展览空间设置独立的出入口，以及餐厅放置在三层的做法，这也造就了公共区与酒店之间特殊的流线关系和多样的游走体验。场地上的百年香樟和梧桐树，与远处的山景成为多角度反复借用的视觉对景。窗景的思考延伸到每个房间，获取了不同的客房平面和窗户视线的关系。结构上采用了砖混与钢木屋架的混合体系，在造价与景观视觉上取得相对较优的平衡。

The Project is Located at the southwest corner of the previously state-owned silkworm farm in Yucun, Moganshan. Concerning the existing rural land property problems as a result of the past, the boundaries of the site are very irregular, interwoven with others' properties. The design is motivated by both of the current dynamic constraints and the unpredictable status of surroundings. The strategy of internalized scenery, is not only to make a defense against the unfavorable exterior, but also turn the saturated inner space into part of the landscape. Meanwhile, the design hopes to provide public space to share with the local, such as café and small exhibition areas with separated entrances, and also the restaurant on the third floor. The spatial arrangement creates an unique circulation between public space and hotel area, and provides special experience along the routes for the guests. Taking advantage of the site's 100-years-old Camphor and the Platanus tree and even the distanced mountains, the design frames scenes to create a sequence of dynamic views from the building. A mixed system of masonry structure and the truss of wood and steel is adopted, maximizing the balance between construction cost and the use of landscape.

大厅休息区
客房室内

咖啡厅入口
餐厅内景

1 咖啡厅
2 公共庭院
3 大厅
4 客房
5 厨房
6 内部庭院
7 凉棚
8 戏水池

一层平面　0　5　10　20m

西南鸟瞰

设计将办公、便民服务、报关、会展、金融等使用需求整合为5个独立的功能实体，将这5个实体相互分离产生不均质的虚空，这些虚空被定义为共享空间。三层通高的中央大厅将这些空间串联起来——空气是流动的，空间也是流动的，建筑与环境之间、人与人之间的隔绝被消解了，创造出具有参与性、开放性的人性化公共服务性建筑。近地楼层坡形屋面绿化与平台结合，营造出多标高、层次丰富的活动空间，与内部共享空间形成良好的延续和沟通。借景的精心组织，展示了令人震撼的芭蕉湖山水长卷。各立面针对不同朝向分别采用最有效的遮阳系统，通过巧妙设计，使其成为了精美立面的一部分。采用大量被动技术，并利用火电厂废热蒸汽形成大"三连供"的节能模式。

Functional spaces such as office, public service, customs declaration, conference and exhibition, and finance are integrated into the building with five independent parts, apart from one another with heterogeneous voids that can be defined as public sharing spaces. The three-storied central hall connects these voids, where air can flow everywhere and the whole space is flowing, too. The isolation between the building and the environment is removed, while the separation of people from each other is eliminated, resulting in a humanized public service building with a more open atmosphere inviting public participatory. The combination of sloping green roofs and platforms creates a multi-leveled space of rich activities, which echoes and extends the public sharing space inside. The scenes afar are borrowed into the building, displaying the breathtaking stroll of painting of the Bajiao Lake and surrounding hills. All façades are covered with effective shading systems according to different directions as part of the unique exquisite ornamentation. The architects utilize passive technologies and the recycling of waste heating steam becomes a large CCHP energy-saving mode.

湖南城陵矶综合保税区通关服务中心
Hunan Chenglingji Free Trade Zone Customs Clearance Service Center

湖南 岳阳　竣工时间 2017年

建筑设计
上海建筑设计研究院有限公司
苏昶　谭春晖

摄影　胡义杰

Architects
Institute of Shanghai Architectural Design & Research Co., Ltd.
SU Chang, TAN Chunhui
Location Yueyang, Hunan
Completion 2017
Photo HU Yijie

鸟瞰
四层屋顶山水长卷

1 共享大厅
2 便民服务中心
3 检验检疫报检大厅
4 海关报关大厅
5 展厅
6 支行营业厅

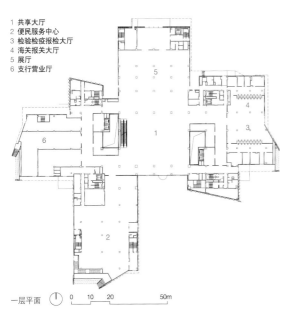

一层平面　0　10　20　50m

中央大厅

开敞错落的南立面

湖上村舍位于苏州阳澄湖畔，是一个集住宿、餐饮、会议、咖啡茶座为一体的建筑。建筑的外立面根据不同的环境有着相对的特征，北侧以实墙面为主，最大限度地隔绝干扰，在入口处形成相对封闭的神秘感；南侧面向水面则是非常开放的姿态。从入口进入建筑的公共空间，再到开放水面旁的餐饮和住宿空间，形成一个开放－封闭－再开放的空间序列。第一层开放的空间由一座长达25米的引桥穿过樱花林带入建筑，形成一个慢慢过滤外界世俗干扰的过渡空间。在封闭的公共空间中，不同高度的高侧光照亮整个建筑室内，形成了具有精神性的光的容器。接待区域被设置在1.8米的高度之上，和一层、二层分别只有半层的高差，视线上得以连续，限定出功能不同却互相贯通的空间。光线随着时间和季节的变化在公共空间中自由变换，完全裸露的混凝土天花和北侧内墙成为接收并反射光线的媒介，空间性和材料性在这里产生了共鸣。

The House by the Lake is located on the lakeshore of the Lake Yangcheng in Suzhou, with functions such as B&B, dining, conference, handcraft workshop, bookshop, café and so on. The facades of the building have distinct characteristics according to the environment. The northern solid wall minimizes interruptions from the outside, developing a relatively closed mysterious atmosphere at the entrance. The southern side is opened up to the water. The spatial sequence that undergoes openness, closedness and openness again is invented as visitors enter the building and pass by the dining and hotel space close to the open waterfront. A 25-meter-long bridge introduces people into the building by crossing the woods. The relatively closed space is lit by the high windows on different levels from the south side. The space is a vessel of light with spiritual quality. The reception was set on the level of 1.8 meters, between the 1st and 2nd floor. The space is continuous while well defined. The sun light is constantly changing during the day and the year. The exposed concrete soffit and the inner walls are reflective medium for the light, creating a resonance between space and material.

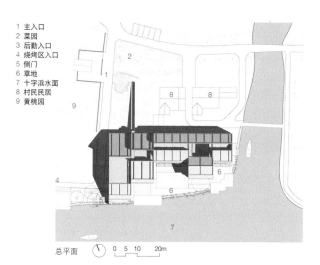

1 主入口
2 菜园
3 后勤入口
4 烧烤区入口
5 侧门
6 草地
7 十字浜水面
8 村民民居
9 黄桃园

总平面

东南方向看湖舍

湖上村舍
The House by the Lake

江苏 苏州　　竣工时间 2016年

建筑设计　苏州个别建筑设计工作室　王斌

摄影　侯之　唐伟　陈灏

Architects Group BE, Suzhou　WANG Bin
Location Suzhou, Jiangsu
Completion 2016
Photo HOU Zhi, TANG Wei, CHEN Hao

从入口前台看向低半层的手作活动区
住宿区一层走廊

住宿区二层走廊

带阁楼的客房

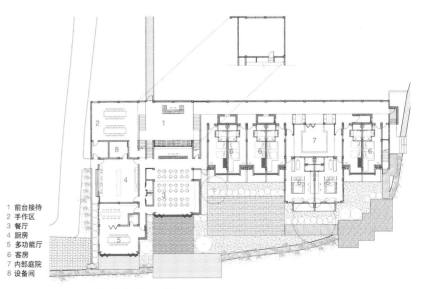

1 前台接待
2 手作区
3 餐厅
4 厨房
5 多功能厅
6 客房
7 内部庭院
8 设备间

一层平面　0　2　5　10m

北侧鸟瞰

东立面晨景

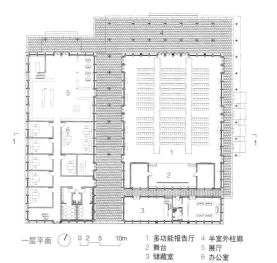

一层平面　0 2 5 10m

1 多功能报告厅　4 半室外柱廊
2 舞台　　　　　5 展厅
3 储藏室　　　　6 办公室

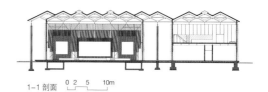

1-1 剖面　0 2 5 10m

石塘互联网会议中心
Shitang Village Internet Conference Center

江苏 南京　　竣工时间 2016 年

建筑设计
张雷联合建筑事务所 / 南京大学建筑规划设计研究院有限公司

钟华颖 张雷

摄影 侯博文 姚力

Architects
AZL Architects; Institute of Architecture Design and Planing Co., Ltd. of Nanjing University
ZHONG Huaying, ZHANG Lei
Location Nanjing Jiangsu
Completion 2016
Photo HOU Bowen, YAO Li

乡村发展的现代化、城镇化趋势不可避免地需要引入新的功能类型，互联网会议中心等大空间多功能建筑是其中之一。江宁石塘村项目以"公社礼堂"及"温室大棚"为原型，尝试重构乡土环境下的公共空间类型。采用工业化快速建造体系，引入预应力细柱结构技术，有选择地以适宜技术消解弱化物化的建筑存在，还原乡村原有的触感，在极短的建造时间内进行了一次乡村复兴的建筑实验。

The trend of modernization and urbanization in rural development inevitably requires the introduction of new types of functions, and the Internet conference center with multi-functional buildings of large span is one of them. Taking the commune auditorium and the vegetable greenhouse as the prototype, the Shitang Village Project in Jiangning tries to reconstruct public buildings in a rural context. A rapid construction system is adopted with the technology of pre-fabricated structure of super-slender pillars. Selected technologies are used to downplay the existence of materialized buildings and to restore the primitive feelings of rural villages, hence a rejuvenated practice in an extremely short construction period in a rural village.

门廊局部

东立面
多功能报告厅室内 ｜ 辅助用房室内

主体建筑高度632米，共127层，是与上海环球金融中心和金茂大厦组成的浦东全球首例三座毗邻的超高层建筑中最高的一栋。大厦采用双层表皮设计，内部由9个圆形建筑彼此叠加构成，其间形成9个垂直空间，外立面与内立面之间的空间则形成了空中中庭。"上海中心"将这一系列垂直社区用于不同用途，每个垂直社区设计为独立的生物气候区进行调节和使用，可改善大厦内的空气质量，双层高透玻璃表皮隔绝了热能的传导，创造出宜人的休息环境。大厦采用多项最新的可持续发展技术以达到绿色环保的要求，是同时符合中国绿色建筑评价体系和美国绿色建筑认证体系的"绿色摩天大楼"。

The main part of the building has 127 stories with a height of 632 meters. It is the tallest of the world's first triple-adjacent super-tall buildings in Pudong, the other two being the Jin Mao Tower and the Shanghai World Financial Center. Adopting double-skin technologies, the tower takes the form of nine cylindrical buildings stacked atop each other to form nine indoor zones. Between inner and the outer layer, these nine areas has its own atrium on air. The vertical communities are used for various purposes, with each being designed as independent bio-climatic zone adjustable, to improve indoor air quality for a better environment for rest. Isolating heat conduction, the highly transparent double-layer glass on facades is a unique design feature. The Shanghai Tower incorporates numerous green architecture elements. Its owners received certifications from the China Green Building Committee and the U.S. Green Building Council for the building's sustainable design.

外景
鸟瞰

上海中心大厦
Shanghai Tower

建筑设计
美国 Gensler 建筑设计事务所
同济大学建筑设计研究院（集团）有限公司

Architects Gensler; Tongji Architectural Design (Group) Co., Ltd.
Location Shanghai
Completion 2016
Photo ZHANG Yong

上海 | 竣工时间 2016 年 摄影 章勇

公共建筑　办公／商业

地下通道
大堂

幕墙
会议厅入口

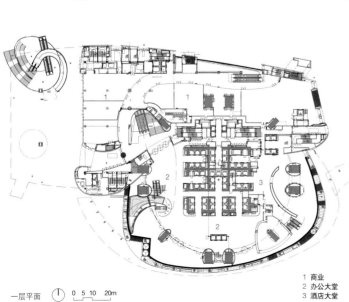

一层平面　0 5 10 20m

1 商业
2 办公大堂
3 酒店大堂

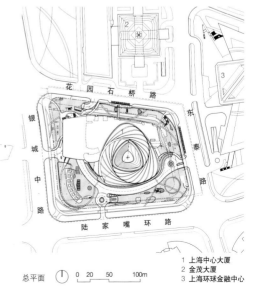

总平面　0 20 50 100m

1 上海中心大厦
2 金茂大厦
3 上海环球金融中心

建筑南侧外景

天赐新能源企业总部是一个功能复合型建筑，集生产和办公于一体。建筑一至三层是生产性用房，四、五层是研发办公和管理用房。建筑体型方正，可使生产用房高效使用空间。办公楼层设计了屋顶庭院，最大化引入自然采光和通风。建筑外观采用U型玻璃幕墙，半透明的表皮使建筑与园区环境建立了一种微妙的张力关系。建筑既内敛又开放。项目位于城市新开发区，周围环境具有一种荒野的场所属性，建筑物意图创造一种内聚的空间性以抵抗外部的无地方感。此外，建筑内特别设计了一道连续的直跑楼梯，从底层门厅转折向上直达四、五层的共享空间。这样一条路径不仅提供了联系各楼层的便捷通道，也让使用者获得一种更紧密的融入建筑的方式。

The building of Headquarters of Zhejiang TCI Ecology & New Energy Technology is a complex including manufacturing factories on the first, second and third floors, and research lab and management offices on the top two floors. The plan is designed with regular shape towards efficient usage for production program. Two roof gardens are set to bring daylight into office spaces and facilitate ventilation. Translucent U-shaped glass curtain walls are used as architectural surfaces, with which a delicate spatial tension emerges between the building and its environment. The building is introversive but open toward the outside, resisiting the sense of nowhere on the site. A continued staircase begins from the entrance hall and goes toward the communal space on the top. It is not only designed as a convenient linkage, but also helps people integrate into the whole space.

五层平面

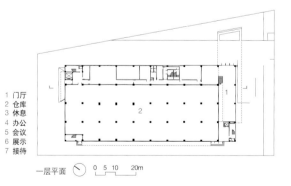

1 门厅
2 仓库
3 休息
4 办公
5 会议
6 展示
7 接待

一层平面

天赐新能源企业总部
Headquarters of Zhejiang TCI Ecology & New Energy Technology

建筑设计
无样建筑工作室
嘉兴联创建筑设计有限公司
冯路

Architects
Wuyang Architecture; Jiashan Lianchuang Architectural Design
FENG Lu
Location Jiaxing, Zhejiang
Completion 2016
Photo HU Yijie

浙江 嘉兴 | 竣工时间 2016年 | 摄影 胡义杰

建筑东立面
办公层内院
四、五层共享空间｜入口门厅

建筑外观效果

这栋综合用途塔楼开发项目位于中国南昌高新区，塔楼体量呈优雅的几何形，昭示其成为南昌新的城市地标。建筑设计的"城市之窗"占据塔楼顶部三分之一的立面，不仅向游客预示酒店的所在位置，还表明这片新区在城市中的重要性。"城市之窗"朝正西方向，与老城区形成直线联系。建筑通过的朝向，与历史建立视觉上的联系和隐喻联系。倾斜玻璃面构成的"城市之窗"营造独一无二的功能用途和场所感，外墙围护结构采用相似的形态策略与之呼应。塔楼的四个立面都布置了三角形遮阳板，不仅呈现出独特的质感，还有助于实现更佳的建筑性能。遮阳系统使用水平遮阳板和斜肋遮阳板，可以减少热量吸收。裙房的朝向、按功能分布的体量以及优美的梯田式露台，为老城区提供壮观的景致视野。在行人流线组织上，建筑的商业分布不仅尽量扩大零售店的展示面，还打造繁华的城市公共广场。开阔的公共庭院利用不同的空间条件，为休闲的行人、酒店客户、办公租户提供与整个城市互动的不同空间。

The project is located in the Gaoxin district of Nanchang, China, with an elegant geometrical image as the new landmark of the city. The building's "great window of the city", occupying the top third of the tower, not only signifies a visitor's arrival into the hotel, but also suggests the arrival and importance of a new district in the city. The "great window" is oriented towards due west, connecting the old city in a straight line. Through the buildings orientation, a connection to the past both visually and metaphorically is created. The sloping, glazed surface that defines the "great window" is unique to the program and the site, and a similar formal strategy is echoed in the façade enclosure. The triangular sunshade panels are placed on all facades, providing a distinctive texture to the façade and ensuring a better performing building. The use of both horizontal and diagonal fins eliminates a large percentage of the heat gain through solar shading. The orientation and layout of the podium and its gracefully terracing outdoor spaces allow spectacular views out over the old city. For the pedestrian circulation, the distribution of retail space is organized in such a way so as to not only extend the amount of retail store frontage, but also create a bustling urban public plaza. The grand public courtyard, with its variety of spatial conditions, provides different areas of interchange between the casual passerby, the hotel guest, the office tenant and the city as a whole.

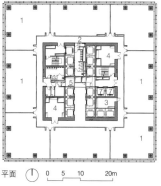

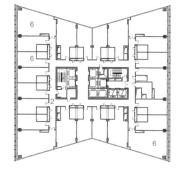

1 办公区
2 走道
3 强电间
4 暖通
5 弱电间
6 客房
7 消防电梯

平面 0 5 10 20m

南昌绿地紫峰大厦
Jiangxi Nanchang Greenland Zifeng Tower

江西 南昌 | 竣工时间 2014年

建筑设计
华东建筑设计研究总院
美国 SOM 设计公司
亢智毅
Ross Wimer

摄影 SOM ECADI 提供

Architects
Eastern China Architectural Design & Research Institute;
Skidmore, Owings & Merrill LLP
KANG Zhiyi, Ross Wimer
Location Nanchang, Jiangxi
Completion 2014
Photo SOM ECADI

公共建筑　办公／商业

鸟瞰效果

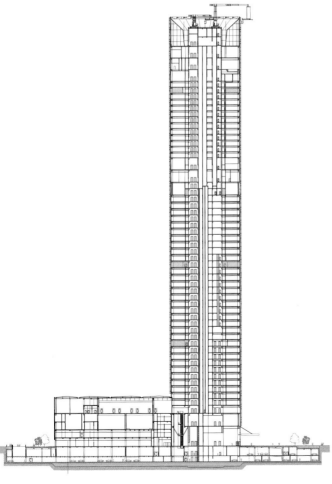

剖面　　0 5 10 20m

塔楼入口效果
酒店空中大堂效果

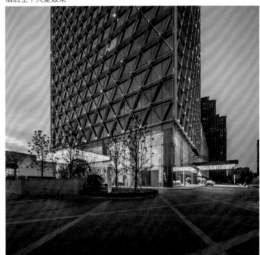

图书区

1 会议室/接待室
2 合伙人办公室
3 模型室
4 陈列室
5 行政办公室
6 厨房和卫生间
7 上空
8 图书区
9 办公区
10 内院
11 海棠院

一层平面　　0 1 2　5m

二层平面

大舍西岸工作室
Atelier Deshaus Westbund

上海　　竣工时间 2015 年

建筑设计
大舍建筑设计事务所
柳亦春　陈屹峰　王伟实

摄影　陈颢

Architects
Atelier Deshaus
LIU Yichun, CHEN Yifeng, WANG Weishi
Location Shanghai
Completion 2015
Photo CHEN Hao

院与树

院与建筑

轴测

上二层的楼梯与天窗 ｜ 合伙人办公室内景
西栋会议室

　　大舍西岸工作室建在原上海飞机制造厂场地内，紧邻保留并临时改造为艺术中心的大厂房。工作室将是一个寿命只有5年的临时建筑，必须低成本且快速建造，但在设计时仍然考虑了与场所气质的协调性。

　　基地原来是一个混凝土地坪的停车场，通过采用原有的墙体结构，将混凝土地坪利用为建筑的基础以节省造价，原有的墙体结构也恰好能对应于工作室所必需的会议室、模型制作、行政办公、储藏等小面积的功能空间。建筑的二层则采用了符合大空间办公使用要求的轻钢结构。在二层的轻钢结构中，根据办公桌的空间布局和使用，确定了结构的间距及其和窗户的关系，因而结构的逻辑里隐含了使用的空间要素，这些结构也同样在氛围上和这个地方所具有的工业气质相吻合。

The office of Atelier Deshaus is located in the former site of Shanghai Aircraft Manufacturing Works, next to a big plant that are temporarily renovated to an art center. The new office building is a temporary structure for 5 years due to limited budget and rapid construction. However, the architets still embraced the spirit of the site in design.

Originally used as a concrete parking area, the site was transformed smoothly into a studio, using the concrete ground as foundation of the building to lower cost, and the original wall structure can be used for small rooms of conference, model making, administration, and storage required in the new building. The upper floor adopts light steel structure. According to the layout of the office tables, the spacing of the structure and its relationship with the windows were hence set. The structural logics informed by the spatial elements also echoes to the industrial temperament of the site.

西北侧临湖实景

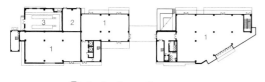

A区一层平面　　0 5 10 20m

B区一层平面

1 商铺
2 变电所
3 开闭间
4 器械健身区
5 接待
6 办公室
7 消防控制室

滨江休闲广场商业用房
Binjiang Leisure Plaza and Commercial Housing

江苏 常熟　　竣工时间 2015年

建筑设计　启迪设计集团股份有限公司
查金荣　程伟

摄影　查正风

Architects
Tus-Design Group Co., Ltd.
ZHA Jinrong, CHENG Wei
Location Changshu, Jiangsu
Completion 2015
Photo ZHA Zhengfeng

滨江休闲广场外景
西南侧实景

临湖小广场

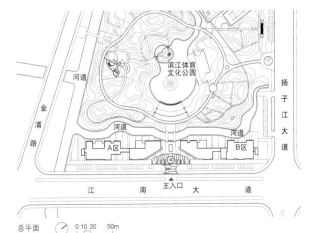

总平面

总体空间布局上,该项目设计采用"漂浮的盒子"的构思,创造了灵活多变的趣味性母体空间。项目临水而建,形成了与公园景观的和谐对话。建筑的体量被划分成为若干较小的形体组合,用白色穿孔板在建筑立面上抽象地展现树的形态,形成水墨画的意境。同时,设置在西侧立面的穿孔板也遮挡了西晒的阳光。

For overall layout, the project adopts the idea of "floating box" to create the prototype of flexible and interesting space. Close to water, it has been in a harmonious dialogue with park landscape. The building bulk has been divided into several smaller parts, using white perforated plates on the facades demonstrating the shape of trees in an abstract way, with an imagery of traditional Chinese painting. In the meantime, the perforated plates set in west effectively reduce western sunburn.

东侧沿外滩

外滩SOHO位于上海老外滩的最南端,人民路和新永安路之间。本项目由多栋错落排列的高层办公楼和商业裙楼围合而成,巧妙地形成了楼宇中央独特的小型广场。贯穿项目东、西向的户外步行街一直延伸至外滩,沿街分布着各种不同形态的公共休闲空间,塑造了活泼生动的商业步行街区。从狭长的"里弄"、星罗的街道网和竖向石材的老建筑中获得灵感,漫步外滩SOHO,能细细品味出对传统文化及历史建筑的深刻理解与继承,同时也能清晰地感受到极具未来感及前瞻性的"国际化街区"的开放和包容。

The Bund SOHO is located at the northern tip of the bund, between Renmin Road and New Yong'an Road, consisting of a group of high-rise office buildings and commercial podiums that enclose a small unique square at the center. The outdoor pedestrian street winds throughout the buildings from east to west, extending to the bund. Various shapes of public leisure spaces are dotted along the street, making the area lively. Inspired from narrow "Shanghainese alleys", scattered streets and old buildings decorated by vertical stone strips, the Bund SOHO reflects an understanding and a precise expression of traditional culture and historical buildings, and the diversity and inclusiveness is clearly revealed by the futuristic idea of "international block".

外滩 SOHO
Bund SOHO

建筑设计 德国 GMP 国际建筑设计有限公司 华东建筑设计研究院有限公司华东建筑设计研究总院
施特凡·肖茨 施特凡·瑞沃勒 乔伟

上海 | 竣工时间 2015 年 | 摄影 阴杰

Architects GMP International GmbH; East China Architectural Design & Research Institute
Stephan Schütz, Stephan Rewolle, QIAO Wei
Location Shanghai
Completion 2015
Photo YIN Jie

公共建筑　办公/商业

黄浦江对岸夜景
南侧外观

内庭院

1　AB栋
2　C栋
3　D栋
4　E栋
5　F栋
6　G栋

总平面　0 10 20 50m

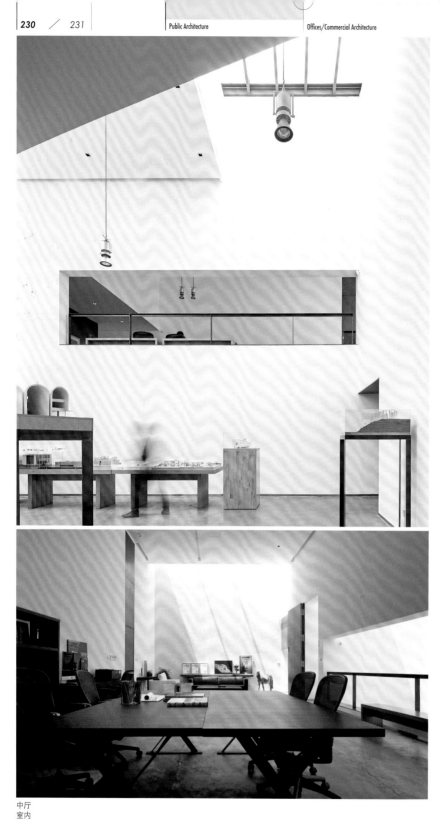

中厅
室内

该项目是建筑师周恺的工作室。基地周边多是居住区,没有多少可以观赏的景色,加之建筑师本身也希望营造一种内敛、安静的氛围,因此在设计中弱化对建筑外部的表现,着重强化内部空间的表达。

设计之初,建筑师最先想到的就是光,随时间变化的光线总是能给人带来惊喜。于是,建筑中有16道大小不一、或长或方的天窗,用最简洁的白墙,以最朴素的形态承托光影的变化。建筑内部主要有公共和私密两个分区。建筑中心是一个两层通高的厅,平时用于展示项目的工作模型,也可以举办规模不大的会议或者聚会。一层南侧主要是休息和休闲空间,压低的空间尺度令人舒缓、放松。建筑南侧还有一个小院子,种着海棠和竹子,四季变化的自然触手可及。建筑二层是办公空间,分别在中厅的南北两侧,并借助两个较大的洞口向展厅打开,两侧办公空间由连廊连接。

It is a design and research studio of architect Zhou Kai. The studio is surrounded by residential areas with little sceneries. More importantly, the architect prefers a introversive and quiet working environment, hence the weakening of external expressions and strengthening internal expressions.

At the beginning of design, the first thought that came to the architect's mind is light, as the change of light can always be surprising. As a result, the building has sixteen rectangular or square skylights of different sizes. White walls in their simplest forms are used as the background to cast the change of light. There are two main public and private sectors inside the building. The center at the building sits a two-storied lobby for project models usually or for medium-sized conferences or parties. The south side on the ground floor is mainly used for rest and leisure. The height is reduced for the purpose of soothing and relaxing. A small courtyard is placed south of the building, planted with Begonia and bamboos to reflect the tangible rotation of different seasons. The second floor of the building are offices, located on the north and south sides of the central hall, which are connected by a corridor through two large openings facing the exhibition hall.

水西工作室
Shui Xi Studio

建筑设计
天津华汇工程建筑设计有限公司
周恺

天津
竣工时间 2015 年
摄影 周恺 陈鹤

Architects
Huahui Architectural Design & Engineering Co., Ltd.
ZHOU Kai
Location Tianjin
Completion 2015
Photo ZHOU Kai, CHEN He

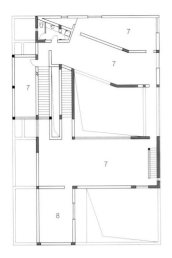

二层平面

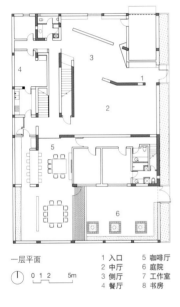

一层平面

① 0 1 2 5m

1 入口	5 咖啡厅
2 中厅	6 庭院
3 侧厅	7 工作室
4 餐厅	8 书房

公共建筑 — 办公/商业

水西工作室
2015

入口
中厅

鸟瞰

北京绿地中心坐落在大望京商务区中心,位于机场高速五元桥边上。260m摩天主楼"中国锦"取自中国传统元素文化符号"织锦"的设计理念,立体编织式炫动外立面的设计灵感,将东方艺术与现代商务建筑文明完美融合。外立面图案采用浮雕雕刻技术,以等腰梯形模块作为棱镜,吸取阳光并将其折射,创造光和影的互动。梯形每个面都由两块竖直的低辐射隔热玻璃板组成。建筑立面由两种梯形棱镜共同组成,梯形窄端朝向天空或是下方街区,增强了建筑的视觉冲击力,为立面提供了自遮阳系统,节约了能源使用。

Beijing Greenland Center is located in the center of the Great Wangjing Business District, on the edge of Wuyuan Bridge on the airport highway. The design idea of the 260-meter-high building comes from the "China Jin ", a traditional and cultural symbol of "Chinese tapestry". The design of the stereoscopically braided facades perfectly combines the Oriental art with modern business and architectural achievements. The patterns on the facades applies the relief carving techniques, using the isosceles trapezoid module as the prism that absorbs and refracts sunlight, creating the interaction of light and shadow. Each trapezoidal surface is composed of two vertical low-radiated insulation glass panels. The building facade is composed of two trapezoidal prisms. The narrow ends of trapezoids face the sky or urban blocks beneath, enhancing visual impact of the building and providing a shading system on the facades to save energy.

北京绿地中心
Beijing Greenland Center

北京 | 竣工时间 2015 年

建筑设计
中国建筑设计院有限公司
美国 SOM 设计公司
汪恒 安澎 孟海港

摄影 张广源

Architects China Architecture Design Group; Skidmore, Owings & Merrill LLP
WANG Heng, AN Peng, MENG Haigang
Location Beijing
Completion 2015
Photo ZHANG Guangyuan

公共建筑 办公／商业

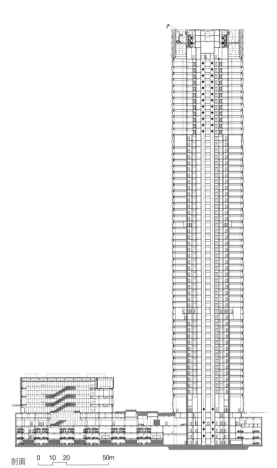

剖面 0 10 20 50m

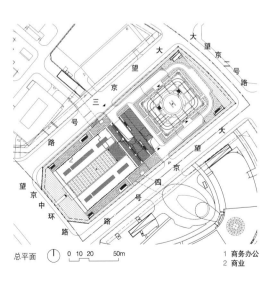

总平面 0 10 20 50m

1 商务办公
2 商业

街景
入口

Public Architecture　　Offices/Commercial Architecture

西立面

本项目位于上海市青浦区，为其东侧商业综合体的临时售楼处。基地所在的条状地是沿道路的退界绿化地带，在区域城市化过程完成后，将交还给城市使用。

特殊的项目条件让甲方选择采用集装箱作为项目的主要建造体系：快速，经济，可部分回收。设计采用交错码放箱子的方法来节约箱子的数量，并围合成一个最大化的中庭空间作为沙盘放置区以及洽谈区。中庭两层通高，面对南面入口，东西对称，简洁有力的虚实关系是理性选择的结果，也因此空间的秩序得到了强化。建构的逻辑无论是室外还是室内，都被清楚地表达出来。交错码放的原则形成了虚实交替的中庭室内外空间界面，虚为洞口，实为隔墙。昼夜接替，定制的照明灯具让空间虚实关系发生互换，但空间整体结构仍然保持完整。

The project is located at Qingpu district of Shanghai, used as a temporary sale office for a mix-use commercial complex on the east. The site of a strip of land spreads along the set-back green areas. With the completion of urbanization of this area, the site will be returned to the city as public green space.

The client decided to adopt container construction system as the main building system regarding specific situations of the site. Adopting this system guarantees speed of construction, efficiency and partial recycling. The design stacked containers jaggedly to minimize the number of containers, and maximizes the space of atrium for placing models and meetings. Symmetrically arranged, the two-storied atrium faces the southern entrance, resulting in the powerful relations between the solid and void reasonably. The spatial order has been reinforced too. The construction logic both for the interior and exterior has been explicitly expressed. The principle of staggered arrangement results in the spatial interface alternating the solid with void for the atrium. The void are openings while the solid are walls. Despite the alternation of day and night that reverses the relation between the solid and void, the integrity of the space remains intact.

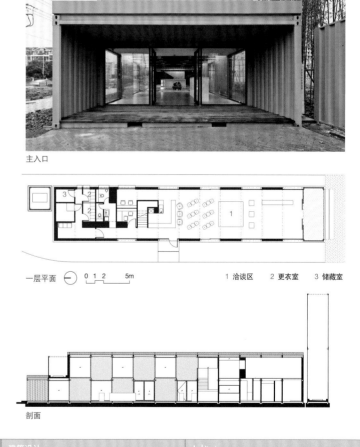

主入口

一层平面　　0 1 2　　5m　　　1 洽谈区　　2 更衣室　　3 储藏室

剖面

宝龙城市广场集装箱售楼处
Baolong Qingpu Plaza Container Sales Office

建筑设计　上海旭可建筑设计有限公司
刘可南　张旭

上海　｜　竣工时间 2014 年　｜　摄　影　苏圣亮

Architects　Atelier XUK
LIU Kenan, ZHANG Xu
Location　Shanghai
Completion　2014
Photo　SU Shengliang

街角透视

接待区空间局部
接待区空间｜吧台

办公楼屋顶合院

西南角鸟瞰
都市夜景下的宿舍楼

项目包含办公和居住两个主要功能，设计依据地段周围城市环境特征，采用了"馆"与"舍"分置的错落布局，在保持与大使馆协调的同时，形成了介于高层酒店和低层民宅之间尺度的"中间"城市景观。馆舍屋顶格栅和北立面可滑动的遮阳板（因处于南半球，住宅楼北立面为受阳面），则分别体现了适应于当地气候条件的"双层屋顶"和"双层立面"的理念，它们在遮蔽当地强烈日照的同时又保证了办公与居住活动的私密性。

设计力图在不同的层次上表达出中国传统文化和当代精神。办公楼入口雨棚、窗格及门厅天窗是中国传统意象的直接体现，而入口窄院及屋顶回廊与北面住宅形成的半围合空间，则是对中国传统院落空间意象和内涵的抽象式表达和意境式再现。

The project contains two functions: office and housing. The design adopts separate halls and houses interwoven according to characteristics of the urban environment. By keeping harmonious with the embassy, the middle urban landscape emerges between high-rise hotels and low vernacular houses. Grilling on the top of the halls and sliding sunshades on the north (as it is located on the southern hemisphere the northern side receives sunlight) represent the ideas of double roofs and double facades accommodating to the local climate. Reducing direct sunlight, these maneuvers guarantee privacy of housing and offices.

The design aims to reveal Chinese traditional cultures and contemporary spiritual quality on different levels. The porte-cochere at the entrance, window panes and skylight of the hallway reinterpret perfectly the Chinese traditional imagery. In addition, the space and its connotation of traditional Chinese courtyards are best represented in a narrowed courtyard at the entrance and the semi-enclosed space surrounded by roof ambulatory and northern houses.

中国商务部驻印尼商务馆舍
Office Building of Chinese Ministry of Commerce in Indonesia

印度尼西亚 雅加达 | 竣工时间 2014 年

建筑设计
清华大学建筑学院单军工作室
清华大学建筑设计研究院
单军 刘玉龙 铁雷

摄影 铁雷

Architects
SHAN Jun Atelier; Architectural Design of Research Institute of Tsinghua University Co., Ltd.
SHAN Jun, LIU Yulong, TIE Lei
Location Jakarta, Indonesia
Completion 2014
Photo TIE Lei

办公楼与宿舍楼之间的内庭院　　办公楼门厅
办公楼入口　　内庭院

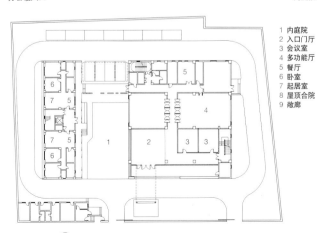
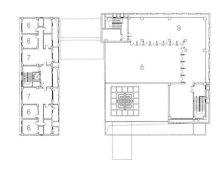

1 内庭院
2 入口门厅
3 会议室
4 多功能厅
5 餐厅
6 卧室
7 起居室
8 屋顶合院
9 敞廊

一层平面　　0 2 5　10m　　　　　三层平面

远处的五龙山与办公楼
雾化玻璃小窗

三层通高的共享中庭

本项目处于整个规划的办公区域最中心的位置，也是最高的一栋建筑。设计的策略是先强调建筑的体量感，把它作为整个办公区的灵魂和标志来设计，通过两个不同形态、不同表皮的体量穿插围合出入口广场空间。简单的体量传达出建筑本身的张力。两层高的多功能厅外表皮采用超白玻璃，而主体建筑是6层高的四合院，外侧采用竖向的白色铝板遮阳，立面形式清晰严谨而富有序列。

The project sits at the center of the planned office area as the tallest building. The design strategy emphasizes the sense of building volume, taking it as the soul and landmark of entire area. Through two interspersed blocks with different forms and skins, the square at the entrance is shaped by various building volumes. The simple volume conveys the tension of the building itself. The 2-storied block with super-translucent glass is a multi-purpose hall. The main building, on the other hand, is a 6-storied courtyard with vertical white aluminum shades outside, and the facade is clear and orderly designed.

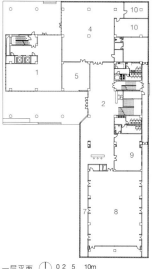

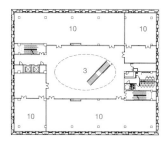

四层平面

1 办公区门厅
2 服务区门厅
3 共享中庭
4 开场查阅
5 封闭查阅
6 前台
7 休息走廊
8 多功能厅
9 贵宾休息室
10 办公室

一层平面 0 2 5 10m

苏州科技城国家知识产权局
苏州中心办公楼
Patent Examination Cooperation Jiangsu Center of the Patent Office, SIPO

江苏苏州 | 竣工时间 2014年

建筑设计
苏州九城都市建筑设计有限公司
张应鹏 王凡

摄影 姚力

Architects
Suzhou 9 Town Studio Co., Ltd.
ZHANG Yingpeng, WANG Fan
Location Suzhou, Jiangsu
Completion 2014
Photo YAO Li

西立面通过架空与围合形成入口广场
入口广场空间

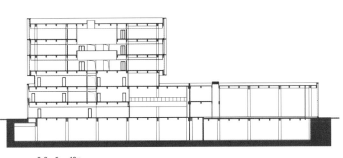

剖面 0 2 5 10m

夜晚的灯光透过雾化玻璃形成朦胧的光晕

中心庭院

规划布局南北两翼是客房及康体部分，中间以餐饮会议为连接，围合出一个对西面山景开放的庭院。此外在入口等处设置了大小院落。建筑设计采用朴素的风格，主体建筑内外墙及地面均铺贴机切毛面灰麻条石，和谐自然。另一大特点是建筑中设置了大量的金属格栅墙，成为塑造和划分空间的重要元素，形成半实半虚的效果。整个建筑的灰色调给予场所浓厚的北方建筑情调。

In the planning scheme, the two wings of the building contain guestrooms and recreational facilities, connected by the dining hall and conference rooms that surround a courtyard opening to the mountains in west. Courtyards of various sizes are placed at the entrance. A simple style is adopted for the building, and the material of machine-cut green quarry stone with rough surface is used for both interior and interior walls, natural and harmonious. Another characteristics is the use of a large amount of metal plates as a crucial element for shaping and dividing space, with an effect seemingly false and real at the same time. The strong appeal of northern architecture is embodied in the grey tone of the building.

建筑局部

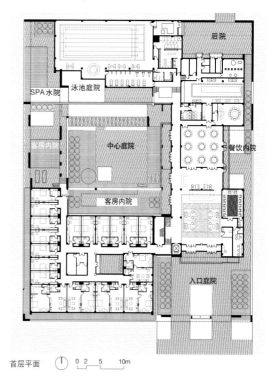

首层平面

北方长城宾馆三号楼
NO.3 Building of Northen Great Wall Hotel

建筑设计
中国建筑设计院有限公司
陈一峰　杨光　尚佳

Architects
China Architecture Design Group
CHEN Yifeng, YANG Guang, SHANG Jia
Location Beijing
Completion 2014
Photo SHU He

北京　　竣工时间 2014 年　　摄影 舒赫

中心庭院俯视

中心庭院西边院
客房院

泳池庭院
入口庭院

玉树州政府南侧内院
玉树州行政中心及北侧普措达泽神山

玉树藏族自治州行政中心
Yushu Tibetan Autonomous Prefecture Administrative Center

青海 玉树藏族自治州 | 竣工时间 2014 年

建筑设计
清华大学建筑设计研究院有限公司
庄惟敏 张维

摄影 姚力

Architects
Architectural Design & Research Institute of Tsinghua University Co., Ltd.
ZHUANG Weimin, ZHANG Wei
Location Yushu Tibetan Autonomous Prefecture, Qinghai
Completion 2014
Photo YAO Li

从玉树州政府远眺南侧加吉娘神山
玉树州委藏式"林卡"景观

玉树州委藏式"林卡"景观

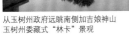

1 州委主楼
2 州政协
3 州人大
4 会议中心
5 办公用房
6 老干部活动中心
7 保留建筑
8 州政府主楼
9 西配楼办公
0 东配楼办公
1 南区办公

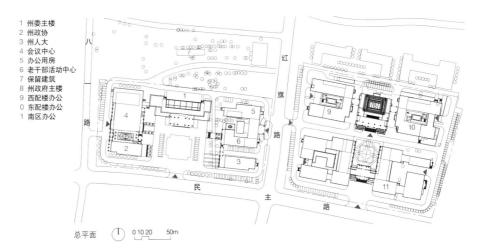

总平面　0 10 20　50m

玉树州政府主楼门廊采用抽象形式的藏式牛角柱

　　建筑设计以藏区历史传统上的雪域宗山为原型,以多样性的藏式院落空间组合为载体,在建筑细部上以藏民熟悉的亲民姿态展示在公众面前。运用藏式园林"林卡"的造景手法,保留基地所有现存树木,围绕树木进行建筑和景观设计,体现对生命的尊重。设计在现代建筑技术和工艺条件下尊重藏区当地的风俗文化,体现当地文脉特色,强调行政建筑的时代特质,赋予建筑群落以场所精神,构筑起神山、景观、建筑融合为一的空间环境。

　　The design takes the Tibetan "Dzongshan" as the prototype to translate it into various Tibetan-style courtyards. Architectural details are designed to be familiar with Tibetan. Based on reservation of existing trees and the typical Tibetan landscape, "Linka", landscape design shows respect to life. Modern architectural technologies are used with respect to local customs and culture, so as to realize genius loci through an integration of the environment of divine mountains, landscape and architecture.

内院

在竹林广场看坝坝电影

内立面——外挂楼梯

西村大院的设计一反中心体量集合的城市综合体常见模式,采用外环内空的布局,环绕街区沿边修建,围合出一个公园般的超大院落,成为一个外高内低,容纳纷繁杂陈公共生活"绿色盆地",呼应了四川盆地的原风景。大、中、小竹林院落层层相套,面向公众开放,自由穿行,再现传统生活环境,激发集体记忆,延续了成都人民热爱的竹下休闲传统生活方式。功能设施式的骨架设计,任由富于个性的世俗生活自由填充,又被大院的巨大尺度所归纳,最终形成"市井立面",传达出群体创造的丰富表现力。一条架空休闲跑道巡回大院并攀升至屋顶又环绕一周,为跑步和骑车人们带来了自由兴奋的超常体验,并赋予建筑具有动态能量的鲜活形象。西村大院的建设使日常休闲方式获得了具有纪念性尺度的集合表现,成为周边社区民众休闲生活的乐园,也为自身带来了丰沛的活力和巨大的成长空间。

West Village is a super-courtyard that adopts an atypical design compared to other commercial complexes in that its centrifugal layout encircles the entire block to maximize the inner area of sports and greenery like a park, echoing to the form of Sichuan Basin and encompassing a diverse public life as a "green basin". Smaller bamboo courtyards are embedded within those of medium and large sizes, open to the public so that visitors are welcome to walk through freely. The spatial design carries on the ever-popular traditional leisure lifestyle under the shade of bamboo among locals in order to refresh their former collective memories. Its functional structure allows a miscellaneous forms of expressions to utilize the customizable spaces at will, while the large scale of the courtyard harbors and displays a "vernacular façade of the commoners". A suspended runway overlooks the courtyards and climbs up to the roof around the building, providing exciting experience for joggers and bicyclists with a lively and dynamic profile of the building. The construction of West Village is a collective representation of daily recreation in a monumental size, resulting in much vigor and room for growth.

西村大院
West Village

建筑设计
成都市家琨建筑设计事务所
刘家琨

Architects
Jiakun Architects
LIU Jiakun
Location Chengdu, Sichuan
Completion 2014
Photo Arch-Exist Photography, Jiakun Architects

四川 成都 | 竣工时间 2014 年 | 摄影 存在建筑 家琨建筑

全景

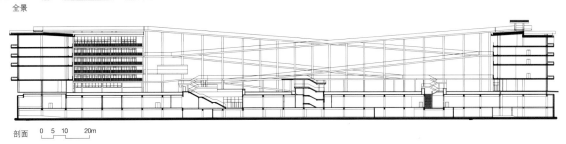
剖面 0 5 10 20m

1 门厅　2 架空入口　3 商业　4 竹园　5 多功能空间

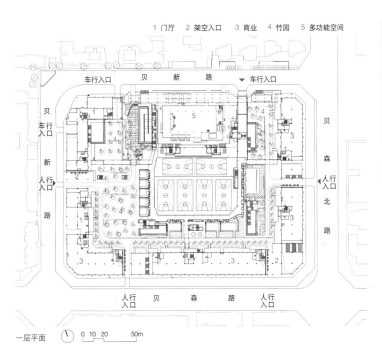
一层平面　0 10 20 50m

沿街夜景

东南向鸟瞰全景

成都远洋太古里是一个开放式、低密度的街区式购物中心。项目以现代的手法演绎传统建筑风格,与散落其间的6座富含历史底蕴的保留院落交相辉映,营造出一片开放的城市空间,而毗邻的千年古刹大慈寺更为其增添独特的历史和文化韵味。项目的建筑设计独具一格,融汇古今,以创新设计理念打造出充满活力的零售休闲社区,以人为本的"开放里"概念贯穿始终,让人感受到遍布里巷、茶馆、商铺和花园的勃勃生机。在深刻理解成都这座城市以及当地民众生活习惯的基础上,项目首创"快里""慢里"概念,并精心选择品牌和业态创造出"快耍慢活"的全新生活方式。

Sino-Ocean Taikoo Li Chengdu is an open-plan low rise mall inspired by the layout of lanes. The architectural design of the project pays homage to traditional Sichuan architecture with an innovative, modern approach. Six traditional courtyards and buildings within the site have been preserved and revitalised, and the adjacent Daci Temple also contributes to the historical and cultural essence of the complex. The urban and architectural design uniquely integrated modern and traditional elements and has created a sustainable and creative retail and community complex, invigorating the lanes, streets, teahouses, shops and gardens. The project features a unique concept of Fast Lane and Slow Lane retails, which is rooted in the culture of Chengdu. Shoppers can enjoy shopping for glamorous international fashion labels in the three Fast Lanes linking the East and West Plazas, or they can enjoy a relaxing moment with a cup of coffee or a stroll through lifestyle stores in the Slow Lanes that weave through the complex.

成都远洋太古里
Sino-Ocean Taikoo Li, Chengdu

建筑设计
欧华尔顾问有限公司
MAKE Architects
Spwaton Architecture Ltd. & Elena Galli Giallini Ltd.

四川 成都　　竣工时间 2014 年　　摄影　存在建筑 陈尧

Architects
The Oval partnership Ltd. ;
MAKE Architects;
Spwaton Architecture Ltd. & Elena Galli Giallini Ltd.
Location Chengdu, Sichuan
Completion 2014
Photo Arch-Exist Photography , CHEN Yao

大慈寺东侧慢里步行街
西广场

漫广场

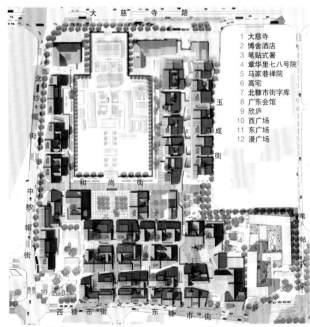

1 大慈寺
2 博舍酒店
3 笔贴式署
4 章华里七八号院
5 马家巷禅院
6 高宅
7 北糠市街字库
8 广东会馆
9 欣庐
10 西广场
11 东广场
12 漫广场

总平面 0 10 20 50m

中里

鸟瞰

1 入口景观广场
2 下沉广场
3 步行广场

总平面 0 20 50 100m

凌空 SOHO
Sky SOHO

建筑设计
上海建筑设计研究院有限公司
扎哈·哈迪德事务所
刘恩芳 Satoshi Ohashi 周燕

上海　　竣工时间 2014 年　　摄影 陈伯熔

Architects
Institute of Shanghai Architectural Design & Research Co., Ltd.;
Zaha Hadid Architects
LIU Enfang, Satoshi Ohashi, ZHOU Yan
Location Shanghai
Completion 2014
Photo CHEN Borong

入口实景

内庭院实景 1
内庭院实景 2

引人入胜的流线型外观,动感十足的曲线造型,流动而丰富的空间变化——这座极具未来感的建筑将成为上海的新名片,也将为所有到访者呈现一场记忆深刻的视觉盛宴。

凌空SOHO的12栋建筑被16条空中连桥连接成一个空间网络,多个楼层彼此互通,融为一体。这些结构叠放在下沉式广场上空,最高处落差约40m,俯瞰或仰望,如置身大自然中的峡谷。占地8.6万余m^2的凌空SOHO,东西长约480m,南北距离约250m,呈现出开放连通的超大"庭院"。骑行或漫步其中,跟随不断变化的流线外观,穿越一个个大小峡谷,别有一番天地。超过7 000m^2的下沉式广场,不仅将地下商业区流畅地连接起来,更提供了一个巨大的公共休闲空间。广场南北两侧的水景楼梯,让下沉式广场与地面相连,与动感的地上建筑相映成趣,带来更具活力的氛围。

The building has an attractive curvilinear appearance, dynamic shape and enriched change of space in flow. The futuristic building is to become the new landmark of Shanghai and to provide visitors and users a profound and impressive visual attraction.

Sky SOHO consists of 12 buildings effortlessly connected by 16 bridges into a network of spaces. Viewed from above or below, these structures are stacked up to 40m above the sunken garden on the ground, resembling the smooth, towering walls on a canyon. Located on the site of 80 000m^2, Sky SOHO is 480m in length by 250m in width, with a super enormous courtyard open and connected to other parts. Walking or biking here provides a dualistic experience characterized by natural and futuristic surroundings. Covering an area of 7 000m^2, the sunken garden offers a blended mix of retail and public spaces. Both waterscapes in South and North gardens provide the connection between the sunken garden and the ground, evoking the vitality along with the building.

主入口与金属文字墙

一层公共空间
二层公共空间与调温通风孔

项目是一个两层高、白顶绿身的双坡体量，坐落在用地北侧，提供办公、营业、培训、会议和员工住宿功能。建筑平面呈矩形，服务核心筒组织了"回"字形的流线，并将内部空间分隔成东西大小不等的两部分。核心筒由家具、集成卫浴构成。房屋悬浮于地面上，6个面设有保温层，并使用了挂板与其后的空气间层形成"集热墙"。它背后的空气被阳光加热后向上流动，可在冬季白天通过三角形檐部构造进入房间，提供二层室内空间的供暖。用户通过控制房间南北两侧翻板的开闭，可达到在冬夏两季调节室内舒适性的目的。房屋地面部分采用了轻质泡沫混凝土板，以达到蓄热、减震、隔声多重效果。新建房屋的设计、生产与建造时间只有8周。建筑主体的构件组装和现场建造仅用时4周。建造该房屋的工人全部系本村村民，先进建造的体验使村民对乡村银行有了更高的认同感。

The 2-storied building with double-pitch white roof and green body sits on the north side of the site. Its functions include office, business, training, conference, and staff housing. The rectangular plan has an ambulatory service core, and the interior space is divided into 2 parts – east and west – with varying sizes. The service core consists of furniture and integrated bathroom. Suspended above ground, the building has 6 sides installed with thermal insulation layers and adopts cladding technique to form a Trombe wall with the air insulation layer behind. The sun-heated air flows upwards, which during daytime in winter can enter the house through the triangular cornice to heat up the space on the 2nd floor. Users may adjust interior comfort during winter and summer by switching the shutters on the south and north sides of the house. Lightweight foam concrete panels are applied to part of the flooring for the purpose of thermal storage, shock absorption, and sound insulation. From design to production and construction, the project took only 8 weeks. Component assembling for the main body and in-situ construction only took 4 weeks. Construction workers are all local people, who accept the rural bank extensively due to the construction experience.

尤努斯中国中心 陆口格莱珉乡村银行
Yunus China Center Lukou Grameen Village Bank

江苏 徐州　　竣工时间 2014 年

建筑设计　深圳元远建筑科技发展有限公司；香港中文大学建筑学院
朱竞翔　夏珩　韩国日

摄影　韩国日　赵妍

Architects Unitinno Architectural Technology Development Co., Ltd.; School of Architecture, The Chinese University of Hong Kong
ZHU Jingxiang, XIA Heng, HAN Guori
Location Xuzhou, Jiangsu
Completion 2014
Photo HAN Guori, ZHAO Yan

南侧广场
北侧院落

西南视角

1 营业大厅
2 厨房
3 储藏间
4 客厅
5 卧室
6 多功能空间
7 储藏间

二层平面

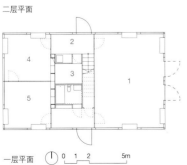

一层平面　0 1 2　5m

二层公共空间

项目位于北京朝阳公园西南角。除媒体办公和演播制作功能之外,建筑安排了大量对公众开放的互动体验空间,以体现凤凰传媒独特的开放经营理念。设计理念是用一个有生态功能的外壳将具有独立围护的使用空间包裹在里面,体现了楼中楼的概念,两者之间形成了许多公共空间。在东西两个共享空间内,设置了连续的台阶、景观平台、空中环廊和通天的自动扶梯,使得整个建筑充满着动感和活力。此外,建筑造型取意于"莫比乌斯环",这一造型与不规则的道路方向、转角以及和朝阳公园形成和谐的关系。

建筑的双层外皮可很好地提高功能区的舒适度和降低建筑能耗。设计利用数字技术对外壳和实体功能空间进行量体裁衣,使二者的空间关系精确地吻合。

Phoenix Center is located in the southwest corner of Chaoyang Park. Apart from the media office, the broadcasting studios and the production offices, the building provides abundant of open spaces for the public to get interactive experiences, which expresses the unique operation concept of Phoenix Media. The logic of the design concept aims to create an ecological environment shell embraces the Individual functional spaces as a building-in-building concept. The two independent office towers under the shell generate many shared public spaces. In the east and west parts of the shared spaces, there are continuous steps, landscape platforms, sky ramps and crossing escalators which fill the building of energetic and dynamic spaces. Furthermore, the building's sculptural shape originates from the "Mobius Strip". The sculptural shape provides the building a harmony relationship with the irregular direction of the existing streets, the sitting corner of the site, and the Chaoyang Park.

The double layer exterior of the building can improve the comfort in the functional areas, and reduce the consumption of energy. Digital technology is applied to tailor the physical space of the exterior shell and the inside volume precisely in order to ensure exact matches.

西中庭
中心广场

凤凰中心
Phoenix Center

北京 | 竣工时间 2013 年

建筑设计
北京市建筑设计研究院有限公司
邵韦平

摄影 傅兴

Architects
Beijing Institute of Architectural Design
SHAO Weiping
Location Beijing
Completion 2013
Photo FU Xing

公共建筑　办公、商业

中心广场局部

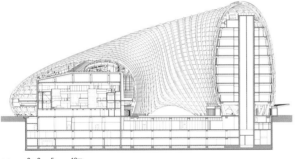
剖面　0 2 5 10m

东中庭
西南鸟瞰

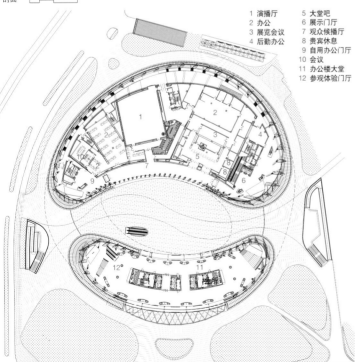

1 演播厅　　5 大堂吧　　　9 自用办公门厅
2 办公　　　6 展示门厅　　10 会议
3 展览会议　7 观众候播厅　11 办公楼大堂
4 后勤办公　8 贵宾休息　　12 参观体验门厅

一层平面　0 2 5 10m

航站楼夜景环视

设计将航站区建筑群体打造成为功能多样的航站楼综合体;外部空间环境与建筑形态一体化设计,集中设置2万多 m² 绿化景观;为提供精致的出行服务,将航站楼功能重新整合,增设交通中心、引入智能化设备、商业重新策划等;对原有航站楼建筑深入解读,汲取既有建筑元素,融汇现代设计手法重新演绎;关注绿色节能,量身定制被动式绿色策略;采用分阶段、置换改造的方式,确保各阶段旅客流线完整。

Design strategies are adopted as follows. The buildings in terminal area would be made into an integrated terminal complex. External space design would be integrated with formal design, with a concentrated area of more than 20000m², for green space. Functions of the terminal are integrated to provide a complete and considerate service for passengers with newly added transportation center, intelligent and smart technology equipment and re-programmed commerce, etc. The design aims to re-interpret the existing terminal and absorb architectural elements for expression in the modern architectural language. Green and energy saving become the focus of the project, and passive green strategy is customized for the building. Periodization of construction and replacement is adopted to ensure complete circulations.

国际出发办票大厅

上海虹桥国际机场 T1 航站楼改造及交通中心工程
Shanghai Hongqiao International Airport T1 Renovation & Traffic Center

上海 竣工时间 2016 年

建筑设计
华东建筑设计研究院有限公司
华东建筑设计研究总院

郭建祥 张宏波 吕程

摄 影 庄哲

Architects
East China Architectural Design and Research Institute Co., Ltd.,
East China Urban Architectural Design and Research Institute
GUO Jianxiang, ZHANG Hongbo, LV Cheng
Location Shanghai
Completion 2016
Photo ZHUANG Zhe

交通换乘大厅
国际候机厅

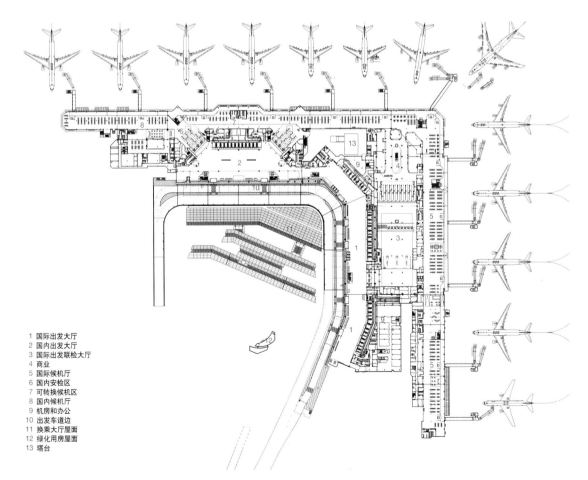

1 国际出发大厅
2 国内出发大厅
3 国际出发联检大厅
4 商业
5 国际候机厅
6 国内安检区
7 可转换候机区
8 国内候机厅
9 机房和办公
10 出发车道边
11 换乘大厅屋面
12 绿化用房屋面
13 塔台

二层平面 0 10 20 50m

住院楼北立面

夜景

在总体布局设计上，本项目结合基地情况将整个医院分为门急诊区、医技区、住院区和行政区4个部分。医技区作为医院的核心，布置在基地的中部，门诊、急诊、住院围绕其周边布置，各医疗功能区之间通过回廊式空间有机衔接，交通组织便捷合理。在立面设计上，本项目明快而具有开放感的设计、清晰而端庄的形象展现了安亭地区新的现代化医院的特征。

In the overall design, this project takes the site situations into account to divide the whole hospital into four parts, including the gate emergency area, the medical technology area, the hospitalization area and the administrative area. As the core of the hospital, the medical technology area is arranged in the center of the site, outpatient, emergency, hospitalization, etc. around its periphery. Each part is connected by a looping corridor organically which organization of circulation guarantees convenient and reasonable. In the design of facades, the lively and open design, clear and dignified image of the project show the characteristics of the new modern hospital in Anting.

上海东方肝胆医院
Shanghai Oriental Hepatic Hospital

建筑设计　上海建筑设计研究院有限公司　山下设计株式会社
陈国亮　唐茜嵘　邵宇卓

上海　｜　竣工时间 2015 年　｜　摄影　邵峰

Architects Institute of Shanghai Architectural Design & Research Co., Ltd.; Yamashita Design Corporation
CHEN Guoliang, TANG Qianrong, SHAO Yuzhuo
Location Shanghai
Completion 2015
Photo SHAO Feng

门诊医技楼共享中庭

住院部共享中庭

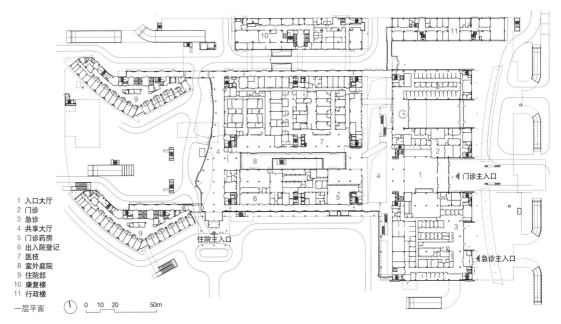

1 入口大厅
2 门诊
3 急诊
4 共享大厅
5 门诊药房
6 出入院登记
7 医技
8 室外庭院
9 住院部
10 康复楼
11 行政楼

一层平面　0　10　20　50m

门诊主入口
住院主入口
急诊主入口

主入口鸟瞰

外景

本案旨在建立一个能够支持并推动农村医疗管理及护理改革的卫生院模范。其理念包括提供现存卫生院所缺乏的基本医疗设施，例如简单的候诊室。此外，设计着力提倡卫生院向社区开放，重新把它定位成一个真正让公众享用的公共建筑。

方案把传统的医院功能重新配置。设计始于一个简单的策略，即利用一条连续的坡道贯穿所有楼层。宽阔的坡道设有休息的地方，也加强了建筑内部的流通性。同时坡道围合构成了一个开放的供村民使用的中央庭院。庭院添设的台阶提供更多休憩场地，使其成为室外候诊区。材料方面，大楼的外墙使用了可循环再用的传统青砖，螺旋式通道的内侧则采用定制的混凝土镂空砌块。这些定制砌块从远处看来跟普通砌块没分别，却是由富弹性的乳胶模具制成。此技术改变了混凝土固有的坚硬感，令庭院在一天的光影变化中表现出柔和及动态的一面。

The design aims to develop a model rural health care building capable of supporting the many progressive reforms on rural hospital management and care giving. This includes providing basic necessities absent in current establishments, some as simple as waiting rooms. Additionally, seeing that most institutions in China, such as schools and hospitals, are walled off and managed as contained programs, the architects were interested in re-introducing the hospital as a friendly facility to the public.

The design re-configures the conventional form of the hospital. The design begins with a simple strategy to provide a continuous ramp access to all floors. A wide ramp allows for seating and improves circulation. This also creates a large central courtyard space open for public use. At the ground level, the courtyard provides additional steps for seating and serves as an outdoor waiting area. Materials consist of both recycled traditional bricks for the exterior façades, and custom designed concrete screen blocks, which flank the interior spiral passageway. Though from a distance they appear like the common type, these custom blocks are cast in a flexible latex mold. The resulting courtyard exhibits a soft and smoothly changing quality, casting variable shows throughout the day.

昂洞卫生院
Angdong Hospital

湖南 湘西　　竣工时间 2014 年

建筑设计
城村架构 / 香港大学
保靖设计院
Joshua Bolchover 林君翰

摄　影　城村架构

Architects
Rural Urban Framework, The University of Hong Kong;
Baojing Design Institute
Joshua Bolchover, John Lin
Location Xiangxi, Hunan
Completion 2014
Photo Rural Urban Framework

庭院

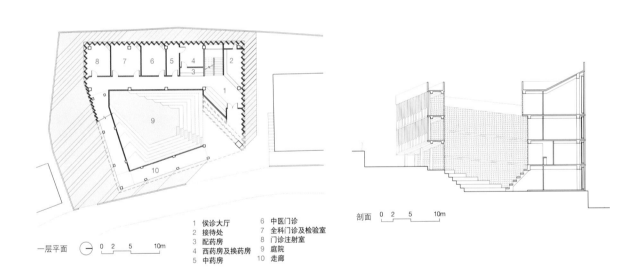

1 候诊大厅	6 中医门诊
2 接待处	7 全科门诊及检验室
3 配药房	8 门诊注射室
4 西药房及换药房	9 庭院
5 中药房	10 走廊

一层平面　0 2 5 10m

剖面　0 2 5 10m

病房　　走廊　　候诊厅

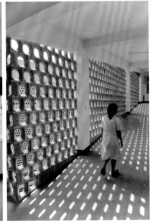

鸟瞰

在山西省境内，太原市周边，中国仅存的几座唐朝木构建筑，向我们传达着唐朝盛世的历史信息。太原南站传承这一历史文脉，体现地域文化特色。站房主体采用建筑结构单元体形成大空间，建筑即结构，结构即空间空间即形式。"树"一样生长的钢结构单元体为建筑空间提供了自由生长的可能性，同时构成极具特色的室内空间。针对太原城市的气候环境条件，太原南站前瞻性地采用建筑遮阳、自然通风采光、墙体保温等被动式节能技术，全面提高站房舒适度，成为具有示范效应的绿色车站。

There are several Tang Dynasty wooden architectures around Taiyuan, Shanxi Province, which are only remained in China. They tell us the story of Tang Dynasty prosperous time. Taiyuan South Railway Station inherits such historical context and reflects the regional and cultural characteristics. The main building of the station uses the construction unit to form a large space and to show the method of architecture as structure, structure as space, space as form. The steel structure unit rises like a "tree" and provides the possibility for the free rise of the building space, at the same time, it forms a unique interior space. According to the climate conditions of Taiyuan City, the project has been using the passive energy-saving technologies such as: building shading, natural ventilation and lighting, as well as wall insulation, etc. This will comprehensively improve the comfort level of the station and becomes a green station with demonstration effect.

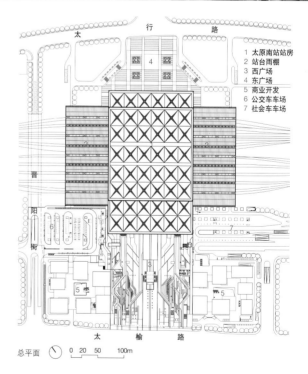

总平面

1 太原南站站房
2 站台雨棚
3 西广场
4 东广场
5 商业开发
6 公交车车场
7 社会车车场

太原南站
Taiyuan South Railway Station

建筑设计 中南建筑设计院股份有限公司
李春舫 王力

山西 太原　　竣工时间 2014 年　　摄影 施铮 丁烁

Architects Central South Architectural Design Institute Co., Ltd.
LI Chunfang, WANG Li
Location Taiyuan, Shanxi
Completion 2014
Photo SHI Zheng, DING Shuo

主入口夜景
候车大厅

室内商业空间

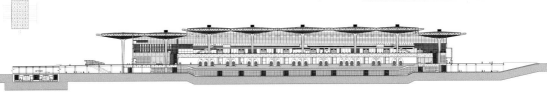

剖面　0 10 20 50m

行政办公区内院

都江堰大熊猫救护与疾病防控中心是汶川"5.12"地震后由香港特区政府投资援建的项目,是世界首个熊猫医院,兼具大熊猫的疾病救治、疗养、疾病研究和科普教育等功能。项目设计贯彻"全寿命周期"内资源消耗最小的绿色建筑理念,已获得绿色建筑三星设计及运营标识。设计以单体建筑"隐""融"的策略获得整体环境品质的提升,表明设计者对人、建筑、环境之间关系的态度,轻触式地还原了场所存在的意义。

Dujiangyan Giant Panda Rehabilitation Medical Treatment and Research Center is funded by Hong Kong government after the 2008 Sichuan Earthquake in 2008. It is world first panda hospital which has functions of giant panda disease treatment, recuperate, disease research and scientific education. The project has a design concept of minimal energy consumption during the full life cycle of the architecture. The project has obtained China's three-star green certification of both design and operation. The design tries to make every artificial single building integrated within the environment and become "invisible to improve environmental quality". This relationship of human, the environment and building forms our central design theme: to recover the meaning of the place through light intervention.

监护兽舍鸟瞰
从花田景观看疾病防控中心

都江堰大熊猫救护与疾病防控中心
Dujiangyan Giant Panda Conservation and Disease Control Center

四川 都江堰 | 竣工时间 2013 年

建筑设计 中国建筑西南设计研究院有限公司
钱方 戎向阳 茅锋

摄影 存在建筑

Architects China Southwest Architectural Design & Research Institute Co., Ltd.
QIAN Fang, RONG Xiangyang, MAO Feng
Location Dujiangyan, Sichuan
Completion 2013
Photo Arch-Exist Photography

科教中心
后区熊猫自然放养兽舍聚落

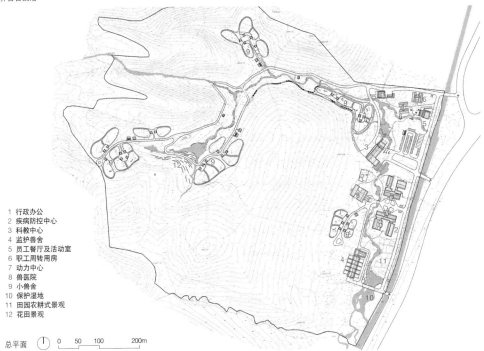

1 行政办公
2 疾病防控中心
3 科教中心
4 监护兽舍
5 员工餐厅及活动室
6 职工周转用房
7 动力中心
8 兽医院
9 小兽舍
10 保护湿地
11 田园农耕式景观
12 花田景观

总平面　0　50　100　200m

Renovation
Heritage Preservation

改造及修复

P264

P341

从溪水往南看七园居

"七园居"是一座容纳着七间客房的旅舍,位于德清的山谷中。基地依山傍溪,四面被景色环绕。基地上两层高的木构民宅被保留下来,客房被安排在这里。原本紧凑的6个生活开间被重新划分为4间。木柱顺势落入空间,暗示着先前的使用。第7间客房是完全新建的。不规则的圆形平面充分利用了基地的轮廓,顺势创造了一种流动、神秘的空间体验。钢筋混凝结构平行于木结构加建,以容纳卫生间和新的公共设施。建筑师舍弃了惯常的走廊模式,通过3组不断分岔的路径与7间客房连接,使居住直接与自然联通。每个客房都依借着场地或屋顶,获得专属的庭园。场地中的高差也被保留下来,纳入到底层环形的动线系统中,在相对狭小的场地中,提供连续、自在的休憩体验。

The Hotel of Septuor is a boutique hotel of 7 rooms located in a valley among the hills of Deqing, surrounded by beautiful sceneries with bamboo hills and a winding stream. The original dwelling of two-storey timber structure and loam walls is preserved. To transform a compacted house into a hotel, the width of each room is redistributed, thus the wooden pillars stand independently from the walls and meanwhile suggesting the previous space. To make the most of the backyard, an irregular rounded-shape-plan guest room is newly built, which have a flowing and mysterious atmosphere. New concrete structure is added in parallel with the old one, to accommodate bathrooms and new public services. To create intimate relationship between living and the nature, the linear corridor is replaced by three sets of paths with continuous turnoffs leading to various rooms which own a private garden each. The natural terrain is preserved and extended to the inner space, forming a open and inviting, yet quiet and intimate experience.

七园居
The Hotel of Septuor

建筑设计
上海博风建筑
王方戟 董晓 肖潇

浙江 湖州　　竣工时间 2017 年　　摄影 博风建筑

Architects Temp Architects
WANG Fangji, DONG Xiao, XIAO Xiao
Location Huzhou, Zhejiang
Completion 2017
Photo Temp Architects

改造及修复

7号客房看露台　　　　　　　　　　　4号客房卫生间

4号客房长窗
二层公共露台

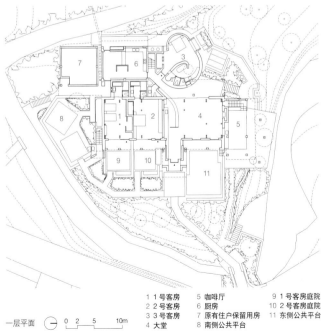

一层平面　　0 2 5 10m

1 1号客房　　5 咖啡厅　　　9 1号客房庭院
2 2号客房　　6 厨房　　　　10 2号客房庭院
3 3号客房　　7 原有住户保留用房　11 东侧公共平台
4 大堂　　　　8 南侧公共平台

从思庭北侧看五龙庙

五龙庙建于唐大和六年（公元832年），是现存最早的道教建筑。这是一次国家专项资金与社会资金合作进行文物保护事业的新尝试，使这个千年古庙的文物本体在获得国家文保资金修葺之后，又获得了环境品质的改善，将一个孤立古庙转换为一座关于中国古代建筑的博物馆，融入到当下生活。

环境整治设计围绕着两条线索展开。明线是以五龙庙为主体，展开一系列有层次的空间序列，并植入相关展陈，从而使观者能够更好地欣赏、阅读、理解文物；暗线则是通过提升五龙庙的环境品质和重新解读五龙庙，加强了这一场所的凝聚力，使村民重新聚集、交往在这一世代相传的公共空间，为当下农村精神价值的重塑创造契机。

Built in 832 A.D. in Tang dynasty, the Five Dragons Temple is the oldest surviving Taoist temple. Converting the one-thousand-year-old temple into a museum of traditional Chinese architecture history as part of contemporary life, the project is an exciting attempt combining the government and the private funds to preserve cultural relics and enhance the quality of their environment.

The design of the environment uplift for the Five Dragons Temple is centered around two themes. An outstanding theme is creating layers of overlapping spaces around the main building to tell the story of the temple history and ancient Chinese architecture. Through this theme, people will learn about the knowledge of traditional Chinese architecture to better understand the importance for the preservation of heritage. The latent theme is restoring the temple into an area of public gathering in the village, offering a lively environment that encourages the harmony between contemporary lifestyles and ancient architecture.

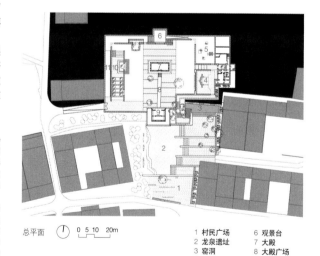

总平面　0 5 10 20m

1 村民广场　6 观景台
2 龙泉遗址　7 大殿
3 窑洞　　　8 大殿广场
4 序庭　　　9 戏台
5 斗栱庭　　10 思庭
　　　　　　11 晋南古建展廊

五龙庙环境整治工程
The Environmental Upgrade of the Five Dragons Temple

建筑设计
URBANUS 都市实践建筑事务所
清华大学建筑设计研究院有限公司
王辉

山西 运城　｜　竣工时间 2016 年　｜　摄　影　杨超英　阴杰

Architects
URBANUS; Architectural Design and Research Institute of Tsinghua University Co.,Ltd.
WANG Hui
Location Yuncheng , Shanxi
Completion 2016
Photo YANG Chaoying , YIN Jie

从西侧思庭看五龙庙山墙

村民广场与五龙泉遗址

框景中的五龙庙

村民与五龙庙

剖面　0 2 5 10m

四叶草之家外观

　　四叶草之家是一座像家一样的幼儿园。由于用地紧张，幼儿园的经营者决定把自用住宅改建，将原有的家庭私立保育所扩建为儿童教育机构。

　　改造从如何处理这幢 105 平方米的二层小住宅开始。原建筑本是从房屋建设公司购买来的装配式成品，为全木结构。MAD 决定保留并利用这座住宅的主体木结构，使它成为新建筑结构的一部分。被保留的坡屋顶木框架不但造就了有趣的室内空间，对房屋的主人，幼儿园的经营者来说，也是对家的一种纪念。新的房子则像一块布一样，包裹着房屋老的木结构，并形成一种全新的混沌的空间。对孩子们来说，原本的木结构好像记录了四叶草之家的传统和故事，而对日常使用来说，这个木结构也是孩子们主要的学习空间，它既通透又有围合感，灵活地适应着不同的教学内容。从四周的窗户射进来的阳光，给它带来不断变幻的光影，追逐着孩子们好奇、天真的想象力。

The Clover House is a home-like kindergarten. The manager of the kindergarten decided to convert his own house into an educational facility due to land shortage.

The transformation began with an investigation of the existing two-stories house of 105 m². The original house was built in wooden pre-fabricated housing components. MAD decided to retain the main original structure as part of the new building. Gabled wooden roof truss preserved not only contributes to the making of interesting interior space, but also plays a role of memorizing the old house for the managers, the owner of the property. Enveloping the old, wood-frame structure with a secondary skin, MAD conceptually and spatially blurs the threshold distinction between the old and the new. The roof forms a continuous surface with the facade, as asphalt shingles clad and waterproof the structure. Punctured through the cladded envelope, distinctively-shaped windows cast sunlight to the interior and create ever-changing shadows, chasing the students' curiosity and innocent imaginations.

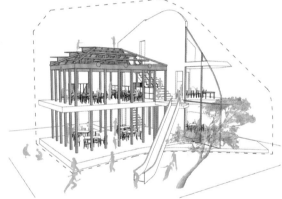

轴测

建筑庭院里的滑梯，孩子可从二层滑滑梯到庭院

四叶草之家 / Clover House

日本 爱知县 冈崎市　｜　竣工时间 2016 年

建筑设计　MAD 建筑事务所　马岩松 早野洋介 党群

摄影　Fuji Ko; Ji Dan Honda

Architects MAD Architects
MA Yansong, Yosuke Hayano, DANG Qun
Location Okazaki, Aichi, Japan
Completion 2016
Photo FUJI Ko; Ji Dan Honda

充满记忆的老木头屋架
孩子们在专属他们的"角落"空间中玩耍

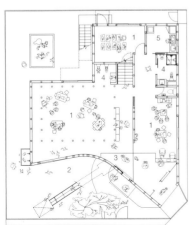

一层平面　0 1 2 5m

1 教室
2 操场
3 入口
4 浴室
5 厨房

房屋新结构包裹住老结构，又稍微脱开
如山洞般的建筑入口

A 栋与 2 号楼之间形成的室外展场

作为乌镇 2016 国际艺术邀请展主展场的北栅丝厂建于 20 世纪 70 年代,后破败荒废。建筑师希望尊重既有格局和空间特征,在此基础上与新的城市环境积极互动并适应新的功能。

改造尽可能保留原有建筑质朴的外观,并进行了结构加固和屋顶翻修处理。只将最高的老建筑用深灰色铝板网进行包裹,在新旧对比中赋予其朦胧而现代的独特气质。扩建的三栋新建筑通过并置、插入、连通等不同的方式,来缝补厂区与城市及未来的割裂关系,营造出室内外一系列具有创意感的艺术交流场。所有现状树都被保留下来,并因地制宜地与建筑产生对话,最终让建筑与环境形成一种更为紧密的新状态。

As the venue of Wuzhen 2016 International Contemporary Art Exhibition, Beizha silk factory was created after the renovation of an abandoned 1970s factory. In the renovation original layouts and spatial features of the site are conformed to adapt new context and new program.

Structure reinforcement and roof refurbishment were generally taken in the reconstruction while keeping the original simple exterior as much as possible. Only the tallest old building was wrapped with dark grey stretched aluminum mesh, which brings an intriguing image of the old and new. Three new buildings were added by different approaches such as Juxtaposition, Insertion and Connection to remedy the gap with urban context and bring creative new indoor and outdoor spaces for art exhibition. All exist trees are reserved and mapped in the reconstruction, specific methods are taken to handle the dialogue between trees and buildings, hence a new status which combines the buildings and the environment emerged.

C 栋东侧二层观景平台

厂房展厅

乌镇北栅丝厂改造
Renovation of Beizha Silk Factory in Wuzhen

浙江 桐乡   竣工时间 2016 年

建筑设计
上海道辰建筑师事务所
陈强 付娜 陈剑如

摄影 艾清 吴清山

Architects
Design Creates Atelier
CHEN Qiang, FU Na, CHEN Jianru
Location Tongxiang, Zhejiang
Completion 2016
Photo AI Qing, WU Qingshan

B栋艺术品商店夜景
主入口夜景

1 门厅
2 序厅
3 展厅
4 艺术品商店/咖啡厅
5 室外展场
6 入口坡桥
7 内院
8 水池
9 室外阶梯展台

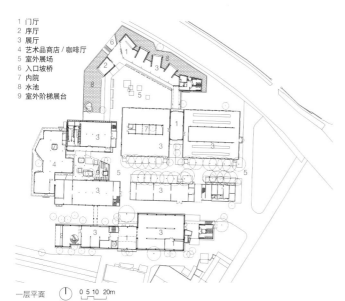

一层平面　0 5 10 20m

C栋大厅天桥

Renovation / Heritage Preservation

北立面夜景

项目是上海典型里弄中的一栋非典型的二层楼房子,房龄仅有十几年。原设计整体带有一点现代主义风格的造型特征,立面贴满米黄色瓷砖,在法租界一栋栋砖石结构的欧式花园老洋房中显得十分突兀。

工作室的改建设计,在保留原有房子的结构基础上,一层南侧加建阳光房,最大化加强室内与室外自然的视觉联系。在忙碌的工作间隙,偶然间的回眸一瞥,感知秋叶退黄、春草渗绿的时光变迁。二楼的天窗切斜角,将阳光温柔地引入室内。在改建细节上,地板采用传统灰色水磨石做法,配以白色涂料墙面与木色门、窗框,让室内空间更加简洁明亮。对建筑材质质感的强调更暗示着建筑师对工作的细腻敏感与精益求精。

Located in a 1990's two-story building within the typical Shanghai li-nong (alley) neighborhood, the original architecture combined the modernist aesthetics with maize-yellow ceramic tiles on the facade walls. The renovation sought to add a sunshine room based on the original construction to build a contact with indoor and outdoor closely. One of the most crucial aspects of the new design is the consideration of construction materials such as gray terrazzo floor, wooden doorframe and wrought-iron glass door panels, which implies the perfection and sensitivity of the architect.

1-1 剖面

弄堂入口

五原路工作室
Wuyuan Rd. Studio

建筑设计
刘宇扬建筑事务所
刘宇扬

Architects
Atelier Liu Yuyang Architects
LIU Yuyang
Location Shanghai
Completion 2016
Photo Eiichi Kano , CHEN Hao

上海 | 竣工时间 2016 年 | 摄 影 Eiichi Kano 陈颢

二层空间与天窗

一层空间
前厅

1 院子
2 展示区
3 储藏
4 工作室
5 阳光房
6 庭院
7 模型室
8 茶水间
9 卫生间
10 会议室

一层平面　0 1 2　5m

阳光房与后院

南向外景

西浜村昆曲学社位于昆山阳澄湖畔，这个小村子曾经是史上玉山草堂的北界，百戏之祖昆曲在此孕育而生，然而时过境迁，如今的西浜村凋敝失落，再无当年歌舞升平之象。为了恢复乡村之生命，让600年前的水磨之音再萦绕于水巷桥头，故将4座坍塌的民房改造成小戏校，取《玉山雅集》中"读书舍"之描述："舍前有修竹，舍后有芙蕖"，依诗词意境，造一座很轻的房子来解读当年之意境。建造过程尽量降低干扰，村民自由耕作于房前屋后。用最小截面的钢构再现江南民居的框架，用最轻的金属瓦、夹板墙降低荷载。竹菊梅兰4小院用竹以《牡丹亭》曲谱韵律做墙，于河边做新牡丹亭为戏台……江南的粉砖的草泥墙，压顶里搁置的废旧小青瓦，共同营造出江南水乡新与旧的篇章，小舍织补在乡村中，与乡村融为一体。

The Society of Kun Opera is located at Xibang Village by the Yangcheng Lake, and the village used to be the north border of Yushan Garden in history where Kun Opera was born. However, with the passage of time the village grows into a poor situation with no sign of prosperity. To regenerate the village and call the Kun Opera back to the watery region as 600 years ago, 4 collapsed houses are rebuilt to be a small school of Kun Opera. As described in the book of *Yushan Collection*, bamboo grows before the house and lotus are flapping behind it, the architects designed a building to show the artistic conception, and to reduce impact of construction on the local people of farming work around the buildings. Using steel structure of minimized section to represent the frame of vernacular houses in the lower Yangtze region, the lightest structure, metal tiles and plywood walls are chosen. Four small yards are designed as bamboo yard, plum blossom yard, chrysanthemum yard and orchid yard, in which bamboo walls are made to show the music score of *Peony Pavilion* by the spacing of the bamboo. Also a bamboo pavilion is built by the river as a Kun Opera stage. The white walls with grass mud, and the old Chinese-style tiles inside the top beams of these walls, both show the new and old life of the place. The new building, which should be a part of the land, is already integrated into the village.

1 入口　6 菊院　11 教室
2 梅院　7 序厅　12 兰院
3 道具间　8 竹院　13 食堂
4 多功能厅　9 化妆间　14 办公室
5 戏台　10 舞蹈教室　15 现状民居

一层平面　0 2 5 10m

西浜村昆曲学社
The Society of Kun Opera at Xibang Village

江苏 昆山　｜竣工时间 2016 年

建筑设计　中国建筑设计院有限公司
崔愷 郭海鞍 沈一婷

摄　影　张广源 蒋彦之

Architects China Architecture Design Group
CUI Kai, GUO Hai'an, SHEN Yiting
Location Kunshan, Jiangsu
Completion 2016
Photo ZHANG Guangyuan, JIANG Yanzhi

在演出的戏台与观演村民

菊院
二层房屋夜景　　　　　　　　　　　　　　　　　　　　　踏影寻歌

改造及修复

1-4号院鸟瞰

1 1-4号院
2 5号院
3 6号院

总平面

1 客房
2 阳台
3 餐厅
4 茶室
5 厨房
6 设备间
7 庭院

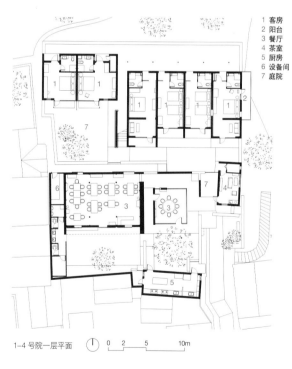

1-4号院一层平面

5号院改造后棚下空间与河道景观

乡宿上泗安
Shangsi'an Cottage

建筑设计
亘建筑事务所
范蓓蕾 孔锐

浙江 湖州　　竣工时间 2016 年　　摄影 侯博文

Architects
genarchitects
FAN Beilei, KONG Rui
Location Huzhou, Zhejiang
Completion 2016
Photo HOU Bowen

6 号院改造后的庭院
1-4 号院二层走廊
1-4 号院公共路径
1-4 号院新建客房

上泗安村位于太湖西岸，泗安塘穿村而过，一座清代石桥连接两岸。桥头的树荫和凉亭构成了村中最重要的公共空间。

业主挑选了散布在村中的六栋房子，希望将它们改造成分散式的乡间酒店。这些房子有建于清末的木屋，有杂物仓库，有贴瓷砖的二层小楼，以及一个刚落成的位于石桥北侧的仿古小展厅。建筑师将最靠近石桥的小展厅，改造成为整个酒店的起居室，提供图书和茶点，并向村民开放，其他几栋则改为客房。

这些房子散布在村子里，没有边界，也没改变村子原有的道路系统，只在铺地上做了一些暗示，村民仍可穿行其中。由于房屋现状复杂，建筑师并没有预设一套统一的手法，而是尽量回答现实所提出的问题。一座危房拆除并重新设计，瓷砖小楼调整了内部功能和内外关系，木屋则只是进行加固。村中心的小展厅，建筑师用植被和坡地替换了原来颇为城市化的花岗岩大台阶，再用轻质的棚架搭起了一个户外歇息的空间，村民和游客都可以在这里喝茶聊天，环绕的溪水缓缓流过。

Located on the west shore of Taihu lake, Shangsi'an village is flown through by Si'antang creek. A historic stone bridge connects two sides of creek. There is a significant public space near the bridge, which consists of evergreen trees and existed pavilions.

Six buildings scattered throughout the village were chosen by the client in order to be transformed into a distributed country hotel. Among these six buildings, there is a traditional style exhibition hall nearby bridge. Due to its unique location, The architects transformed it into a public living space for the entire hotel, which serves as a teahouse and library opened to both guests and villagers. Other buildings including a timber house from late Qing dynasty, a warehouse, as well as three two-story buildings with tile facade has been converted into hotel rooms.

Without new fences, these buildings stand with original local paths and yards, with only some hints on pavement. In this case, villagers might pass through the hotel area in their daily life. The architects try to answer questions with the buildings' distinguishing situation. For example, they redesigned unsteady buildings after demolition, adjusted the tile facade buildings' functions and reconnected them to landscape, simply strengthened the timber house. Moreover, in front of the small exhibition hall located, the urban-flavor granite stairs has been converted back into grassy slope. The architects also constructed an outdoor space by setting up a pavilion, so that both villagers and guests can stay and chat there, feeling the natural breath surrounded by creek.

池舍正立面夜景

池社，作为在破败不堪的老建筑的基础上改造完成的艺术空间，提供了一处精致但丰富的复合艺术空间，在紧凑的建筑内叠合了展示收藏、创作讨论、休息交流等多重艺术活动。

设计保留了原有建筑的外围护墙体，进行基本的性能改善和结构加固后获得最大化的展厅空间；同时在不影响整个园区空间感受的情况下局部将建筑屋顶抬高，获得一处可以享受完整天空的夹层休息空间；屋面结构换为更轻质并富有温暖气息的张拉弦木结构屋顶，并局部抬高获得天光。池社的外墙，将回收自老建筑的古老灰砖与先进的机械臂现场建造工艺相结合，采用一个具有张力的曲面形态，展现了勃勃生机，实现了数字化施工技术在现场完成真实建造的首次尝试。

Renovated from a dilapidated building as an artistic space, Chi She provides a delicate and abundant compound art space, which contains various artistic events, such as curiosities exhibition, creative workshop and unpremeditated communication.

In order to attain the maximum exhibition space, The architects retained the initial exterior walls followed by the elementary performance enhancement and structure reinforcement. Therefore, under the condition of maintaining the space perception of the whole artistic park, part of the roof has been elevated in order to create an interlayer space, where people could enjoy the intact sky view. Furthermore, the roofing structure has been replaced by a lightweight and more efficient tensioning string timber structure, and part of them is lifted to obtain skylight. The external walls of the Chi She were built by the recycled grey green bricks from old buildings and constructed with the help of the robotic arm technology, which generates a cambered surface morphology, showing the vitality of the Chi She and realizing the first endeavor to utilize the on-site digital fabrication.

池社 / Chi She

建筑设计：上海创盟国际建筑设计有限公司
袁烽 韩力 孔祥平

上海　　竣工时间 2016 年　　摄影 苏圣亮 袁烽

Architects Archi-Union Architects
Philip F. YUAN, Alex HAN, KONG Xiangping
Location Shanghai
Completion 2016
Photo SU Shengliang, Philip F. YUAN

改造及修复

池舍雨篷与立面一体化折叠 | 池舍内部楼梯
池舍内部展览空间

池舍内部空间

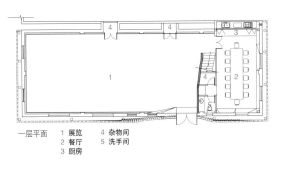

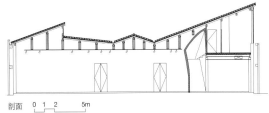

一层平面　1 展览　4 杂物间
　　　　　2 餐厅　5 洗手间
　　　　　3 厨房

剖面　0　1　2　　5m

Renovation / Heritage Preservation

苏州河畔的四行仓库由两座仓库组成，西部的"四行仓库"建于1935年，原高5层，主要为钢筋混凝土无梁楼盖结构体系。这里是1937年淞沪会战中闻名的"四行仓库保卫战"的发生地，是市级文物保护单位。

本次保护与复原设计以尊重历史真实性为原则，用多种方法查明西墙在抗战时的炮弹洞口位置，力求准确复原梁柱边的洞口并采取多种创新技术确保建筑安全；拆除7层后期搭建，6层后退，恢复南北立面历史原貌；恢复原中央通廊特色空间并将其改作中庭，其西侧设立"抗战纪念馆"，彰显抗战遗址历史意义；其余部分提高舒适度，作创意办公等使用。

The project consists of two warehouses on the shore of Suzhou Creek. The west one of five storeys is Joint Trust Warehouse, which was built in 1935. The main structure of this five-storeys building is a reinforced concrete slab-column frame. In Battle of Shanghai, 1937, the well-known "Lone battalion" took a final stand against the Japanese troops before retreat here. Now the warehouse is a Municipality Protected Historic Site to memorize this "Defending the Joint Trust Warehouse Battle".

Historical information is highly respected and authenticity is taken as design principle in preservation and restoration. In the battle the west wall of the warehouse was torn by the explosive shells. Several methods have been taken to locate the exact positions of these damage parts and make sure they are revealed in a way as same as in the battle time. At the same time various new techniques have been used to ensure safety of the building. The 7th floor, an additional construction, was removed. The external wall of 6th floor has been moved inward so that the north and south facades could be restored to the original. The central passage is restored and used as a characteristic atrium, west of which is used as a "Memorial Hall for Anti-Japanese War" to demonstrate the significance of this historic site, while the quality of the rest is highly improved to meet the demand of "office for creativity".

远景鸟瞰
南立面

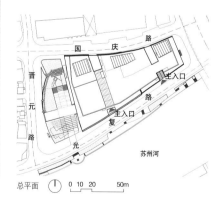

总平面

四行仓库修缮工程
Protection and Restoration of the Joint Trust Warehouse

上海 | 竣工时间 2016年

建筑设计 上海建筑设计研究院有限公司
唐玉恩 邹勋
摄影 邵峰

Architects Institute of Shanghai Architectural Design & Research Co., Ltd.
TANG Yu'en, ZOU Xun
Location Shanghai
Completion 2016
Photo SHAO Feng

南立面主入口
中庭

西立面弹孔墙及北立面
街景

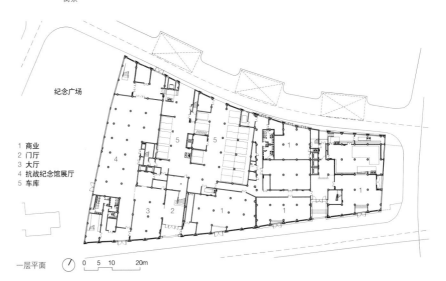

纪念广场

1 商业
2 门厅
3 大厅
4 抗战纪念馆展厅
5 车库

一层平面   0 5 10 20m

村落鸟瞰

谢店村位于湖北省东北部，毗邻山水一色的尾斗湖畔，村落顺应地势，延绵布局；灰墙石基，瓦屋栉比，是一处山水环抱的遗世。村民至今保留传统的生活和生产方式，乡风淳朴。

规划秉持最小干预原则，遵循低冲击、低干预、低消耗、低维护的设计理念，通过最少的场地介入、最低的环境干预与最高的场地环境资源利用来减少对传统村落的冲击；优先选择乡土植物、石材等地域材料；鼓励村民、社会群体参与其中；种植有观赏价值的经济类作物，对外租赁绿地作为家庭花园，在绿地周边设置餐饮休憩服务设施回笼资金，作为传统村落可持续发展的有效途径。规划设计延续村落空间肌理，从修复建筑、梳理道路、整治院落、丰富植被，柔化驳岸五大方面进行改造。

Xiedian Village is located in the northeast of Hubei Province, adjacent to magnificent Weidou Lake. The site is closely connected with mountains and water, conforming to the terrain. There are lots of traditional features the architects would like to keep such as stone wall base, grey wall and roof tiles. Xiedian Village is like a wonderful lost world waiting to be discovered. Villagers keep original styles of living and production, simple but rich in cultural meanings.

Adhered to the principle of "minimal intervention", the design concept of low impact, low intervention, low consumption and low maintenance, through the minimum site intervention, lowest environment intervention and the highest use of environment resources is adopted to reduce impact on the ancient village. Priorities are given to local plants and materials, and villagers are encouraged for participation . It is an effective way for sustainable development of traditional villages and making profit by cultivating economic crops and leasing the green space as a family garden with catering facilities to return investment, as an effective way to the sustainable development of traditional villages. The design principle aims to continue with the local context in terms of building restoration, road construction, courtyard repairing, plant diversification, and softening the revetment.

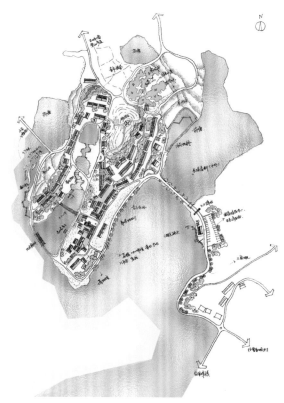

总平面

谢店村传统村落保护与再生规划设计
Planning of Xiedian Traditional Village Protection and Regeneration

湖北 麻城　　竣工时间 2016 年　　摄影 李阳

建筑设计　中信建筑设计研究总院有限公司
肖伟　何欣然　李伟强

Architects CITIC General Institute of Architectural Design and Research Co., Ltd.
XIAO Wei　HE Xinran　LI Weiqiang
Location Macheng Hubei
Completion 2016
Photo LI Yang

改造及修复

沿湖栈道
驳岸景观
巷道外景

无边水池

天目湖微酒店位于天目湖旅游度假区，东南西三面被湖水环抱，景观得天独厚。酒店只有35间客房，充分发挥景观优势，让下榻至此的客人感受四季变化和自然妙趣是设计的重点。设计采用半岛式布局，使酒店坐拥南山，视野开阔，望见整个天目湖。又用庭院串联起空间，精致的细部演绎出东方文化特有的优雅之美。

Located in the Tianmu Lake Tourist Resort, Tianmuhu Lake VIP Club is surrounded by the lake with fantastic views. While this Hotel only features a total of 35 guest rooms, it features a panoramic view, allowing the guests to fully immerse themselves within the change of seasons and delight of nature. A peninsular style layout is adopted for the hotel, allowing the hotel to embrace the southern mountains and overlook the whole Tianmu Lake. Space is connected by courtyards with exquisite ornaments of unique Oriental elegance.

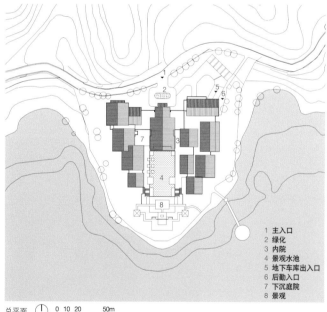

1 主入口
2 绿化
3 内院
4 景观水池
5 地下车库出入口
6 后勤入口
7 下沉庭院
8 景观

总平面 0 10 20 50m

天目湖微酒店
Tianmu Lake VIP Club

建筑设计 上海建筑设计研究院有限公司
程明生 张萌 王维

江苏 溧阳 | 竣工时间 2016年 | 摄影 由业主提供

Architects Institute of Shanghai Architectural Design & Research Co., Ltd.
CHENG Mingsheng, ZHANG Meng, WANG Wei
Location Liyang, Jiangsu
Completion 2016
Photo By client

改造及修复

临湖方向外景
酒店入口

标准客房　　　　　　　　　　　　　庭院内景

街道

酒店位于杭州大兜路历史文化街区,改造前为两幢已破败的四层安置房建筑。本项目旨在无形中向人们传递"空间才是居住体验的灵魂"的理念,保留原有建筑的基本形态,重新梳理建筑的空间、体量、流线。一栋建筑呈I型,直面运河;另一栋建筑呈L型,安静雅致。在两单体间置入玻璃盒子作酒店大堂,连接底层空间,对外为主入口。酒店外部增加围墙,三面围合,界定空间,自然而成的中庭内向收敛。形体设计从"加、减"两个角度切入。先做"减法",拆除多余建筑体量,再做"加法",填平立面。

Seclusive Jiangnan Boutique Hotel is located in Dadou Road Historic District, Hangzhou. Before the renovation, the existing structures used to be two dilapidated affordable apartment buildings. The project aims to convey the idea to the public that "spatial quality is the key for living experience in residence". The new design keeps the original form of the buildings, but reorganizes the circulation and spatial divisions. The I-shaped building faces the canal while the other L-shaped building is quiet. Designers insert a glass box as the hotel lobby that connects the two separate buildings as the main entrance. In this way, it forms a courtyard as a space enclosed on 3 sides. Excessive parts are demolished and some added, so as to keep the facades clean and straight.

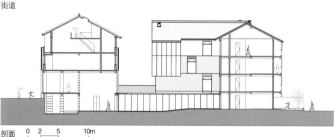

剖面　0　2　5　　10m

隐居江南精品酒店
Seclusive Jiangnan Boutique Hotel

| 浙江 杭州 | 竣工时间 2016 年 |

建筑设计
gad·line + studio
浙江绿城建筑设计有限公司
孟凡浩　李昕光

摄　影　范翌

Architects
gad·line + studio; Zhejiang Greentown Architectural Design Co., Ltd.
MENG Fanhao, LI Xinguang
Location Hangzhou, Zhejiang
Completion 2016
Photo FAN Yi

改造及修复

入口

巷道
庭院

改造示意

室内

1 入口门厅　5 陈列展示
2 大堂接待　6 会议
3 图书阅览　7 厨房
4 办公　　　8 客房

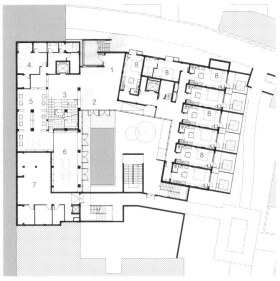

一层平面　　0 2 5 10m

Renovation / Heritage Preservation

一层空间

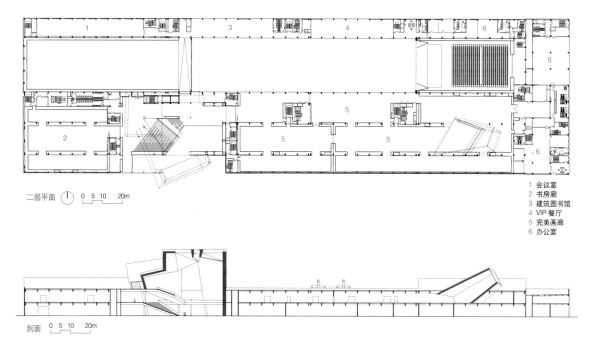

二层平面　0 5 10 20m

1 会议室
2 书房廊
3 建筑图书馆
4 VIP 餐厅
5 完美画廊
6 办公室

剖面　0 5 10 20m

北京民生现代美术馆
Minsheng Museum of Modern Art

建筑设计
朱锫建筑设计事务所
朱锫

Architects
Studio Zhu-Pei
ZHU Pei
Location Beijing
Completion 2015
Photo FANG Zhenning

北京　　竣工时间 2015 年　　摄　影　方振宁

剧场阶梯
二层中庭

一、二层空间

外观
庭院

民生现代美术馆由一座20世纪80年代的工业建筑改造而成。改造设计塑造了大小不一、尺寸各异、层高显著不同的展览空间，它们有机地组织在一个充满张力的中心空间周围，结合美术馆前装置公园、屋顶展览平台、中心院落等开放式展览空间，构成一组尺度不同形态各异的空间组群。未来美术馆不再是成功艺术家呈现辉煌的圣殿，而是激发公众和艺术作品及艺术家互动、交流的艺术场所。空间不再是为呈现作品而作，更是为艺术创作而生。一些灵活可变、功能不明、有用无用的空间，却可激发艺术家和公众的创作激情。为特定环境和场地而创作，让艺术品、公众和美术馆融为一体。

The gallery is converted from an industrial building built in the 1980s. The renovation design results in a set of exhibition halls of various sizes and heights, all of which are organized in a central space full of spatial tension. Combined with open exhibition spaces including the sculpture park in front of the gallery, a roof deck for exhibition and central courtyard, etc., the group of spaces is in place. The gallery of the future is no more a sacred temple to exhibit masterpieces of successful artists, but rather an artistic place to encourage interactions and exchanges between the public and artists with their works. Space is no more a product for representation but rather the genesis of artistic creation. Some space of flexibility and ambivalence is potential to inspire the creativity of artists. Designs hence are made as a result of specific surroundings and locations to integrate the artworks, the public and the gallery as a whole.

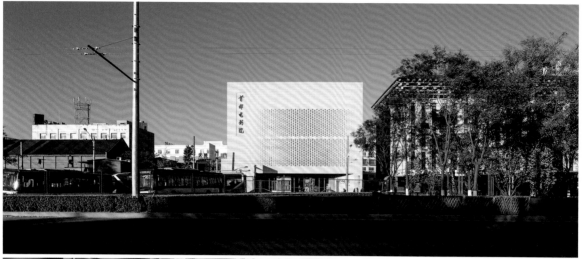

东立面
室内门厅

中华巨幕影厅

首都电影院作为天桥地区重要的文化节点，具有悠久的历史，并于1987年到2014年期间，进行过多次修缮改造。此次在保持原有建筑结构的基础上，对内部空间进行改造，将二层改造为中华巨幕影厅，首层及地下设置普通小厅，并结合公共空间布置零散商业空间。

建筑立面主体风格与场地两侧的天桥艺术大厦和天桥艺术中心相呼应，周身装饰淡黄色石材，局部采用石材错位砌筑的形式，形成镂空的表皮肌理，既与天桥艺术中心表皮肌理相契合，又体现一定的时代特征，保证了北京城南中轴线建筑风格的连贯性。

As the central point in Tianqiao district, the Capital Cinema has undergone many repairs and renovation from 1987 to 2014. On the basis of preserving the original structure of the building, the architects transform its internal space, clean up the second as a auditorium, which is the biggest IMAX auditorium in Beijing, reform the first and the underground floor, set up two small general halls, and arrange the scattered stores in the public space.

On the facade of the building, the architects choose yellow stone which is randomly constructed intentionally in some degree, forming hollow skin texture, which coordinates with the Tianqiao arts center, and shows a strong characteristic of the times, guarantees the consistant architectural style on the south of Beijing Central Axis.

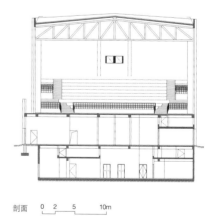

剖面　0　2　5　10m

首都电影院装修改造
Preservation and Reconstruction of the Capital Cinema

北京 ｜ 竣工时间 2016 年

建筑设计
中国中元国际工程有限公司
郭骏 刘鼎纳 刘洪涛

摄　影　楼洪忆

Architects
China IPPR International Engineering Co., Ltd.
GUO Jun, LIU Dingna, LIU Hongtao
Location Beijing
Completion 2016
Photo LOU Hongyi

改造及修复

入口门廊

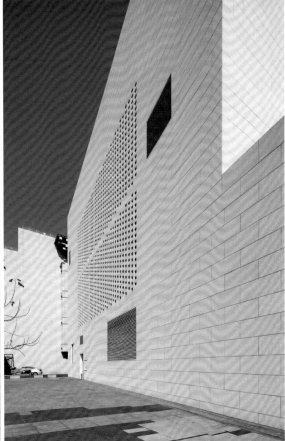

西南角外景

东北角外景

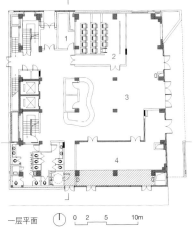

二层平面

一层平面　0　2　5　10m

1 放映间
2 VIP 影厅
3 门厅
4 商业
5 中华巨幕影厅

南向外景
公共窗

阳台

1 居住单元
2 屋顶阳台
3 公共休闲平台
4 新增电梯厅
5 电梯厅
6 消防楼梯
7 原有建筑物外墙

五层平面　0　2　5　10m

居住集合体 L
Housing L

建筑设计
汇一建筑设计咨询(北京)有限公司
徐千禾 梁幸

Architects
in:Flux Architecture
Chien-Ho Hsu, LIANG Xing
Location Laiyang, Shandong
Completion 2015
Photo SU Shengliang

山东 莱阳　　竣工时间 2015年　　摄　影 苏圣亮

居住层公共走廊

这是一个将带有明显时代特征、空置十多年的烂尾楼改造为城市住宅的项目。在不调整原有结构柱网的原则下，分割出的244户居住单元被置入在15 000平方米的混凝土构造物中，尺度和功能各异的公共空间穿插其间。各楼层公共露台的设置消化了建筑平面不适合居住的区域，以弱化对于一个地级市来说原有过于巨大的建筑体量对城市空间的影响，同时在垂直向度上为各楼层住户提供一个能发生多样活动的公共空间。单元的室内规划避免了单一功能非停留空间的存在。必要的食、住和清洁等生活机能，结合交通空间，再通过必要的储物空间的配置串联成为一个高弹性的生活综合体，而这些单一功能空间边界的模糊化处理，使得居住空间整体具备了更多使用功能的自主性和更高的效率。

"Housing L" is a renovation project of an office building that had been abandoned for more than ten years. Without modifying existing concrete structure, three types of 244 residential units (from 22 to 38 m²) are placed on eight floors, and various functional parts of different sizes are arranged in between the grid. Terraces on various levels accommodate areas unsuitable for living, reducing negative impact of enormous building bulk on urban space of a medium-sized city, and offering a public space containing various kinds of events on the vertical dimension. The interior design avoids exclusiveness, and combines necessary functions such as eating, living, cleaning, transportation, storage, etc. as a highly flexible living complex. Functional boundaries are blurred so that living space attains multi-functional autonomy and high efficiency.

阳台
36m² 居住单元
入口大厅

微园外观

白厅

微园
Wei Yuan Garden

建筑设计
东南大学建筑设计研究院有限公司
葛明

江苏 南京　　竣工时间 2015 年　　摄　影　赖自力 孔德钟 耿涛 贾安明

Architects Architects & Engineers Co., Ltd. of Southeast University
GE Ming
Location Nanjing, Jiangsu
Completion 2015
Photo LAI Zili, KONG Dezhong, GENG Tao, JIA Anming

墨池东　　　　　　　　　　　复厅

项目是在并无关系的一组老房子之间经营扩建加建，改造为书法馆，以期小中见大，自成微园。将原单层厂房坡顶各向两侧接续一跨，应对书法展示，并将坡顶桁架包覆，使原厂房进深空间转向而与院子水平连通。借力于结构制造起落，降低视点，以回溯宋时曾经具有的特殊视高的空间观法。置石理水，植树培土，均以连接内外为要，以期扩大空间容量（房），以期眼前有景（园）。

By extension and addition, a group of unrelated old buildings are renovated into a calligraphy museum as a whole, i.e. Wei Garden, in the hopes of imagining the big from the small. The one-storied building extends one bay of the roof truss on either side, used as a calligraphy exhibition room. By wrapping the trusses, the former direction of space in depth veers to connect with the courtyard outside. Structure helps create the undulating profile and lowers the eye level, in order to recall the spatial viewpoints once possessed in Song Dynasty. The layout of stone and water, tree planting and earth cultivation are well designed in accordance with the principle of continuity of inside-outside, in the hope of expanding the spatial capacity (like the house) and offering sceneries (like the garden).

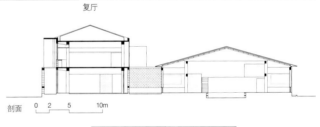

剖面　0　2　5　10m

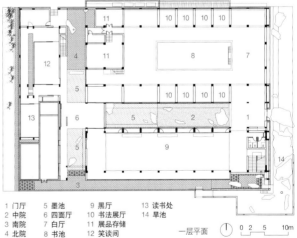

1 门厅　5 墨池　9 黑厅　13 读书处
2 中院　6 四面厅　10 书法展厅　14 旱池
3 南院　7 白厅　11 展品存储
4 北院　8 书池　12 笑谈间

一层平面　0　2　5　10m

白厅　　　　　　　　　黑厅

纪念馆东南角
入口内院与东廊

陈化成纪念馆移建改造
Removal Renovation of Chen Huacheng Memorial

建筑设计
上海阿科米星建筑设计事务所有限公司

上海 | 竣工时间 2015 年 | 摄 影 唐煜

Architects Atelier Archmixing
Location Shanghai
Completion 2015
Photo TANG Yu

陈化成纪念馆原址是上海宝山临江公园内一座巍峨的孔庙,因为孔庙恢复需要移建,业主要求利用公园内一处折尺形的小型附属用房做立面改造。设计师说服业主,在相同的造价下,改用空间整理的方式来重新组织流线及氛围。设计主要引进了钢木结构的连续开敞围廊,塑造出端庄有序的外观和空间序列,并采用精准的构造和低沉的调性处理,实现了历史纪念馆应有的纪念性。同时开放的边界跟公园周边环境互动,形成了积极而融合的日常公共活动场所。

Originally located in a magnificent Confucian Temple, now relocated to elsewhere, in Riverside Park of Baozhan District, Shanghai, Chen Huacheng Memorial Hall uses an adjacent commercial building of a zigzag plan with renewed surface. Atelier Archmixing persuaded the client to change this facade renovation program into that of a spatial reorganization. By introducing continuous open galleries with steel-timber structure, the architect produces the design of orderly appearance and spatial sequence with accurate details and modest color, contributing to the monumentality of a historical memorial. The design achieved an appropriate sense of remembrance, and also created a positive everyday public space integrating with the surrounding landscape.

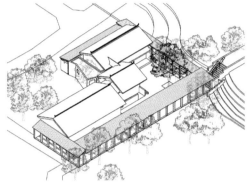

增设游廊轴测

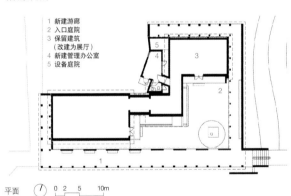

1 新建游廊
2 入口庭院
3 保留建筑（改建为展厅）
4 新建管理办公室
5 设备庭院

平面　0 2 5 10m

从南廊望东廊
南廊内景

内外空间的渗透

外立面改造后的夜景
从二层办公室看中庭

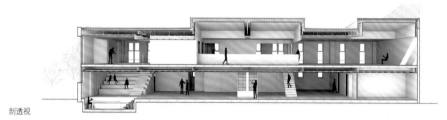

剖透视

竞园 22 号楼改造
Jingyuan No.22 Transformation

北京　　竣工时间 2015 年

建筑设计
C+ Architects 建筑设计事务所
程艳春

摄　影　夏至

Architects
C+ Architects
CHENG Yanchun
Location Beijing
Completion 2015
Photo XIA Zhi

玄关　　　　　　　　　　　　二层西侧开放办公区

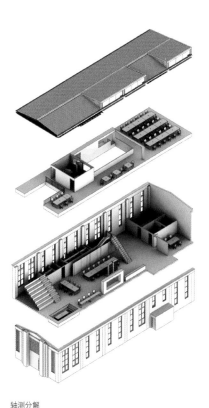
轴测分解

一层下沉讨论区及大台阶
中庭及滑梯

竞园22号楼是一个将旧棉麻仓库改造成联合办公空间的项目，业主为两家年轻的互联网金融公司。建筑师利用互联网式思维和体验式空间的共通之处，以空间趣味性和灵活性为出发点，创造出具有多重办公可能性、充满互动的叙事空间。

植入了钢结构夹层的竞园22号楼共有两层，每层平面交通均呈"回"字形。中庭不仅构成了整栋房子的视觉中心，其下方的大型会议区和投影墙同时也成为了功能中心，是公司集会的主要场所。围绕中心生成的办公空间具有很大的开放性，保证员工之间能够随时进行无障碍交流。改造后的主入口是一个盒式暗空间，为自行车爱好者安装有悬挂自行车的装置。这个黑盒子与室内的主楼梯紧密联系并形成了一条主要交通动线。次入口位于建筑物西侧，与大阶梯相连，是员工进出的辅助线路。此外，一条隐藏在隔墙里的滑梯可以让人快速从楼上到达楼下的洗手间和休息室，更丰富了室内垂直交通。屋顶打开了天窗，为室内空间引入更多自然光线。

Jingyuan No.22 project transforms an old cotton warehouse with a ground floor area of 330 m² to a joint office owned by two young Internet financial companies. The design concept, originated from the concentration of space interest and flexibility, aims to connect the experiential space with Internet thinking to create the space of multiple possibilities and narratives.

Inserted with a steel-structured interlayer, each level of this two-storied has an independent looping circulation system that is organized along the central courtyard, as the visual focus of the building. The main conference space beneath and a white wall specially for projection within courtyard become the functional core as the main venue for company meetings and entertainment. The open working space around the courtyard ensures efficiency of teamwork and communication between employees. The dark box-style main entrance is also equipped with bike wall mounts particularly benefit to bicycle enthusiasts. The stairs in black connects to the entry box leads employees directly to the coworking space and formed the major interior circulation. The secondary entrance located in the west of building along with the grand staircase is an auxiliary route. Furthermore, in order to enrich the vertical circulation, the "ninja channel", a slide hidden between partition walls from where programmers can quickly reach the restroom and bedroom downstairs is subtly built. Newly opened skylights on the roof invite more sunlight inside.

改造后的主楼远景
改造中的顶层空间

项目位于上海杨浦区凤城路，该厂建于1985年。这幢建筑此前多次被改造的痕迹层层叠叠。直到清理掉四处堆放的杂物、局部的吊顶、临时的棚架，这座老建筑的骨骼才缓缓站了出来。厂房的结构是装配式的，预制混凝土构件像积木一样，简简单单地垒了上去，柱头上扛着梁，梁则搁着槽板，一层接着一层。建筑师意图保留这种明了的感觉，也让走在马路上的人同样能看到它清晰的骨架。

设计把朝向城市的南面打开，换上玻璃，并让玻璃幕墙后退，露出了一排排结构，这也让建筑里的人能够走出来，在阳台上吹吹风。南边是一片在城市中难见的开阔景色。立面打开之后，室内也变得明亮。到了夜晚，预制的槽板被灯光打亮，混凝土的结构从透明的正面展示出来，远远就能看到。为了这份简单，建筑师小心地组织空调设备、消防、管线、雨水，因为它们全都暴露在外。在一层，实墙也被打掉了，整个场地连成一个通透的整体，将来这里会容纳商业，为上面的办公人员和周围的居民服务。

Located on Fengcheng road in Yangpu district of Shanghai, stands the former Fuli clothing factory, which was founded in 1985. It was not the first time that the factory was renovated. After the garbage surrounding the building is cleaned up, and suspended ceilings and temporary scaffold removed, the structural frame of the old building slowly appeared. The factory was assembled with prefabricated concrete components, simply piled up as toy blocks: columns supporting the beams, beams carrying the slabs, and one layer atop another. The architects wanted to keep this distinct feeling for pedestrians walking on the city streets to see this frame.

The southern facade is opened up with changed glass. Rows of structural components are exposed due to the set-back glass curtain wall, which forms a balcony. It invites people to enjoy the rarely broad southern view towards the urban area. By opening up the facades, natural light is also introduced inside. Prefabricated grooved panels are lighted up in the evening, displaying the concrete structure form the transparent facades that can be seen from afar. For this purpose, conditioning equipments, fireproofing and piping are well organized as they are exposed. On the ground floor, solid walls are removed so that the site is connected as a whole, open to accommodate shops that serve people above and around.

富丽服装厂改造 Renovation of Fuli Clothing Factory	建筑设计 亘建筑事务所 范蓓蕾 孔锐		Architects genarchitects FAN Beilei, KONG Rui Location Shanghai Completion 2015 Photo HOU Bowen, CHEN Hao
上海	竣工时间 2015 年	摄 影 侯博文 陈颢	

改造及修复

改造后的室外场地

改造后主楼南向凹阳台
沿街的弧形透明界墙

改造后的东立面——清晨的街景

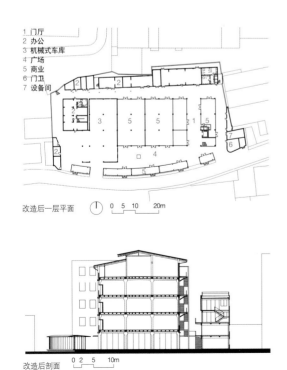

1 门厅
2 办公
3 机械式车库
4 广场
5 商业
6 门卫
7 设备间

改造后一层平面　　0 5 10 20m

改造后剖面　　0 2 5 10m

报告厅和咖啡厅（原宾馆游泳池）

通过功能置换的改造与再生把宾馆建筑变成一个能够满足日益多元的使用需求的创意设计园区。

改造设计的原则是尊重原有建筑的总体布局，保留原有树木；由于功能的需要而不得已的加建，在形体上和原有建筑尽可能地呼应，采用低调的设计手法，消隐建筑体量。立面的改造遵循"新旧混成"的原则，既保留历史的痕迹和记忆，同时也加入新时代的元素，以及自主研发的"离瓦"的运用，体现创意型企业的文化和时代的特征。充分利用基地原有的三个庭园并新增加了一个庭园，围绕建筑群形成了四个不同主题的庭园空间，营造了高效率、高舒适度和高附加值的办公空间。

Through the transformation and regeneration of functional replacement, the hotel is transformed into a creative design park.

The principle of renovation design pays homage to the overall layout of the original building and retain original trees on site. Due to the functional needs, the architects added parts echo to the original building formally at best. The AR use low-key design methods to downplay the building volume. The renovation of the facades follows the principle of "old and new mixed", preserving both the traces and memories of history, as well as the elements of a new era. The application of the "offset tile" as independent research achievement reflects the creative enterprise culture and the characteristics of the times. The architects make full use of the original three gardens and add a new garden, which formed four different themed garden spaces around the complex, creating an office space of high level of efficiency, comfort, and value added.

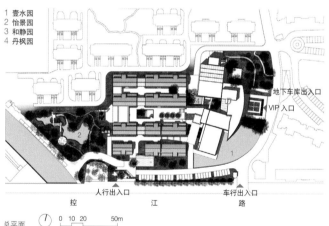

1 壹水园
2 怡景园
3 和静园
4 丹枫园

总平面

咖啡厅室内（原宾馆泳池）

上海联创国际设计谷
Shanghai UDG International Design Valley

上海 | 竣工时间 2015 年

建筑设计
联创国际设计集团
东南大学
钱强

摄影 姚力 钱强

Architects
United Design Group Co., Ltd. ; Southeast University
QIAN Qiang
Location Shanghai
Completion 2015
Photo YAO Li , QIAN Qiang

改造及修复

小楼办公区鸟瞰（原宾馆客房区）
壹水园主楼东南侧外观

主楼东侧外观
和静园中会议室

从C楼看A、B楼及花园

A楼南立面

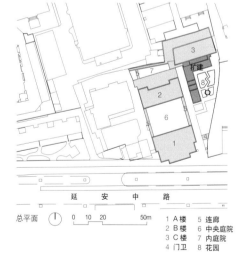

总平面　0　10　20　50m

1 A楼　　5 连廊
2 B楼　　6 中央庭院
3 C楼　　7 内庭院
4 门卫　　8 花园

上海延安中路816号修缮项目——解放日报社
Renovation Project of 816# Middle Yan'an Road: the Jiefang Daily Office

上海　　竣工时间 2015 年

建筑设计　同济大学建筑设计研究院（集团）有限公司
原作设计工作室
章明　张姿

摄影　章勇

Architects Original Design Studio, Tongji Architectural Design (Group) Co., Ltd.
ZHANG Ming, ZHANG Zi
Location Shanghai
Completion 2015
Photo ZHANG Yong

改造及修复

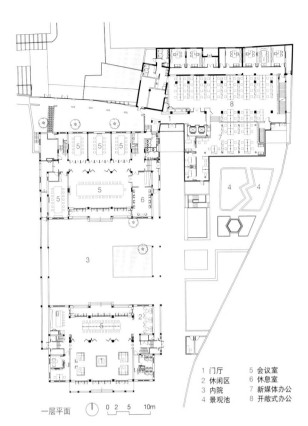

1 门厅　　5 会议室
2 休闲区　6 休息室
3 内院　　7 新媒体办公
4 景观池　8 开敞式办公

一层平面　0 2 5 10m

入口门厅
休闲讨论区

本项目为保护性修缮工程，其中，严同春住宅（A、B楼）为优秀历史建筑。设计包括外立面、建筑内部的天棚、墙面、地面、室外场地及花园、电梯等。同时根据解放日报的日常办公功能对室内空间及功能进行改造设计。将历史建筑与解放日报的历史、人文相结合，体现历史街区的当下社会价值及文化内涵。

The design is a protective repairing project. Buildings A and B which are dwellings of Yan Tongchun in this project are excellent historic buildings. The design objects involves facades, interior ceilings, walls, the ground, exterior space and garden, and elevators and so on. At the same time the architects redesign the interior space and related functions according to the demand of daily office work inside *Jiefang Daily* Office. Combining historic buildings with the history and culture of *Jiefang Daily* Office, what they want to present is the social value and cultural connotation of the historic district.

C楼下沉庭院

下沉书屋与庭院

项目位于洱海畔，是一个基于原有农宅的改扩建项目。建筑师对设计建造的全过程——选址策划、建筑方案及施工图、室内、家具及景观设计、施工现场等环节进行着全程的把控。原有农宅建筑高度为二层，为了让建筑形式能融入村庄肌理，改扩建时在总体策略上将建筑体量化整为零。两个新建体量则引入石头墙的设计元素，石头墙作为边界存在使得新建建筑与周围邻居之间既有所区别又有所关联。用地与洱海之间隔着公路，这个设计难点在于使用者要跨过面前的公路才能欣赏到前面的洱海水景。建筑师采用了一个半下沉的公共空间以塑造一个双重的联系，同时为客人创造了多层次的公共空间体系。

Located beside the Erhai Lake, "Munwood Lakeside" is expanded and reconstructed originally on a farm house. The architects took overall control of the whole process including site planning, architecture and construction drawing design, interior, furniture, landscape design and construction site follow-up, etc. The original building was 2 stories in height. To integrate architectural form into its rural context, the overall expansion strategy follows the idea of "breaking the whole into parts". Thus two new buildings are created along with the element of stone wall, which functions as a boundary making the relationship between the new building and the surrounding neighbors distinct yet related. The building site is separated from Erhai Lake by roads. Accordingly, the challenge lies in a view of the lake before crossing the road. The architects adopts a semi-sinking public space to create a double connection, while providing a multi-level public space system.

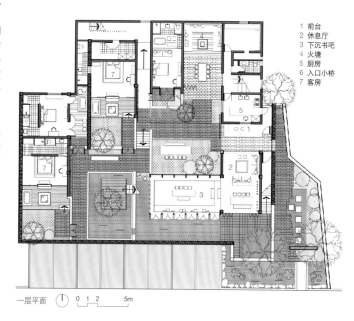

1 前台
2 休息厅
3 下沉书吧
4 火塘
5 厨房
6 入口小桥
7 客房

一层平面

大理慢屋·揽清度假酒店
Dali Munwood Lakeside Resort Hotel

建筑设计
元象建筑
重庆合信建筑设计院有限公司

云南 大理　　竣工时间 2015 年　　摄 影 存在建筑

Architects
Init Design Office; Chongqing Hexin Architectural Design Institute Co., Ltd.
Location Dali Yunnan
Completion 2015
Photo Arch-Exist Photography

建筑外景
下沉书屋内部

前台接待

建筑与周边

北侧外观

徐家汇观象台位于徐家汇天主堂南侧,与天主堂隔草坪相望。观象台经过多次改造及加固后,主体结构已成为砖、木、钢、砼的混合结构。

改造将观象台于1930年代的历史风貌作为修缮恢复的目标,因为这一时期的历史风貌较为完整地体现了该建筑的综合历史价值,且与建筑的现状形体基本一致。本着真实性、可逆性和可识别性的原则,观象台的保护修缮工作主要为:原有建筑立面风貌的保护和修复、保留基于功能要求的历史改造痕迹、使用功能调整、空间格局恢复、交通流线调整、室内历史风貌展现与当代氛围塑造的结合和建筑物理性能的恢复、建筑设备的提升。

L'Observatoire de ZI-KAWEI is located at the south of the Xujiahui Cathedral, separated by a lawn, and the Tome of Xuguangqi is just not far away at its west. After many times of renovation and reinforcement, the main structure of the building has turned to be a hybrid one of masonry, timber, steel and concrete.

The architects decide to renovate it to look like what it was in 1930s, because the building in this period has presented the comprehensive historic value and appeared to be most similar to what it looks like today. Based on the principle of authenticity, reversibility and recognizability, the refurbishment will be carried out in the following aspects: protection and restoration of the original building facade, preservation of the traces of previous renovations based on functions, adjustment of program, restoration of spatial distribution, adjustment of circulation, combination of the historic indoor style and contemporary atmosphere, restoration of physical performance of building and upgrading of mechanical equipment.

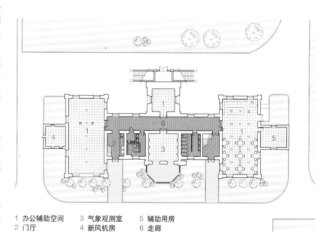

1 办公辅助空间　3 气象观测室　5 辅助用房
2 门厅　　　　　4 新风机房　　6 走廊

一层平面　0 2 5 10m

徐家汇观象台修缮工程
Refurbishment of L'Observatoire de ZI-KAWEI, Xuhui, Shanghai

上海　　竣工时间 2015年

建筑设计 致正建筑工作室
张斌　周蔚　金燕琳

摄影 胡义杰

Architects Atelier Z+
ZHANG Bin, ZHOU Wei, JIN Yanlin
Location Shanghai
Completion 2015
Photo HU Yijie

气象科普展示间
展厅

二层走廊
二层西端大展厅

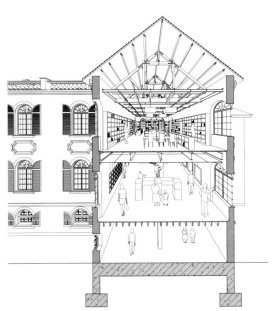

西端剖透视

南向鸟瞰

2010年落成的上海油雕院美术馆如一块棱角分明的巨石座落于金珠路旁。具有标志性的斜墙源于场地中原有的界墙，却巧妙地形成了具有导向性的入口和一个雕塑庭院。3米宽2米高的混凝土预制大板如斑驳的巨石方，累叠成严整的立面。而内部空间则以完全自由的大空间策略回应多样的展览需求。

2015年底，美术馆标志性的不锈钢网斜墙中开启了一扇隐秘的门，大烟囱咖啡馆在雕塑庭院内落成。虽然庭院本身的空间非常狭窄，最宽处也只有5米，却仍然通过园林式的迂回路径营造出了丰富有趣的空间序列。一道富于动感的弧墙引导着人们饶有趣味地通过长院来到咖啡厅入口；缀以翠竹的天井，光线明亮而生动。阳极氧化铝材料的应用使入口长院的空间骤然扩大，稍有模糊的镜面效果，给人以朦胧虚幻的视觉感受。室内白墙为小型展览和艺术沙龙创造了最有利的条件。而追随弧线的光带加强了墙的流动感，使空间更富有动态。小建筑在克服限制条件的同时，努力营造出自身的空间性格：柔软而不乏张力，自然而富有情趣。

In 2010, SPSI Art Museum was built as an abstract stone on Jinzhu road. An iconic wall leads to the entrance and encloses a sculpture courtyard. Prefabricated concrete panels of 3m in length by 2m in height are stacked over one another. In order to supply a flexible space for different kinds of exhibitions, the interior space has been created as large as possible.

In the end of 2015, a mysterious door was opened up on the symbolic steel mesh wall of the gallery, and Chimney Café was added to the internal courtyard. Although the sculpture courtyard space itself is very narrow, only five meters at its widest point, a rich and interesting spatial sequence has been successfully created. A dynamic curved wall guides to the café entrance through an interesting winding path, and behind the arc wall is café space. The patios decorated with green bamboo are bright and vivid. Application of anodized aluminum expands the width sense of courtyard space; the slightly blurred mirror effect gives people a hazy unreal visual experience. The interior white walls create the most favorable conditions for small exhibitions and art salons. And lighting bands following the arc walls strengthen the dynamic effect of the curve wall. The small cafe space overcomes all constraints and finds its own identity: soft but full of tension, natural and full of fun.

美术馆入口空间
咖啡馆

上海油雕院美术馆及咖啡厅
SPSI Art Museum & Chimney Cafe

建筑设计 大象建筑设计有限公司
王彦 王一博

Architects GOA
WANG Yan, WANG Yibo
Location Shanghai
Completion 2015
Photo LV Hengzhong

上海 | 竣工时间 2015年 | 摄影 吕恒中

改造及修复

美术馆室内
咖啡馆与天井

咖啡馆入口
咖啡馆入口廊道

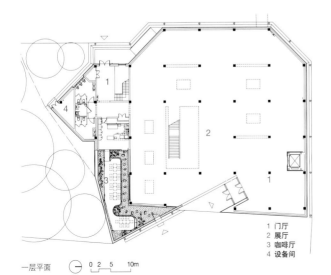

展厅楼梯

1 门厅
2 展厅
3 咖啡厅
4 设备间

一层平面　0 2 5 10m

二层空间
改造后北侧街景

桐庐莪山畲族乡先锋云夕图书馆
Tonglu Librairie Avant-Garde, Ruralisation Library

建筑设计 张雷联合建筑事务所
张雷

浙江 杭州 | 竣工时间 2015 年 | 摄影 姚力

Architects AZL Architects
ZHANG Lei
Location Hangzhou, Zhejiang
Completion 2015
Photo YAO Li

桐庐先锋云夕图书馆位于浙江桐庐县莪山乡戴家山村，是先锋书店开设的第11家书店。项目凭借"先锋和书店"的文化传播理念，以及独特的"畲族"山村的地域自然人文景观背景，成为当地村民和"异乡读者"的公共生活纽带，成为地方文化创意产业的一个聚焦点。图书馆的主体是村庄主街一侧闲置的一个院落，包括两栋黄泥土坯房屋和一个突出于坡地的平台。建筑设计保持了房屋和院落的建筑结构和空间秩序，将衰败现状修整还原到健康的状态，新与旧的关系强化了"时间性"，土坯墙、瓦屋顶、老屋架这些时间和记忆的载体成为空间的主导，连同功能再生的公共性，共同营造文脉延续的当代乡土美学。

Located in Daijiashan Village, Eshan She Nationality Township of Tonglu, Zhejiang Province, the Avant-Garde Ruralation Library is the 11th Bookstore run by the Librairie Avant-Garde. With the culture-spreading idea-Avant-Garde and Library, as well as the unique regional natural and human landscape of the "Shes" village, this project has become the public life bond among local villagers and alien readers, and also become a focal point of local cultural and creative industries. The main body of the library was an idle yard lying at one side of the village's main street, including two yellow mud adobe houses and a platform projecting from a slope. While maintaining the structural and spatial sequence of the buildings and courtyards, the architectural design restores the current declined status to a healthy state. The relationship between the new and the old strengthens the "timing". The carriers of time and memory-the adobe walls, tile roofs and roof trusses have become the spatial dominance, and have jointly created context-continuing contemporary local aesthetics, together with the publicity of function regeneration.

门厅

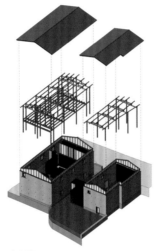
轴测分解图

1 图书馆门厅
2 阅览室
3 卫生间
4 咖啡厅门厅
5 咖啡厅
6 室外露台

首层平面

咖啡厅
入口庭院

从嘉陵江上远眺

为了慎重实施全国重点文物保护单位——四川广元千佛崖石窟的整体保护设施建设项目，2014年首先针对一小部分石窟建设了试验段工程，要求能有效保护石窟，使其免遭风雨侵蚀。

项目延续中国四川地区为石窟建设窟檐建筑的传统，以监测数据为基础，利用现代建筑技术为石窟提供更有利于保护的物理环境，并使建筑融入整体景观环境。建筑采用悬挑的异型钢框架结构作为支撑系统，按照最小干预原则与可逆原则，建筑结构与崖壁完全脱开，并可以在必要时拆除。墙身和屋面的围合体系采用专门设计的透空瓦幕墙，构件尺寸、排布方式和连接细部都经过特殊设计和试验，以同时满足建筑对通风与防水排水的高标准要求。建筑材料选用钢、瓦和木3种材料，其中后两者是当地传统建筑材料。建筑内部在尊重摩崖造像周边物理环境的设计原则下，只设计了最基本的避雷、夜间照明、安防和消防系统。

To ensure compatibility of the approved overall protective project for Thousand Buddha Cliff Grottoes in Guangyuan City, an experimental protective project covering a small portion of the site was authorized to be constructed in 2014 in advance.

The project continues the tradition to build shelters for grottoes in Sichuan Province to create a more efficient protecting environment for the cultural property by adopting modern architectural technologies on the basis of monitoring data, which harmonizes itself into the immediate landscape. A cantilever structure, independent from the cliff and only with its foundation bearing the load of weight, is employed to minimize the intervention so that the cliff was intact during and after construction works, obeying the principle of irreversibility. The specially designed and computer-calculated tile-screen facade and roof features the structure of high quality, which keeps out rainwater and reduce wind force while letting in filtered light and air. Only steel, terra-cotta tiles and wood are employed in the structure, as the latter two materials are amongst local building materials. Accessory equipment installed in the structure is minimized to retain the natural environment intact as possible, including an intelligent equipment system for illumination and multi-media, and several post fire hydrants.

广元千佛崖摩崖造像保护建筑试验段工程
Experimental Protective Structure for Thousand Buddha Cliff

四川广元 | 竣工时间 2015年

建筑设计 清华大学建筑设计研究院
崔光海 安心默 马智刚

摄影 五季

Architects
Architectural Design and Research Institute of Tsinghua University Co., Ltd.
CUI Guanghai, Andrea Gianotti, MA Zhigang
Location Guangyuan, Sichuan
Completion 2015
Photo WU Ji

改造及修复

室内空间
通透的墙体

建筑与崖壁关系

表面肌理
屋顶细部

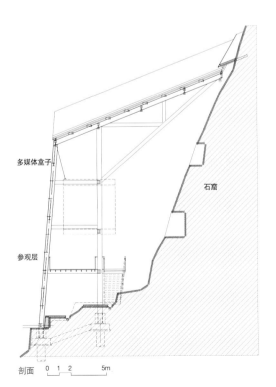

多媒体盒子

参观层

石窟

剖面　0　1　2　　5m

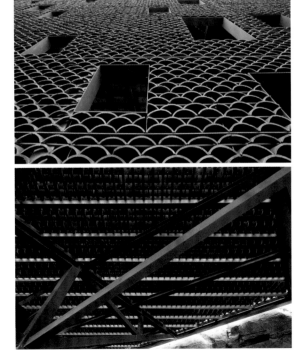

村内环境

摩梭文化，是世代生活在泸沽湖畔的摩梭人所独有的母系氏族文化。随着"全球化"的发展，摩梭传统文化面临逐渐消亡的威胁。

设计基于对特有的母系氏族文化和传统生产方式的全域保护更新原则，充分尊重摩梭人传统生产生活方式，同时改善村落的公共环境，完善基础设施，提升生活水平，实现对文化和村落的共同保护。应对多样性传统居住空间，设计采用了模块化策略，将传统摩梭民居"祖母屋、草楼、经堂、花楼"4部分细化设计，提供不同的模块标准。居民可自由组合，形成多样性自然生长的建筑群。设计强调建筑材料的乡土性和建造的原真性，在设计和建造的过程中，保留大量摩梭人原始的建造手段和方式，整旧如旧，减少建筑师创作的痕迹。针对偏远地区的特殊条件，项目采取了"统一规划设计+村民参与自建+建筑现场控制指导"的建设模式，摩梭人充分参与其中，实现文化的传承和创新。

Mosuo culture is a matriarchal clan culture of ehtnic Mosuo people living beside the Lugu Lake. Under the onslaught of globalization, Mosuo culture is facing the threat of gradual demise.

Based on the unique maternal clan culture and their traditional production methods, the design fully respects Mosuo traditional production methods and lifestyle. This project aims to improve their standard of living by improving the infrastructure and public environment of this village. In response to a diversity of traditional living spaces with a modular method of design, this project makes the four parts of the traditional Mosuo residential – Grandmother's Room, Thatched Cottage, Buddha Hall and Flower House, for design of precision with different standard modules. The design emphasizes local building materials and authenticity. The project retains a large number of Mosuo original construction methods during the design and construction process, aims to reducing the traces of architects' design. Taking the special conditions of remote areas into account, the project uses special construction modes, which are "unified planning", "native villagers constructing" and "architects guiding on the site". Mosuo residents participated in the construction and design processes, realizing cultural inheritance and innovation.

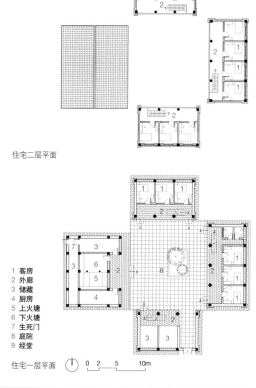

住宅二层平面

1 客房
2 外廊
3 储藏
4 厨房
5 上火塘
6 下火塘
7 生死门
8 庭院
9 经堂

住宅一层平面

摩梭家园
Homeland of Mosuo

四川 泸沽湖 | 竣工时间 2015年 | 摄影 存在建筑

建筑设计
中国建筑西南设计研究院有限公司
张远平 郑欣 马俊

Architects China Southwest Architectural Design and Research Institute Crop., Ltd.
ZHANG Yuanping, ZHENG Xin, MA Jun
Location Lugu Lake, Sichuan
Completion 2015
Photo Arch-Exist Photography

改造及修复

村口牌坊

村内道路
民居改造

民居改造

建筑东立面

上海电子工业学校由20世纪80年代建造的上海声美无线电厂改造而来。六号楼从前是厂区内混合用途的两层建筑，现在被改造为学生浴室，一层为男生用，二层为女生浴室及洗衣房。六号楼原本为混凝土框架结构，设计基本保留并加固了主体结构，重新布置了室内空间并新建了外墙。建筑的北侧增加了新的入口空间，通过小尺度的空间变化，在使用者之间以及使用者和环境之间建立了一种细腻的对话关系。建筑与东侧紧邻的室外活动场地之间利用一片混凝土空心砌块墙创造了一个半透明的空间界面。空心砌块旋转45°砌筑，使整个建筑在不同空间方向上呈现出不同的通透性。它改变了浴室建筑通常的封闭性，使建筑与环境的关系变得柔和并且生动。

The campus of the Shanghai Electronic Industry School is a project transformed from factory buildings which belonged to Shanghai Shengmei Wireless Factory built in 1980s. It used to be a two-storied building of mixed uses, and is converted into a student bath house, the first floor of which is made for male students and the second floor for females. The original concrete framed structure of the building is retained in the design with reinforcement, and the interior space is rearranged with newly added walls. New entrances are added to the northern side of the building, and subtle dialogues between different users and between users and the environment are established through small-sized spatial change. An opaque interface is created between the building and an outdoor playground on the east of the building. Hollowed blocks are laid rotated by 45 degrees, resulting in transparency of various levels on different directions, which breaks normal closeness of bath houses, softening the relation of the bath house and the environment vividly.

二层平台入口转角

上海电子工业学校六号楼/学生浴室
Block 6 of Shanghai Electronic Industry School/Student Shower Block

上海　　竣工时间 2014 年

建筑设计
无样建筑工作室
上海市建筑科学研究院（集团）有限公司
冯路

摄影 苏圣亮

Architects Wuyang Architecture; Shanghai Research Institute of Building Sciences Group
FENG Lu
Location Shanghai
Completion 2014
Photo SU Shengliang

改造及修复

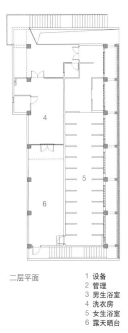
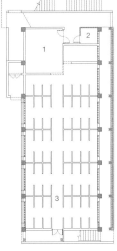

二层平面

1 设备
2 管理
3 男生浴室
4 洗衣房
5 女生浴室
6 露天晒台

一层平面　0 1 2　5m

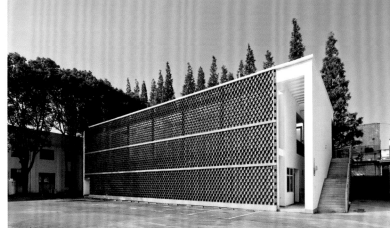

建筑南侧新建白墙
入口空间及空心砌块墙
二层平台空间局部

正立面外景
可以完全打开的外墙

牛背山志愿者之家
Cattle Back Mountain Volunteer House

建筑设计 dEEP 建筑事务所
李道德

四川 甘孜藏族自治州 | 竣工时间 2014 年 | 摄 影 dEEP 建筑事务所

Architects dEEP Architects
LI Daode
Location Ganzi Tibetan Autonomous Prefecture, Sichuan
Completion 2014
Photo dEEP Architects

这是一次公益设计，为一群志愿者们在大山里盖一座房子。为了保证公益实践的开支，这个房子有一定的青年旅社的功能。改造策略是在完善基本使用功能的前提下，让这个建筑更具有开放性与公共性，可为更多的人群服务。从建筑空间与结构上，创新的同时又具有中国传统建筑的记忆与灵魂，使其与村落、与环境相协调，融为一体。尽可能地使用当地村民作为主要的劳动力，用最常见、最基本的建筑材料和传统的搭建方式，比如当地石块的砌墙方式、坡屋顶与小青瓦。在加建的构筑部分建筑师采用了数字化的设计方法与生成逻辑，房子逐渐由传统转变到了现代，甚至是对未来的探索。一个和背后大山、云海相呼应的有机形态的屋顶呈现了出来。内部看似是传统的木结构，但又是一种数字化的全新表现。

It is a social welfare project dedicating to building a house for young volunteers in the mountain of Pumaidi Village, Sichuan Province. The house will be used as a youth hostel to keep the financial balance of the social project. The design strategy is, while improving the basic programs and functions, to make the building more open and public, serving more people. Innovative maneuvers represent memory and soul of traditional Chinese architecture both spatially and structurally, integrated with the village and environment as a whole. Local villagers are employed as construction workers, and most normal and basic materials and traditional construction techniques are used, such as local masonry, gabled roof style and small-sized tiles in green grey, etc. Modern digital design method and generation logic are also used for the added part of the building, making the transition of the building from the traditional to modern, and even explorations of the future. The roof of organic style emerges echoing to the background mountains and the sea of clouds. As for the internal space, it's a brand new expression of wood structure in digital times.

竹木结构营造的灰空间

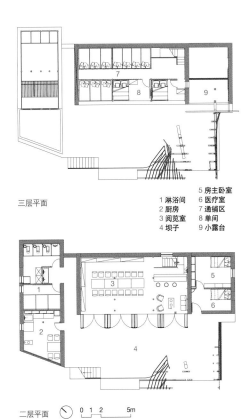

三层平面

1 淋浴间　5 房主卧室
2 厨房　　6 医疗室
3 阅览室　7 通铺区
4 坝子　　8 单间
　　　　　9 小露台

二层平面　0 1 2　5m

水吧
二层书吧

改造后的村民活动中心北立面
场地东侧新建的连廊成为新的界限

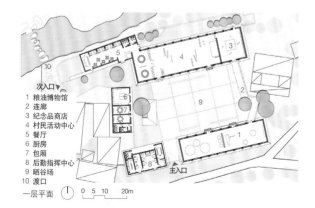

次入口
1 粮油博物馆
2 连廊
3 纪念品商店
4 村民活动中心
5 餐厅
6 厨房
7 包厢
8 后勤指挥中心
9 晒谷场
10 渡口
主入口
一层平面　0　5　10　20m

博物馆南向外景

西河粮油博物馆及村民活动中心
Xihe Cereals and Oils Museum and Villagers' Activity Center

河南 信阳　｜　竣工时间 2014 年

建筑设计　中央美术学院建筑学院　何崴 陈龙

摄　影　何崴 陈龙 齐洪海

Architects China Central Academy of Fine Arts School of Architecture
HE Wei, CHEN Long
Location Xinyang, Henan
Completion 2014
Photo HE Wei, CHEN Long, QI Honghai

项目处于大别山里的小村落，有着丰富的乡村农耕景观，但是经济欠发达而处于衰败状态。项目的场地原为上世纪五六十年代的粮管所。场地内现存建筑5座，均为双坡顶砖木结构，其中有两座体型硕大的粮仓，整组建筑保留状况完好，处于闲置状态。改造设计的目标是为旧建筑植入新功能，为振兴村庄提供支点。两个大粮仓分别改为微型博物馆和村民活动中心，另外一个附属建筑改建为特色餐厅。首先对两个粮仓的大空间做了整体保留，并按新的功能重新布置。两个大建筑立面处理方式是在对外的一侧打开封闭墙面增加较大窗口，引入阳光。而朝向内院的立面则保留了原来的样子，它们和修缮后的内院一起，完整地传递建筑的历史信息。

The project is situated in a small village in the Dabie Mountain, with rich rural landscape but in lack of economic development. The site used to be an office of grain management in the 1950s and 1960s, with five remaining buildings in wooden structure of gabled roofs. Two of the buildings are huge granaries, and all the buildings are preserved well with no specific use. This is a public welfare project whose aim is to revitalize the village through design and agriculture branding, which has been regarded as a new model for rural construction development. It has provided the villagers with a public activity center and has implanted a small-sized museum that reflects the local agricultural history and a restaurant of local uniqueness. The large space within the granaries is preserved as a whole but rearranged according to new functions. For the design of facades, relatively large windows are added to the closed wall on one side to bring in sunlight, while the facade facing the inner courtyard retains previous character, conveying complete historical messages along with the repaired inner courtyard.

村民活动中心室内
西侧山墙室内光影效果
微型博物馆室内

箭厂胡同文创空间紧邻国子监西墙，该项目是在原空间格局的基础上进行的介入性改造，空间内部为了充分利用高度而设置了夹层，并且在夹层上植入了多个具有不同内容的木头阁楼（或盒子）作为激活性的装置。空间中的事件通过不同的流线围绕或进入这些盒子而展开，而走到这些盒子里面又会发现很多有趣的取景窗，对着特定的场景，重新串起多条叙事性的线索，它刺激人通过自身的体验去感知一些预料之外的东西。

As an intervened regeneration project, Jianchang Hutong Creative Cultural Space is right next to the western wall of the Imperial Academy. Fully exploiting the height of the space, the architects add an interlayer inside, and leave several voids connecting the double height space. Furthermore, they have inserted various wood attics (or boxes) into the space so activities can now follow different trajectories, unfolding around or within those attics. Once people are inside those small enclosed spaces, many interesting openings like view-finders can be found, targeted at specific views to string up various spatial narratives, inspiring visitors to explore the unexpected scenarios.

首层门厅

剖透视

箭厂胡同文创空间
Arrow Factory Media & Culture Creative Space

建筑设计 META-工作室
王硕 张婧

北京　　竣工时间 2014 年　　摄影 陈溯

Architects　META-Project
WANG Shuo, ZHANG Jing
Location　Beijing
Completion　2014
Photo　CHEN Su

改造及修复

首层四合院
夹层全景

首层平面　0 2 5 10m

1 门廊　　4 展台
2 内院　　5 会议
3 门厅/接待

盒子装置

入口

上海国际时尚中心前身为始建于1921年的裕丰纱厂，解放后更名为国营第十七棉纺厂，随着经济转型，纺织厂房空置衰败。建筑师们梳理了工业遗产的价值，赋予它新的符合现代生活需求的功能。遵循最小干预原则，采用修旧如旧的手法，突出呈现建筑原有的质感和肌理。新建、改建和修缮部分建筑的风格尊重并保留建筑特有的韵味，成功地将这片上海市区最完整、最具规模的锯齿厂房建筑群完整保留下来，既保存了珍贵的历史记忆，又为城市注入新的空间元素。

The former building of Shanghai International Fashion Center was Yufeng Cotton Mill built in 1921, renamed as No. 17 State-owned Cotton Textile Mill. Accompanied by economic transformation, the textile factories declined. The architects appraise the value of the industrial heritage and endowed it with new functions that meet the needs of modern life. Following the minimum intervention principal, the architects preserved the heritage as much as possible, emphasizing the representation of its original texture. The new, and renovated parts valued the lasting charm of the previous buildings, preserving the factory buildings with the largest and most complete saw shaped roofs as an integrity. The design injected new space and elements into the urban environment, activated the deserted area, and at the same time, kept the collective urban memory.

1 会所
2 时尚秀场
3 时尚精品仓
4 办公

总平面 0 10 20 50m

上海国际时尚中心
Shanghai International Fashion Center

建筑设计
华东建筑设计研究院有限公司华东都市建筑设计研究总院
法国夏邦杰建筑设计咨询（上海）有限公司
邢同和 Pierre Chambron 袁静

Architects
East China Architectural Design and Research Institute Co., Ltd. East China Urban Architectural Design and Research Institute; Arte Charpentier Architects Architectural Design Consultant Co., Ltd. (Shanghai)
XING Tonghe, Pierre Chambron, YUAN Jing
Location Shanghai Completion 2014

上海 | 竣工时间 2014年 | 摄 影 业主提供 | Photo By client

秀场外的空间
水塔

工业感的秀场前厅
局部立面

局部立面
成片的锯齿厂房屋顶

改造后整体鸟瞰及周边
塔吊下方的市民活动

展厅与围合庭园

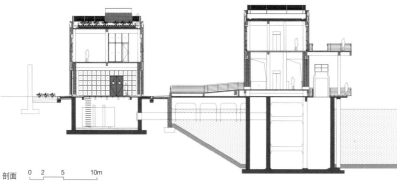

剖面 0 2 5 10m

南京下关电厂码头遗址公园
Relics Park for the Coal Dock of Xiaguan Power Plant

江苏 南京 | 竣工时间 2014 年

建筑设计
华东建筑设计研究总院
杨明 俞楠 于汶卉

摄影 邵峰 庄哲

Architects
East China Architectural Design & Research Institute
YANG Ming, YU Nan, YU Wenhui
Location Nanjing, Jiangsu
Completion 2014
Photo SHAO Feng, ZHUANG Zhe

本项目位于南京市长江沿岸防汛墙外侧。改造前这里曾是下关电厂的出灰运煤码头，厂区整体搬迁地块更新后被废弃。下关电厂是江苏省内第一个百年电厂，很有历史意义，因此确定置换为向公众开放的工业历史主题的城市公园，成为市民观江游憩、品读历史的休闲场所。

基地内现存一些小尺度工业厂房与机械设备，通过筛选保留其中部分建筑的主体结构，保存较有意思的工业空间和工业机械作为展示，增加部分楼面，形成新的功能空间并组织了全新的游览流线。外立面遵循大区域风格特征，与基地旁边的中山码头和下关红楼统一。设计为红砖白墙的民国风格，又融合了现代主义通透的金属边框幕墙，形成新与旧的对比。

The project is located outside the flood control walls of the Yangtze River in Nanjing. The site used to be a coal wharf for Xiaguan Electric Plant, the first power plant of one-hundred-year history, and had been deserted after the relocation of the whole plant. As a historical significant site, it aims as an urban park with the theme of industrial history open to the public, used as a recreational place for the citizens for sightseeing and experiencing.

Factory buildings and mechanical equipments are preserved, and the main structure of part of the buildings are kept as interesting industrial and mechanical exhibits. Part of the old floors become new functional space with newly designed circulation. The design of facades follows architectural characteristics of the region, in harmonious relation with Zhongshan Wharf and the Red House of Xiaguan, all in the Republican style featuring red bricks and white walls. Modernist curtain walls with metallic frames are combined in sharp contrast to the old elements.

沿江展厅与餐厅
亲水观景平台

1 主题展厅
2 保留机器展示
3 机电设备
4 玻璃地面地下展示
5 公共空间
6 餐饮茶室

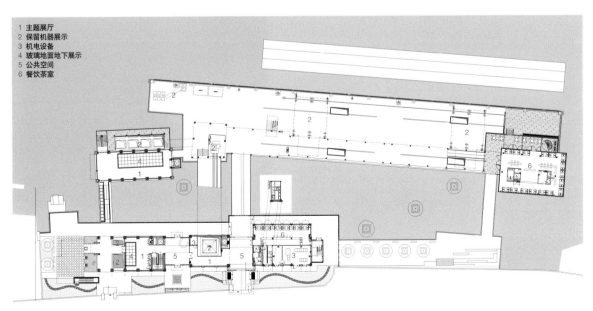

一层平面　0　5　10　20m

东南透视

　　天拖融创市民中心是天津拖拉机厂更新改造项目的第一栋旧厂房改造。设计保留工业建筑遗留下来的十二榀钢架结构，以生成简单高效的展示及办公空间，建筑整体骨架形成一个M型的100米长的腔体。建筑主材采用了陶土烧结砌块，通体只采用一种规格的陶土砌块钢龙骨干挂。为消解尺度巨大的干挂陶土砌块带来的板滞感，建筑师研究了3种陶土砌块的组合方式：错缝搭接；砌块扭转45°上下错搭，如入口处的片墙拼花就用这种方式拼出天拖厂史上1956的重要纪元；错缝平砌。另外，外皮用3种同一色系深浅不同的陶土砌块随机混合使用，模拟原有砖墙斑驳而自然的效果。在经过细部的推敲确定合理的构造逻辑的前提下，陶土砌块丰富的组合语汇延续了老厂房的红砖表皮的时间记忆的广度和深度。

　　The center is the first renovated building of the regenerated project of Tianjin Tractor Factory. The design preserves the 12 sets of steel roof truss in the previous industrial building, generating exhibition and office space of simplicity and high efficiency. The whole structural frame is a M-shaped cavity of 100m in length. Bonded clay is used as the material for the main part of the building, hanging the blocks of the same size on the steel structure. To downplay the rigidity of the building bulk due to materials selected, three types of connection of joints are adopted: overlapping with staggered joints, overlapping with a rotation of 45 degrees as seen in the walls of the factory signaling the crucial epoch starting from 1956, and flat laying with staggered joints. In addition, three types of clay bonded blocks in different colors which belong to the same tone are used on surface, simulating the natural and mosaic effect of the original brick walls. Based on a reasonable construction logic with careful studies of details, the rich combination of bonded clay blocks continue with the temporal memory both in width and breadth.

入口局部

天津拖拉机厂融创中心
Sunac Center of Tianjin Tractor Factory

天津　竣工时间 2014 年

建筑设计　上海日清建筑设计有限公司
任治国　杨佩燊
摄　影　张英琦　苏圣亮

Architects La Cime International Pte. Ltd.
REN Zhiguo, YANG Peishen
Location Tianjin
Completion 2014
Photo ZHANG Yingqi, SU Shengliang

东南广场
东水池

1 市民广场
2 水面
3 天拖历史文化广场

总平面

南入口局部

西北向鸟瞰

　　吉兆营老清真寺是南京城北仅存的一座清真寺，也是一座中国回族传统院落式清真寺。原寺历经沧桑，已不能满足正常使用要求。在城市更新背景下，新寺总体布局采取因地制宜的策略，突破了退让与密度的通常规定，除必要的退让道路红线，谨守旧寺原有边界。外紧内松的空间形态布局表现出对城市街区空间形态的适应。在调适邻里的同时，成功化解了功能需求增长与用地减少的矛盾，整合了城市更新遗存的空间"碎片"。为避免落入伊斯兰风格样式的窠臼，新寺采用了空间重塑和旧物新用的设计策略。设计首先将传统清真寺水平组织的院落进行竖向叠加，构成形态多样、渐次有变的院落系统。其次，设计利用原有树木和新的院落，将礼拜空间由过去的内向封闭转变为内外融合，从而提供新的礼拜空间体验。高大院墙将城市空间划分为世俗与神性，最大限度地保障了祈祷空间的私密性，分隔了世俗的喧扰，形成穆斯林愉悦的精神家园。新寺的设计最后进行建造材料重组，恰当地保护和合理地利用了老建筑遗存物件和老材料，以本土化的设计语言延续了历史的记忆，展现了伊斯兰文化与中国江南地域文化的有机结合。

The old Jizhaoying Mosque, the only mosque remaining in the north of Nanjing, is a mosque of traditional courtyard-style built for Chinese ethnic Hui people. At the level of urban planning, the overall layout of the new mosque is a breakthrough from common setback and density requirements. The design carefully follows site conditions and faithfully retains original boundaries of the old mosque except for necessary road setbacks. While reconciling relations with neighboring environment, the design successfully resolves the conflict of increased demand for usage and decreased site area, integrating "fragments" in urban space. At the level of architectural design, the design of the new mosque hinges on reformation of space and innovative reuse of old construction materials. The design stacks up courtyard spaces to organize different functional spaces. Utilizing trees existing on the site, the architects transform the prayer space from an inward, closed hall to a space that connects the interior and the exterior, creating a new experience of the prayer space. Another highlight of the project is the reorganization of construction materials, as the design appropriately preserves and utilizes remaining objects and materials from the demolished old mosque, sustaining memories of history with a localized design language. Overall, the design is an organic integration of Islamic culture with local cultures in the lower Yangtze region.

吉兆营清真寺翻建工程
Jizhaoying Mosque Renovation

建筑设计　东南大学建筑设计研究院有限公司　马晓东 韩冬青

江苏 南京　　竣工时间 2014 年　　摄　影　耿涛 陈帅

Architects Architects & Engineers Co., Ltd. of Southeast University
MA Xiaodong, HAN Dongqing
Location Nanjing, Jiangsu
Completion 2014
Photo GENG Tao, CHEN Shuai

沿街透视

老龛 – 老树 – 新院

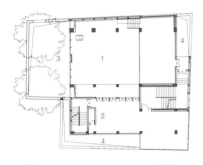

二层平面　　0　2　5　　10m

1 大殿
2 休息活动平台
3 院落
4 边院

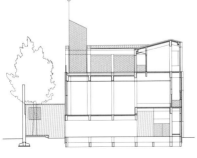

剖面　　0　2　5　　10m

四层经堂檐廊｜老龛新殿
礼拜空间

展厅

天仁合艺美术馆选址于杭州市老复兴路旁、钱塘江畔的白塔公园内，设计是将一栋旧有工业建筑改造成当代美术馆。设计师希望在表达出新旧建筑时代特性的同时在设计元素上使新建筑与老建筑有所关联，因此从木桁架中抽取三角形作为整个设计的几何原型。在门厅空间及接待空间的设计中，设计师利用三角体分形的原理，对于不同尺度的设计对象，采用不同的细分等级，以满足不同的功能需要，形成一个具有清晰几何逻辑的公共功能空间，空间的形象与天仁合艺美术馆的当代艺术气质相吻合。在展厅中则设计隐形灯光，照亮屋顶桁架系统，强调出老建筑的固有特性。

The museum is located in the Baita Park along the historic Fuxing road on the riverside of the Qiantang River. One of the old industrial building was converted to a modern art museum, invigorating the old building.

The architect tries to convey epoch characteristics of both the bold and new buildings, and connect design elements of the two. As a result, the triangle from the wooden truss has been selected and developed into the main geometric prototype of the project. Fractal algorithm of the Triangle is used in spaces like the lobby, VIP reception, etc. Regarding different objects of various sizes, a subdivided hierarchy is adopted to meet different functional requirements, resulting in the public space of clear geometrical logics. The spatial image matches the modern artistic temperament of the museum properly. The hidden light in wooden frame of the exhibition space shaped a nice profile of the traditional building structure.

天仁合艺美术馆
T_Museum

建筑设计
华汇设计（北京）
王振飞 王鹿鸣 王凯

Architects
HHD_FUN
WANG Zhenfei, WANG Luming, WANG Kai
Location Hangzhou, Zhejiang
Completion 2014
Photo WANG Zhenfei

浙江 杭州 | 竣工时间 2014 年 | 摄 影 王振飞

南侧外观
入口
门厅

门厅

改造及修复

门厅

平面 0 5 10 20m

1 展厅
2 门厅
3 多功能厅
4 开敞办公室
5 独立办公室
6 VIP 接待厅

剖面 0 5 10 20m

北立面

设计任务是将华强北旧工业厂房改造为前卫、时尚的设计酒店。基地四面的景观资源差异甚大：北面为最重要的迎街面，改造策略为以标准客房单元为造型元素，窗户向外偏心凸起并做倒角处理，与墙面平滑相接，形成4种基本模数，再通过重复、旋转、镜像、变异等手法排列出动态变化的三维立体效果。西面面向优美的城市中心公园，扩大客房的开窗面积，设置最高级套房。而南面为酒店的背面，面对陈旧的厂房宿舍，存在严重的对视问题，因此采取了侧向开窗的造型单元，避免看到对面杂乱景象又满足自然采光的要求。

This project transformed an old factory building remained in Huaqiangbei into an avant-garde boutique hotel. The landscape conditions around the site differed significantly. The most important facade which defines the hotel image is on the north, facing the main city road. Using the typical hotel room opening as designing module elements, the windows are projected away from the facade to provide maximum viewing possibilities, and the frames are off-centered then filleted smoothly back into the solid wall panels. This modular element is then repeated, rotated, mirrored and deformed to array a dynamic three-dimensional effect. The same module on the west facade is repeated and enlarged; the ensuites units are placed on the west end to provide the best viewing direction towards the city park. Contrarily, the south facade, facing a factory dormitory with no viewing pleasure at all, was ensured with privacy and natural lighting, in order to facilitate the hotel program.

室内

回酒店 / Hui Hotel

广东 深圳 | 竣工时间 2014 年

建筑设计
URBANUS 都市实践建筑事务所
刘晓都 王俊 姜轻舟

摄 影 吴其伟 陈冠宏

Architects
URBANUS Architecture & Design Inc.
LIU Xiaodu, WANG Jun, JIANG Qingzhou
Location Shenzhen, Guangdong
Completion 2014
Photo WU Qiwei, Alex Chan

西北向外景

局部

建筑转角

南立面

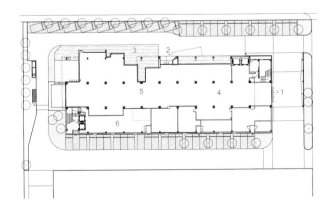

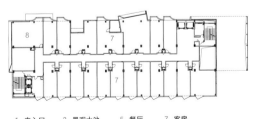

1　主入口　　3　景观水池　　5　餐厅　　7　客房
2　次入口　　4　酒店大堂　　6　厨房　　8　总统套房

一层平面　　0　5　10　20m

二层平面

建成愚园鸟瞰

南京愚园作为一个修缮与重建的项目，目标是通过重建历史上的名园，改善该地长期以来搭建棚户带来的生活环境脏乱差的问题，同时为密集的居民区提供"绿肺"，也藉此传承南京的园林文化。

尊重历史、寻找依据，为第一要则。通过收集照片、比对文献、采访人物、参照周围传统建筑等，修复仅存的既有建筑，形成愚园南区自然质朴和北区精致密集的历史格局。科学定位、有效利用，为第二要则。通过局部考古，对淤塞的水体边界进行拟定，对长期养成的树木进行有效利用，对改变了的山体以山石进行巧妙补形等，使得大局形态自然浑成。需求调整、持续发展，为第三要则。随着将历史上的私园改善为对市民开放的城市园林的需求，重视流线、调整功能成为必须，同时通过采用地源热泵技术解决了室内环境的使用要求。

The Yu Garden in Nanjing was architecturally designed as a restoration and reconstruction project to improve the surrounding living environment, taking advantage of the opportunity of rebuilding the historical site to create a "green lung" for the extensive residential areas in the vicinities, in an attempt to carry forward the city's garden culture.

The design respects history and rebuilds the garden based on proven evidences as the first principle in this project. The architects did this by gathering old photos, comparing literature notes, interviewing relevant people, taking the traditional buildings in the vicinities as reference, and repairing the existing buildings, hence a southern section of the garden that looks natural and simple while keeping the exquisite and a sophisticated northern section. The second principle lies in scientific orientation and effective utilization. Local archaeological excavations gave us a clue to the previous water body boundaries. The architects retained and utilized the trees that have grown on the site for a long time, while making up the changed hillsides using rocks, so as to present a picture of natural blending. The third principle is readjustment of demands and sustainable development. When changing a historical private garden for the purpose of public entertainment, streamlining the garden's layout and readjustments of its functionalities are necessary. Meanwhile, they addressed the needed indoor temperature taking advantage of the ground heat pump technology.

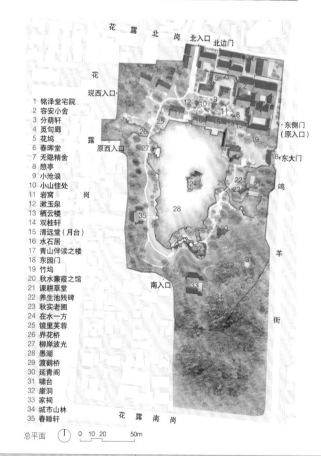

1 铭泽堂宅院
2 容安小舍
3 分萌轩
4 觅句廊
5 花坞
6 春晖堂
7 无隐精舍
8 憩亭
9 小沧浪
10 小山佳处
11 岩窝
12 漱玉泉
13 栖云楼
14 双桂轩
15 清远堂（月台）
16 水石居
17 青山伴读之楼
18 东园门
19 竹坞
20 秋水蒹葭之馆
21 课耕草堂
22 养生池残碑
23 秋实老圃
24 在水一方
25 镜里芙蓉
26 界花桥
27 柳岸波光
28 愚湖
29 渡鹤桥
30 延青阁
31 啸台
32 崖洞
33 家祠
34 城市山林
35 春睡轩

总平面

南京愚园修缮与重建
Restoration and Reconstruction of Nanjing Yu Garden

江苏 南京　　竣工时间 2013 年

建筑设计
东南大学建筑设计研究院有限公司
陈薇

摄　影　高琛 姚力 陈薇 王建国

Architects
Architectural Design Research Institute Co., Ltd.,
Southeast Unicersity
CHEN Wei
Location Nanjing, Jiangsu
Completion 2013
Photo GAO Chen, YAO Li, CHEN Wei, WANG Jianguo

愚园春景
愚园秋景

愚园夏景

南北向剖面

愚园冬景

Housing

居住
建筑

P342

P369

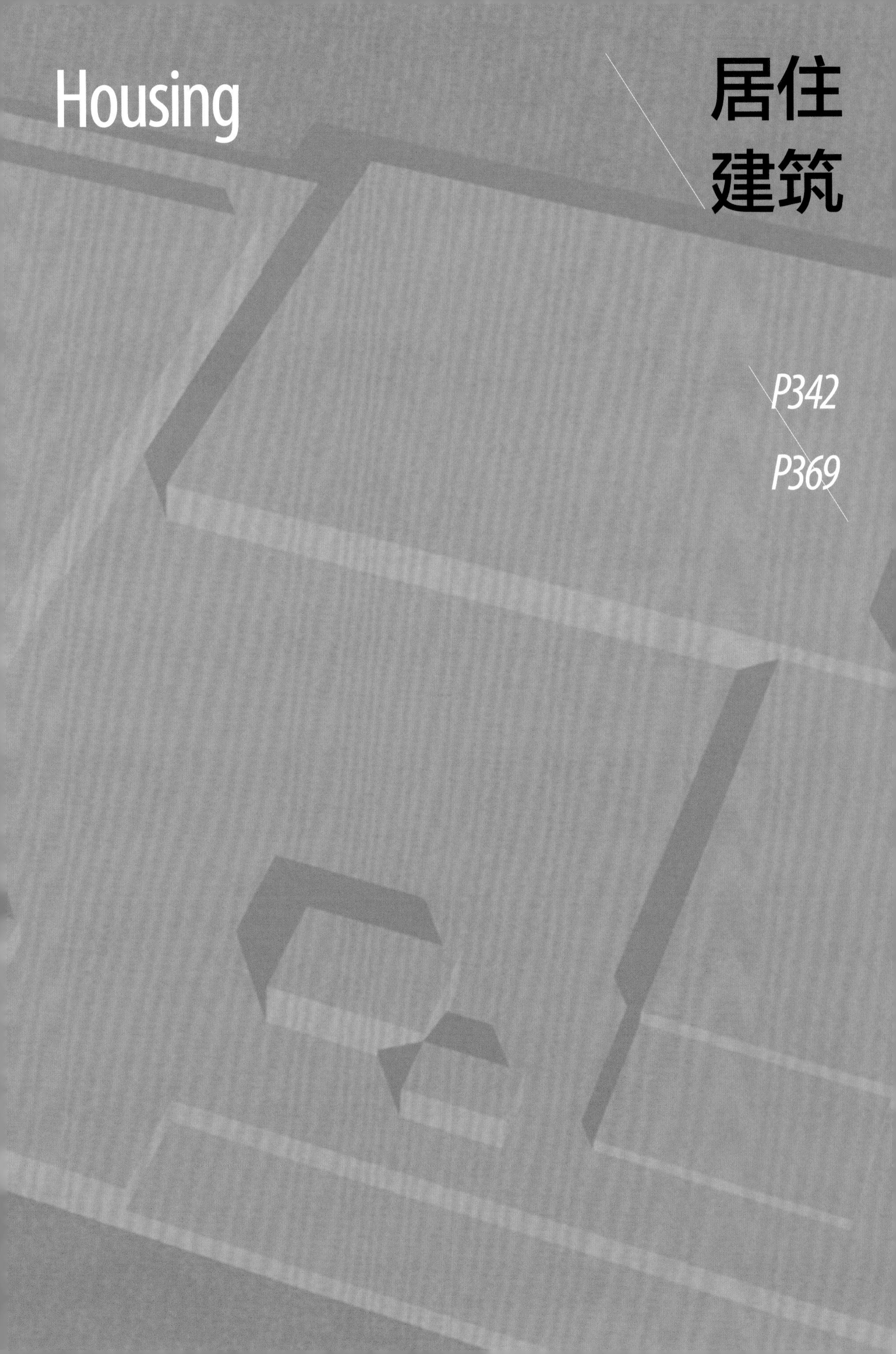

颐乐学院运动馆

雅园总平面

颐乐学院静态教学楼

乌镇雅园是学院式养老社区的一次实践，是一个以健康老人为主要对象的退休活力社区，类型包括老年公寓、合院为主的居住建筑以及幼儿园和颐乐学院等配套设施。

居住建筑的设计重点在于小户型与担架电梯、无障碍走廊、老年人楼梯、轮椅回转、全明式通风等适老化要求之间的协调，以及与中国式建筑立面之间的匹配。全园区的连廊系统连接所有的居住单元及园区配套设施，其间设置休憩和活动空间，方便老年人全天候的使用。同时宅间被连廊自然分成公共庭院、绿篱前庭以及宅廊间的后院。颐乐学院包含老年大学、园区食堂、社区商业及运动健康等功能，丰富的配套服务功能保障了学院式养老模式的成功实现，让每一个老年人享受到"颐、乐、学、为"四位一体的晚年生活。

Wuzhen Graceland is a retirement community, aiming at the elderly of healthy status, including elderly apartments, courtyard-based residential buildings and other facilities like kindergartens and geriatric colleges.
In order to coordinate with the aging requirements and match the Chinese architectural facade, the design of apartments focuses on small-sized apartment, medical elevator, barrier free corridors, elderly stairs, wheelchair turn and effective ventilation. The corridor system of the park connects all living units and supporting facilities for the elderly to rest or exercise under all climatic condicitons. Besides, the corridors also divide apartments into three parts: public courtyards, hedgerow vestibular and backyards. Geriatric College, including the elderly college, park canteens, community commercial and athletic space not only guarantees the college-style caring model successful, but also ensures each old person a restful, joyful, educational and active life.

乌镇·雅园
Wuzhen Graceland

建筑设计
浙江绿城建筑设计有限公司
上海丁周建筑环境艺术设计有限公司
蒋愈 孟骐 周勤

浙江 桐乡　竣工时间 2017年　摄影 祝良基

Architects Zhejiang Greentown Architecture Design Co., Ltd.; Shanghai Dingzhou Architecture Environment Artistic Design Co., Ltd.
JIANG Yu, MENG Qi, ZHOU Qin
Location Tongxiang, Zhejiang
Completion 2017
Photo ZHU Liangji

住宅局部
住宅楼之间的救护通道

颐乐学院运动馆

颐乐学院社区食堂

社区室外实景

泰康之家·燕园是位于北京市昌平区的大型复合型养老社区，由CCRC持续照护退休社区与AAC活力老人社区组成，既为高龄老人提供自理、介护、介助一体化的机构养老服务，也为年龄较低的活力老人提供居家式养老服务。项目根据不同养老模式和服务对象分区规划，通过围合与半围合的建筑布局，形成不同尺度的院落空间，为老人营造具有安全感和归属感的邻里交往场所。采用人车分流的交通组织方式，合理组织适合老人的道路体系。设置弹性空间，采用灵活的结构形式，为老人提供舒适高适应性的生活居住单元。利用采光天窗将自然光线引入室内，形成充满阳光和绿色的共享中庭，创造生动而充满活力的老人公共交往空间。

Taikang Yanyuan Community is a large-scale retirement community located in Changping District, Beijing. It is composed of continuing care retirement community (CCRC) and active adult retirement community. It provides not only institutional care for the elderly people at an advanced age, but also home-based care for the elderly in healthy status. According to different patterns of nursing care and service, courtyards are of different scales formed, surrounded by enclosure and semi—enclosed building layout, creating a neighborhood with a sense of security and identity for the elderly. The traffic is separated for vehicles and pedestrians, suitable for the elderly. It also provides a comfortable and adaptable living units for the elderly by setting flexible space and adopting flexible forms of structure. Skylights introduce natural light inward forming a shared atrium full of sunshine and green, creating vivid and vigorous public communicative space for the elderly.

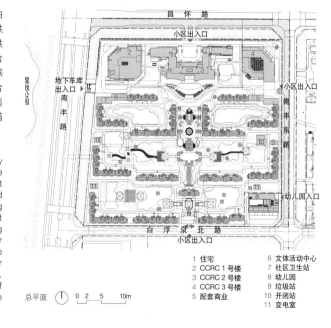

总平面

1 住宅
2 CCRC 1号楼
3 CCRC 2号楼
4 CCRC 3号楼
5 配套商业
6 文体活动中心
7 社区卫生站
8 幼儿园
9 垃圾站
10 开闭站
11 变电室

泰康之家·燕园
Taikang Yanyuan Community

建筑设计　北京市建筑设计研究院有限公司　张广群　石华

Architects　Beijing Institute of Architectural Design　ZHANG Guangqun, SHI Hua
Location　Beijing
Completion　2017
Photo　Arch-Exist Photography SHI Hua

北京　　竣工时间 2017年　　摄影　存在建筑 石华

AAC 活力老人组团室外环境
CCRC 独立生活楼实景

护理楼四季花厅

CCRC 生活单元

黄昏外景

　　设计是从针对结构加固方式的思考开始的。经过一系列的对比，建筑师最终选择了在保留原有砖墙结构的基础上添加一层12厘米混凝土墙的方法。这一策略不仅满足了加固和加建的基本需求，同时也提供了更多可以改善空间品质的可能性。新的结构介入使房屋的空间格局得以重新调整，原有的砖墙可以适当地拆除或移位，在改善通风和采光情况的同时，也使客厅、餐厅与主卧室获得了更好的观赏海的视线。窗的位置和形式得以相对自由地梳理。突出墙面的新混凝土窗套一方面为开启扇提供遮蔽，隔绝沿外墙流动的雨水；另一方面洞口的厚度被转而设计成窗系家具。拱是两边承重墙向上汇聚自然生成的结果，它的形态有利于屋顶的迅速排水，最大程度地降低了漏水和渗水的可能性；它天生具有方向性，在空间两端连接了两片性格迥异的海。人置身其中，视线一边是喧嚣的港口，一边是宁静的大海。

　　The design work starts with the study of structural reinforcement. After a series of careful comparisons, the architects decide to add a layer of 12cm concrete wall to the original brick masonry walls. This strategy brings extra potential to make a better quality of space. The intervention of the new concrete wall allows them to re-manipulate the layout to some extent. The living room, dining room and master bedroom get not only better view but also more natural light and fresh air. The locations and forms of openings also get carefully reconsidered. The new concrete window frame sticks out from the outside wall, which prevents excessive rainwater from seeping into the window from the wall surface. The vault reduces the possibility of water leakage to an extreme for it barely allows any rainwater to stay. Vault is directional. It connects two sides of sea with dramatically different characters: one being the serene sea whereas the other one being the noisy port.

女儿卧室

船长之家改造
Renovation of the Captain's House

福建 福州　　竣工时间 2017 年　　摄影 夏至 陈颢

建筑设计　直向建筑　董功

Architects Vector Architects
DONG Gong
Location Fuzhou, Fujian
Completion 2017
Photo XIA Zhi, CHEN Hao

居住建筑

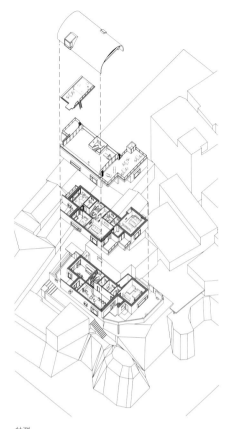

轴测

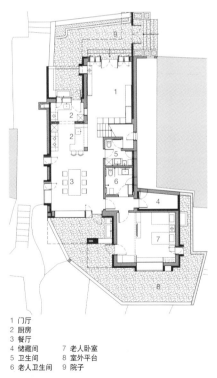

1 门厅
2 厨房
3 餐厅
4 储藏间　7 老人卧室
5 卫生间　　8 室外平台
6 老人卫生间　9 院子

一层平面　0 1 2　　5m

西立面
西南外景
多功能生活空间

5号楼庭院
从8号楼顶看屋顶平台

上海龙南佳苑
Shanghai Longnan Garden Social Housing Estate

建筑设计 上海高目建筑设计咨询有限公司
上海中星志成设计有限公司
张佳晶

上海 | 竣工时间 2017年 | 摄 影 清筑影像

Architects Atelier GOM ; Star Z & C Architecture Design Co., Ltd.
ZHANG Jiajing
Location Shanghai
Completion 2017
Photo Creat A R Images

龙南佳苑共有8栋建筑，其中5栋为成套小户型住宅(户型建筑面积为40~60平方米)，2栋为成套单人宿舍(户型建筑面积大部分为35平方米)，1栋为独立商业建筑。龙南佳苑以不同高度的住宅围合出不同形制的院落空间，逐级跌落的屋顶平台创造了丰富的屋顶活动空间，并且让更多的阳光照入院落。北侧多层区有大量架空两层的半室外空间和处在一二层的公共活动室，而住宅北廊每隔一两层都会有一个凸出的公共露台来迎接一些东西向的阳光，这些公共空间在小户型背景下，是廉租房建筑取得平衡和高效的一种策略。连廊穿行于院落之间，忽明忽暗，它联系了北侧住宅的二层活动室，也为人的活动提供了除了地面之外的另一个维度。五号楼设计采用了混凝土框架内嵌钢结构两个错跃层小户型的框架结构，除了在当下空间不大的住宅及公租房规范里挑战一下超小户型的空间变化以外，也为后公租房时代其使用的多样性提供了无限可能。

Longnan Garden Estate consists of eight buildings, of which five are sets of small apartment unit of the floor area of 40~60m², and two are dormitories with most units of 35m², and the rest building for independent commercial function. Apartments of various heights enclose different courtyards, and the cascading roof created enriched space of activities and bring in more sunlight. There are semi-outdoor space with the stilted first two floors and several public activity rooms in the northern sector of multi-storied houses, and a projecting public terrace on every other floor is set to welcome some sunlight from east and west. These public spaces in the context of small apartments, is a balanced and efficient strategy for affordable housing. The linear corridor runs through courtyards, and the sunlight changes from dark to bright. It also connects those community rooms on the second floor, providing another level for human activities other than the ground. The No.5 building design adopts large concrete frame as the structure, accommodating small units of spring layers and split levels, meeting the challenges of the space variation of minimal apartments of affordable housing standards and allowing unlimited possibilities of functions for the post social housing time.

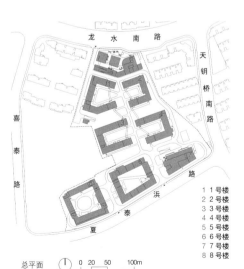

总平面　0 20 50 100m

1 1号楼
2 2号楼
3 3号楼
4 4号楼
5 5号楼
6 6号楼
7 7号楼
8 8号楼

4号楼庭院
4号楼屋顶西向鸟瞰

森林中的树屋鸟瞰

　　齐云山树屋，地属齐云山风景区，簇拥在一片红雪松林区中。建筑从中段玻璃展廊进入玄关后向上向下分为两户。住户通过攀爬中央的螺旋楼梯抵达每个房间，树屋每1.6米升高一层，随着高度的提升，攀爬者可以感受360°视角不同高度带来的森林景观体验。树屋的房间面积不大，但房间端头的玻璃窗却占据了房间的整个景观面。居住者虽居住在小室，但其注意力被引导去了窗外广阔的森林。建筑采用钢结构，在建造完中央旋转楼梯的筒体之后，房间结构部分以悬挑的形式连接在核心筒上。树屋立面材料为当地的红雪松木条。

The Qiyun Mountain Tree House is sited on the area as part of Qiyun Mountain Scenic Area, surrounded by a sea of red cedars. The house invites the guests in from the glass glazed gallery in the central layer, leading to the entrance hall before they go upstairs/downstairs to the two sets of rooms. The guests get to each room by climbing 1.6m up/down the central spiral staircase. As they move to different parts of the house, a panoramic view of the forest is guaranteed, offering a unique visitor experience of enjoying the forest scene at different heights. The rooms in the tree house are not big at all. In fact, they were designed to be extremely small. Yet, the glass window at one end of the room is as big as the entire wall that faces the view. The house is a steel structure. The structural core of the central spiral staircase was first built, joined by the cantilevered room structure. Given the concern that wood from other regions might cause plant disease and insect pest to this area, red cedar wood was chosen as the facade material for the tree house.

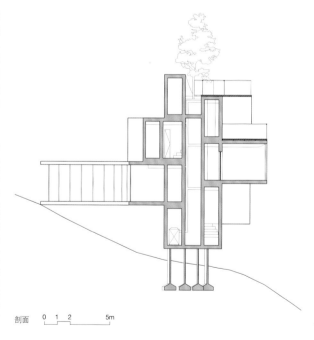

剖面　0　1　2　　5m

齐云山树屋
The Qiyun Mountain Tree House

建筑设计　本构建筑工作室　相南　姚中

安徽 休宁　｜　竣工时间 2016年　｜　摄影 陈颢

Architects Bengo Studio
XIANG Nan, YAO Zhong
Location Xiuning, Anhui
Completion 2016
Photo CHEN Hao

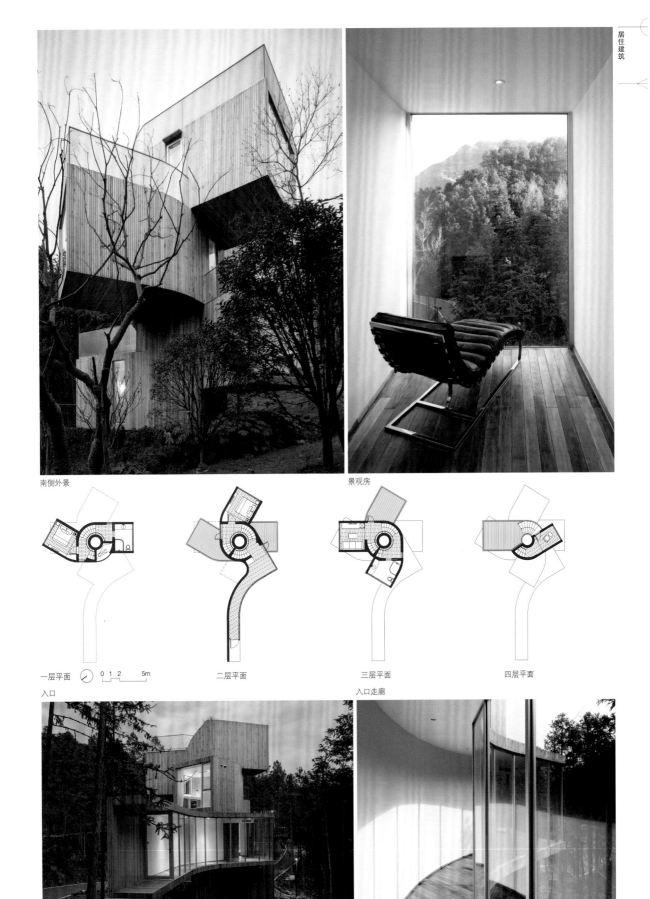

景观步道及入口

　　随园嘉树位于杭州良渚文化村，顺应山地缓坡，采取台地布置，形成丰富自然的建筑形象。该建筑群包括16栋长者公寓、护理楼以及随园会（老年活动中心）。流线处理上，入口处通过两侧密林限定空间，缓缓行至入口广场，进而通过位于建筑群中心位置的随园会通达至各个单体，护理楼则偏居一隅。随园会以十字庭院为核心，辐射整个地块，提供老人健康管理文化娱乐等活动空间。设计鼓励老人走向户外，相互交流，整个地块设置了无障碍的风雨连廊联系随园会及各个公寓楼，并在沿线布置了休憩空间，配合每栋公寓楼前的公共客厅，促进老人间的交流。景观与室内方面也进行了充分的适老化设计。

　　The Suiyuan Jia Shu Project is located at the Liangzhu cultural village in Hangzhou, conforming to the gentle slope on a hill. Its terraced layout formed a naturally enriched architectural impression. The center includes 16 senior apartment buildings, one health center, and one community center. The driveway and entrance are defined by the dense woods regarding circulation. This driveway slowly leads up to the community center square, through which the entrance connects to each individual apartment building. The health center was placed at a peaceful corner of the property. The core of the community center is a cross-shaped courtyard, radiating to the entire community. This community center provides the space for management, health care, culture, and communication for senior citizens. The entire design encourages outdoor activities and communications among residents. The entire property is covered with rainproof hallways with wheelchair accessibility, along with resting areas. In addition, there is a public living room in each apartment complex to facilitate social activities among residents. The landscape and indoor space is also rich adaptive to the elderly.

随园会地下庭院

随园嘉树养生中心
The Health & Longevity Center of the Suiyuan Jia Shu Project

浙江 杭州　　竣工时间 2015 年

建筑设计
上海中房建筑设计有限公司
陆臻 谢强 王静伟

摄影 姚力

Architects
Shanghai Zhongfang Architectural Design Co., Ltd.
LU Zhen, XIE Qiang, WANG Jingwei
Location Hangzhou, Zhejiang
Completion 2015
Photo YAO Li

入口庭院俯瞰

2号楼南侧阳台

随园会大堂南望

1 随园会
2 颐养楼
3 户外活动场地
4 健康步道

白鹭郡北多层住宅

总平面　0 10 20　50m

居住建筑

水园

项目基地位于中国江南地区，这个地区也是孕育中国传统园林的地方。建筑师希望通过这个设计探讨在现代建造技术和保证生活舒适度的前提下，如何来满足传统中国人的精神需求：寄情山水、归隐田居。因此，在本身就已经狭长的基地中，建筑师通过空间的转折和递进拉长了流线、视线和时间，最终使空间序列形成一种深宅深园的效果。建筑的原型来源于传统中国园林建筑中的"亭"，采用钢结构，同中国传统建筑的木构建筑一样，都是杆件受力体系。巨大的悬挑屋顶，使得建筑的边界同庭院融为一体。

The site of project locates in the region south of Yangtze river, which is the motherland of Chinese traditional landscape art. The architect aims to reflect the modern construction technology and living comfort, though this project trying to discuss about how to satisfy the spiritual needs for those Chinese who still really treasured Chinese traditional cultures of living and abandoning themselves to the nature, to be retired from the noisy world. Therefore, on the existing long and narrow site, architect purposely enlarged the longitudinal scale of circulation, scenic views and even timeline though the strategic transition and composition of spaces, in order to form a spatial pattern of wide depth spaces. The architectural prototype of this project originates from the traditional Chinese architectural pavilion element named "Ting", which normally constructed from wooden beams. This project was constructed from steel structural system similar to a bending structural system as wooden beams. The huge overhanging roofs extend the boundaries of architecture into the courtyard and gardens. Finally, this project delivered a consequence, which is a garden house with three different hierarchies of deep depth spaces, deep³ courtyard.

进园

深深·深宅
Deep³ Courtyard

建筑设计
素建筑设计事务所
马科元

Architects
SU Architects
MA Keyuan
Location Nantong, Jiangsu
Completion 2015
Photo SU Shengliang

江苏 南通　　竣工时间 2015 年　　摄　影 苏圣亮

居住建筑

水园　　　　　　　　　　前园
　　　　　　　　　　　　石园

1 泊园
2 进园
3 前园
4 侧园
5 水园
6 望园
7 石园
8 静园
9 动园
10 山园

总平面　0 5 10 20m

东南鸟瞰

东立面外景

新青年公社是一座混合型青年居住社区，位于万科松花湖度假区的边缘，与自然村落接壤。建筑可容纳800人，上层居住着万科员工，中层租赁给镇上个体经营者，首层供学生等营地活动使用，并向周边村民开放。

设计突破了一般集体居住的模型，将建筑体量折一为四，形成外表错落、内在连续的整体。廊桥、楼梯、阶梯座位介入通高中庭，塑造出一条如开放街巷般丰富多变的公共流线，串联起中庭及两侧的各级邻里共享空间，提供了当代年轻人衣食住行可能利用到的空间形式。如此，一个简单而灵活的框架中，新的空间理念与日常生活相融，人们出入于不同层级的个人—共享—集体区域，时有不期而遇，社群自然生长。项目有效回应了当代现实的混杂性，提出一种社群共生的新范式：在个人平等自足的基础上，经由空间的多元共享，促发互助合作、与周边环境积极互动，进而生长出一种当代"新青年"的社群原型。

New Youth Commune, a mixed youth community on the edge of Vanke Songhua Lake Resort bordering natural villages, accommodates 800 people on the upper floors for Vanke staff, while the middle floors are rented to self-employed townspeople, and the ground for camping students and open to surrounding villagers.

The residential stereotype is broken and the building bulk is divided into a quartet, with external parts undulating and internal continuous areas. Bridges, stairs and tiered seating around the full-height atriums compose an open-street-like circulation articulating various public spaces. A simple and flexible framework blends innovative spaces into daily lives, encouraging encounters among private, shared and collective zones and finally the community growth. The project responds to the hybrid contemporaneity and proposes a new community symbiosis: mutual cooperation and positive environmental interaction through inter-spatial sharing based on equality—a prototype community, in the hopes of cultivation of the social prototype of the new youth.

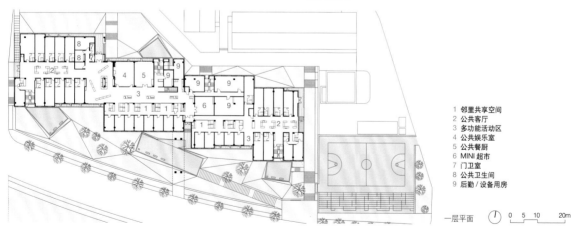

1 邻里共享空间
2 公共客厅
3 多功能活动区
4 公共娱乐室
5 公共餐厨
6 MINI 超市
7 门卫室
8 公共卫生间
9 后勤/设备用房

一层平面

新青年公社
New Youth Commune

建筑设计
META-工作室
王硕 张婧

吉林 松花湖 ｜ 竣工时间 2015 年 ｜ 摄　影 苏圣亮 方淳 陈溯

Architects
META-Project
WANG Shuo, ZHANG Jing
Location Songhua Lake Resort, Jilin
Completion 2015
Photo SU Shengliang, FANG Chun, Chen Su

居住建筑

入口桥下下沉庭院

主入口大阶梯与自发活动
中庭公共空间

中庭公共空间
中庭内立面及住户的日常

老宅夜景

祖屋的存在是设计的源头,它是一层的三开间小屋,毫无特色。但对主人而言,这就是他的童年记忆,设计的主旨则是唤起他儿时的记忆。

新的建筑以一个客厅和祖屋发生关系,退让祖屋并顺应祖屋的屋顶坡度,一方面获得了和祖屋的联系,另一方面客厅也获得了较高的空间。客厅的基调确定了整个新建筑的基本尺度,所有的功能都以此为起点顺应山势自然而成。不同形态的庭院的介入,削减建筑尺度的同时,也增加了建筑和自然的链接,让每一个主要房间都拥有了一个接触自然的场所,一个完整的屋顶把庭院之外的建筑统一覆盖,清晰完整地表达了一个家的场所意义。介于立面和屋顶之间的表情也弱化了建筑通常意义上的存在,更多地把建筑转化为了立体合院的空间氛围。

The existence of an ancestral house is the source of design, which is a plain cabin of one floor with three bays. But to the client, this is his childhood memory, and the theme of the design aims to arouse his childhood memories.

The new building places a living room in relation to the ancestral house, seting back from the ancestral house and conforming to the roof's sloping gradient. On the one hand it responds to the ancestral house, and on the other it produces a living room of higher quality with distinct space characteristics. The tone of the living room determines the basic scale of the entire new building, with all the functions conforming to the contours of mountains. The intervention of courtyard of different sizes reduces the building bulk and enhances the connection between the building and nature, allowing each main room a place in direct contact with nature. An intact roof covers all buildings inside and outside the courtyard, explicitly conveying the message of the home-like place. The expression between the facade and the roof weakens the usual existence of buildings and transforms it into a three-dimensional space of stereoscopic courtyards.

浙江山地老宅
The Old Curtilage in the Mountains of Zhejiang

浙江 嵊州 | 竣工时间 2015 年

建筑设计 江苏中锐华东建筑设计研究院有限公司
荣朝晖 刘敏毓

摄 影 胡义杰

Architects Jiangsu Zhongrui East China Architectural & Design Institute Co., Ltd.
RONG Zhaohui, LIU Minyu
Location Shengzhou, Zhejiang
Completion 2015
Photo HU Yijie

居住建筑

南立面

落院内景 1

落院内景 2

廊道内景

室内空间

剖面　0　5　10　20m

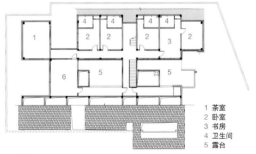

三层平面

1 茶室
2 卧室
3 书房
4 卫生间
5 露台

总平面　0　10　20　40m

一层平面　0　5　10　20m

1 储藏间
2 门厅
3 厨房
4 餐厅
5 客厅
6 灶头
7 佣人房
8 车库
9 卫生间

全景

此项目为网龙公司新总部的员工宿舍，基地位于距海边不远的一片未开发的处女地，既没太多的周边环境，也没有明确的边界。建筑师希望通过创造一种内向的、相对独立的"集体公社"，来形成强烈的社区意识。

根据周边不同的景观和建筑之间的相对关系，三栋房子各自朝不同的方向退台，为居住者提供一系列共享的屋顶平台。同时也将本来完全封闭的内院朝四周的自然景观开放，既可观山也可望海。交通流线设置在内院，并与所有共享平台相连。为了方院内外空气的流通和居住者的便利穿行，三座建筑被架空在地面之上。这里，地面高低起伏，形成复杂几何形态的土丘，既支撑起上面的建筑，又容纳宿舍配套设施。

This project is the staff dormitory on NetDragon Websoft's new campus. Not far from the ocean, the open site is virtually a tabula rasa, with neither much context nor fixed physical boundaries. The idea is to create a collective living space with inward-facing and quasi-autonomous forms that foster a strong sense of commune with strong identities.

According to the relations of various landscape and buildings, three square-shaped buildings with large central courtyards are arranged on the site at different angles. Stepping down in different ways according to wind and views, these three buildings each contain a series of carefully landscaped open terraces to be shared by the residents. This also opens up the otherwise enclosed central courtyards to the surrounding nature, be it mountain or ocean. Circulation is arranged on the courtyard side, and connects to all the shared terraces. These three stepped courtyard blocks are lifted off the ground allowing for free air circulation and foot traffic from outside through the courtyards. The ground has a complex folded geometry, forming different mounds to support the buildings above and accommodate ancillary facilities for the dorms.

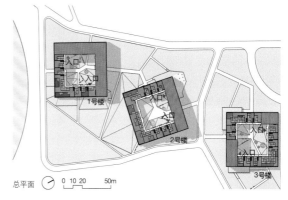

总平面　0 10 20　50m

2号楼屋顶退台

退台方院
Stepped Courtyards

福建 福州　竣工时间 2014年

建筑设计
OPEN 建筑事务所
时代建筑设计院（福建）有限公司
李虎 黄文菁

摄　影　金波安 陈诚

Architects
OPEN Architecture ; Times Architectural Design Institute (Fujian) Co., Ltd.
LI Hu HUANG Wenjing
Location Fuzhou , Fujian
Completion 2014
Photo JIN Bo'an , CHEN Cheng

3号楼内景
1号楼面向内庭院的公共走廊

2、3号楼外景
3号楼屋顶退台

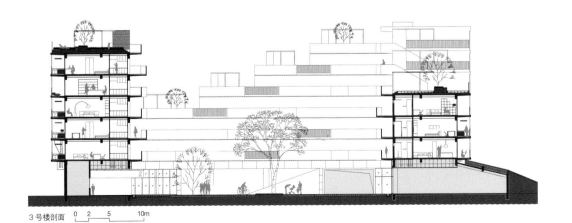

3号楼剖面　0　2　5　10m

建筑中3组院落通过加宽的走道和放大的交通节点连成有机的整体，内置了6条东西、南北不同方向的"街""巷"空间，以满足并强化老人们的公共活动和彼此之间的交流。位于中间的"主街"在尺度与形式上都着墨较重，所有公共性功能都沿着"主街"布置。建筑高度3层，掩映在绿树丛中，空间尺度亲切宜人，通过巧妙设计的天窗引入外界的自然光线。建筑的基本风格是以黑瓦白墙为主，呈现着江南建筑特有的传统气质，但在局部的细节与构件上引入色彩的跳跃与点缀，洋溢着一种晚年时的"青春"色彩；建筑外墙也一改江南建筑原有素雅单纯的白色基调，加进了毛石、U形玻璃、杉木等多种材质，传递出建筑的温暖情怀。

The three courtyards of the building are joined by widening walkways and enlarged traffic nodes into an organic whole. The building has six built-in "streets" and "alleys" in different directions to meet the requirements and strengthen the public activities of the elderly to communicate with one another. The "main street" located in the middle focus on the scale and form in design, and all public functions are arranged along the "main street". The three-storied building is embedded in green trees, and the space scale is kind and pleasant. Natural light is introduced into inside space through the smartly designed skylights. The style of the building is based on the black tiles and white walls, showing the typical traditional temperament of architecture in the lower Yangtze region. Fresh and interspersed color in the details and components is brimming with some kinds of "youth" at an old age. The facade is different from the simple and elegant white keynote of the original architecture in the lower Yangtze region, using rubble, U-shaped glass, fir and other materials to convey a sense of warmth.

室内主街

庭院

南立面局部

苏州阳山敬老院
Suzhou Yangshan Nursing Home

建筑设计　苏州九城都市建筑设计有限公司
张应鹏　黄志强

江苏 苏州　竣工时间 2014年　摄影 姚力

Architects Suzhou 9 Town Studio Co., Ltd.
ZHANG Yingpeng, HUANG Zhiqiang
Location Suzhou, Jiangsu
Completion 2014
Photo YAO Li

居住建筑

东侧立面——建筑与山体的关系
杉木板及天窗——主街阳光的引入

室内与庭院

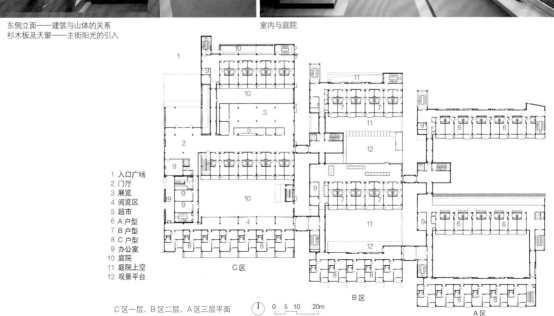

1 入口广场
2 门厅
3 展览
4 阅览区
5 超市
6 A户型
7 B户型
8 C户型
9 办公室
10 庭院
11 庭院上空
12 观景平台

C区一层、B区二层、A区三层平面   0 5 10 20m

D区高层外景
高层住宅顶层复式屋顶泳池

广州南湖山庄 C、D 区
Villa South Lake in Guangzhou-District C and D Residence

广东 广州　　竣工时间 2014 年

建筑设计
华南理工大学建筑设计研究院
倪阳 林毅 张敏婷

摄影 战长恒 陈小铁

Architects
Architectural Design & Research Institute SCUT
NI Yang, LIN Yi, ZHANG Minting
Location Guangzhou, Guangdong
Completion 2014
Photo ZHAN Changheng, CHEN Xiaotie

C区鸟瞰
高层入口花园

高层住宅顶层复式内景

　　南湖山庄项目位于广州著名的风景区南湖东侧，依山傍水。规划建筑布局结合山地特征，采用台地开发模式，因地制宜进行布置，形成顺应地势的小聚落组团空间和层层跌落的山地景观。住宅采用了逐层抬高的布置方式，别墅都有较好的视线通廊。高层布置在半山高台，为住户提供了更加独特的景观体验，将景观资源最大化。高层住宅户型设计将台地的概念引上空中，在空中植入平台，采用层层90°错位，形成6米高的空中院落。客厅结合平台设计在楼宇尽端转角处，为住户提供270°的景观。项目强调住户与自然的互动，别墅的半开敞天井院落、高层住宅的环绕大平台和空中院落、高层住宅下部的半开敞阳光车库都体现了这一理念。

　　The project of villa is located in the east of the South Lake, a famous scenic spot in Guangzhou, close to mountains and the lake. The planning combines mountainous features, and uses the platform development model according to local conditions to form a small adaptation of the terrain and cascading mountainous landscape. Residential use the layout with gradually elevated levels, so most villas will have a good view. High-rise residential houses are arranged in the hillside that do not block the villas and provide the residents of the high-rise buildings a unique landscape experience, maximizing landscape resources. The design of high-rise residential use the concept of the platform cited in the air. With layers of 90° dislocation, the 6-meter-high courtyard on air is created. The living room combines the design of the platform with 270° invincible landscape for the tenants at the corner of the building. The design emphasizes on the interaction between residents and nature, the villa's semi-open patio courtyard, ring-shaped large platform and sky courtyard of the high-rise residential and bottom semi-open sun garage of the high-rise residential all can reflect this concept.

1 C区
2 D区
3 文化活动中心

总平面　　0　25　50　100m

20号楼(A户型)中心庭院1

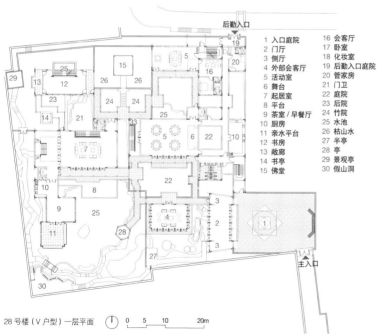

1	入口庭院	16	会客厅
2	门厅	17	卧室
3	侧厅	18	化妆室
4	外部会客厅	19	后勤入口庭院
5	活动室	20	管家房
6	舞台	21	门卫
7	起居室	22	庭院
8	平台	23	后院
9	茶室/早餐厅	24	竹院
10	厨房	25	水池
11	亲水平台	26	枯山水
12	书房	27	半亭
13	敞廊	28	亭
14	书房	29	景观亭
15	佛堂	30	假山洞

28号楼(V户型)一层平面

拙政别墅打破了现代建筑格局，把本来集中的建筑体量分散布置，让园林可以融入其中而非包裹在外，创造出游园式居住体验。走在家中便是走在园林之间，移步换景，重组空间产生深宅大院的大感官—这就是所谓的中式居住体验，也是拙政别墅最大的特点。

The layout of Humble Administrator's Villa breaks down modern architectural layout, dispersing the building bulk and integrating the garden into the villa instead of surrounding it with garden style residential experience. Wandering in the house is like wandering in the garden, changing the sceneries along the route accordingly, reassembling the space while creating the feeling of a compound of connecting courtyards. That is what is being called Chinese garden style residential experience, which is also the most spectacular feature of the Humble Administrator's Villa.

拙政别墅
The Humble Administrator's Villa

江苏 苏州　　竣工时间 2014 年　　摄影 周宁

建筑设计
华东建筑设计研究院有限公司
华东都市建筑设计研究总院
俞挺 傅正伟 丁顺

Architects East China Architectural Design and Research Institute Co., Ltd.; East China Urban Architectural Design and Research Institute
YU Ting, FU Zhengwei, DING Shun
Location Suzhou, Jiangsu
Completion 2014
Photo ZHOU Ning

居住建筑

20号楼（A户型）南庭院
庭院

20号楼（A户型）中心庭院2

东大门

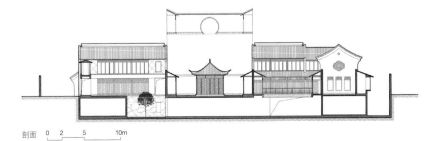

剖面　0　2　5　10m

5m

Production Facilities

生产设施

P370

P383

鸟瞰

项目位于唐山古冶城区边缘的一片农田之中，周边零散分布着村落和房屋。建筑的基本功能是有机粮食加工作坊。设计的思路是创造一个放大的四合院，营造充满自然氛围和灵活性的工作场所，并与周围广阔平坦的田野产生对应性关系。整体建筑由四个相对独立的房屋围合而成，分别是原料库、磨坊、榨油坊、包装区。作为粮食的晒场，便捷的工作循环流线围绕内庭院形成。建筑的边界是联通四个分区的外部游廊，这是参观粮食作坊的流线。空间、结构、材料以及室外庭院共同塑造出这个农场温暖、自然、内外连续的工作场景。

The project is located in the farmland on the fringe area of Guye district, Tangshan city. Villages and houses are scattered around. The site is a rectangular flat land covers an area of 6000 m². The basic function of the building is a processing workshop of organic food. Inspired by traditional courtyard building, the initial idea of design is to build a magnified courtyard house, a workplace full of natural atmosphere and flexibility, and a workplace which is self-contained and forms a corresponding relation with the surrounding broad and flat field. The entire building is enclosed by four relatively independent houses, including storage of raw material, the mill, oil pressing workshop and packing area. The inner courtyard is the drying yard of grain, and convenient cycling procedures are designed around the inner courtyard. The boundary of the building is the external corridor that connects the four areas, and it is the route to visit the food processing workshop. Space, structure, materials and the multi-layered exterior courtyards together creates a warm, natural and continuous working atmosphere for this farm.

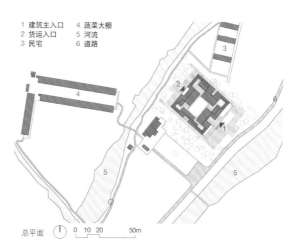

1 建筑主入口　4 蔬菜大棚
2 货运入口　　5 河流
3 民宅　　　　6 道路

总平面

唐山乡村有机农场
Tangshan Rural Organic Farm

河北 唐山 ｜ 竣工时间 2016 年 ｜ 摄　影 金伟琦

建筑设计
建筑营设计工作室
韩文强

Architects
ARCHSTUDIO
HAN Wenqiang
Location Tangshan, Hebei
Completion 2016
Photo JIN Weiqi

生产设施

主立面夜景
室外游廊

竹院

剖透视　0 1 2　5m

室内游廊　　　　　小麦磨坊

生产空间

村委办公

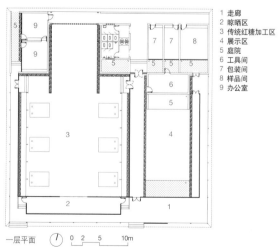

1 走廊
2 晾晒区
3 传统红糖加工区
4 展示区
5 庭院
6 工具间
7 包装间
8 样品间
9 办公室

一层平面　0 2 5 10m

松阳樟溪红糖工坊
Songyang Zhangxi Brown Sugar Workshop

浙江 丽水　　竣工时间 2016 年

建筑设计
北京多维度建筑设计咨询公司
徐甜甜

摄　影　王子凌

Architects
Beijing DnA＿Design and Architecture.
XU Tiantian
Location Lishui, Zhejiang
Completion 2016
Photo WANG Ziling

远景

展示区
参观廊道

工坊兼具红糖生产厂房、村民活动和文化展示的多重功能，是衔接村庄和田园的一处重要场所。作为村庄与四周田园的过渡延伸，村民在生产活动的同时能够欣赏田园风光，工坊也成为村民田间劳作之余休憩的场所。红糖工坊同时也是展现红糖生产活动的舞台。视线开放的空间设计，让生产现场如同戏剧表演现场。不同于传统的二维立面，外部廊道空间实际形成了多层的文化长廊，展示着红糖工坊内外、乡村文化、田园诗意和游客观景的叠加。

As a complex place of brown sugar production, villager activity, and cultural presentation, the workshop in Xing Village is a significant connection between village and rural landscape. As the transitional space between village and surrounding rural site, local villagers can enjoy the landscape outside when they are working and the building also provides a lounge place after work. The Brown Sugar Workshop is also to show production activities. Spatial design with broad vision makes the production a live drama. Unlike traditional two-dimensional elevation, the multi-layered corridors exhibit the inside and outside of the workshop, rural culture, pastoral poetry and tourist scene.

剖面 0 2 5 10m

鸟瞰

基地毗邻3条高速公路，这决定了建筑形体应以一种完整和连续的姿态，与大尺度的城市基础设施之间形成对话。一方面，在保证完整和连续表面的前提下，裂解出实体面之间精致的开阖关系，为地面行人创造近人尺度的视觉感受；另一方面，在基地一端退让出一块三角形区域，改善周边的城市公共环境，从这个视角，建筑在每个侧面的表情各不相同。在建筑内部，盘旋环绕的建筑形体围合出两个风格独具的庭院，是美术馆、印刷工厂、办公空间、底层公共空间等多个部门共享的空中花园。其中美术馆从公共属性的功能部分分离，与企业总部并列，独立地悬浮在景观资源最好的顶部，形成空间效果独特的独立艺术空间。在办公与厂房之间形成的竖向空间被设计成一个垂直的图书馆，用来收藏与展示雅昌日益增加的藏书。一部飞扶梯转折地穿过整个空间，连接了不同楼层，与穿行的人一同形成了这个建筑最富有戏剧性的片段。

Artron Art Center is adjacent to three highways, the massing is determined to be continuous and integrated to dialogue with the surrounding large-scale urban infrastructure. On the one hand, on the premise of integrity and continuity, differentiation of various volumes were created to bring people a comfortable visual impression when walking closer. On the other hand, a triangular plot was reserved as a public park on the corner of the site to improve the urban environment. Observing from this small park, each side of this building is different. For the inner space of this building, the wreathed volume creates two courtyards on the air with different characteristics and shared by the art center, printing factory, office and the ground public space. The art gallery was isolated from the public functions on the ground floor and placed in parallel with the corporate headquarters. Levitating on the top of the building, the art gallery has the chance to create an independent art space with special effect. The vertical space between office and factory was designed as a vertical library to display Artron's growing book collection. An open stair folding through the whole space in the air connects different floors, creating the most dramatic episode of the building.

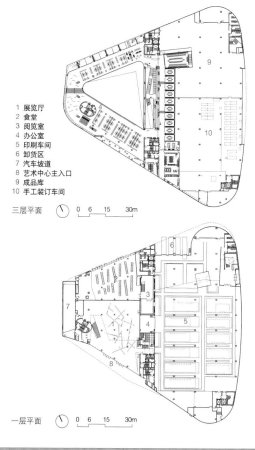

1 展览厅
2 食堂
3 阅览室
4 办公室
5 印刷车间
6 卸货区
7 汽车坡道
8 艺术中心主入口
9 成品库
10 手工装订车间

三层平面

一层平面

雅昌艺术中心
Artron Art Center

广东 深圳 | 竣工时间 2015年

建筑设计
URBANUS 都市实践建筑事务所
孟岩 饶恩辰 周娅琳

摄影 陈冠宏 杨超英

Architects URBANUS Architecture & Design Inc.
MENG Yan, RAO Enchen, ZHOU Yalin
Location Shenzhen, Guangdong
Completion 2015
Photo Alex Chan, YANG Chaoying

生产设施

东南向街景

内庭空间
办公空间   中庭

由廊桥看厂房

基地是沿河展开的一个狭长场地。受工艺流程和地形的双重限制，厂房设计为长条形，在不影响工艺流程和满足设备净高的前提下，把约160m长、13m宽的厂房折成一Z字型。屋顶采用对角线式双坡顶，最高处10m，最低处5m，以降低建筑体量对环境的压迫感。建筑材料以旧青砖、旧瓦、旧门板为主。除生产工艺要求以外，内外主墙面均为清水砖墙，木板均为原色。大部分旧砖与门板均留有时间积淀形成的表面痕迹，整个建筑内外呈现出丰富的细部质感。建筑檐口均为斜面，砖墙檐口封顶时采用和墙身相同的砌法，并顺势作锯齿状退台处理，同时也取得了另类视觉效果，称之"像素檐口"。

The site is a long and narrow space stretching out along the river. Because of the dual restriction from production process and the terrain, the factory building can only be a strip form. In order to reduce the pressure to the environment, the building, 160m long and 13m wide, it is folded as the shape of "Z". The roof top uses a diagonal double crest style, and the height varies from 10m to 5m. Building materials are mainly old bricks, old tiles and old door boards. No painting is made to bricks and wood boards used for both outside and inside walls. Most surface traces as a result of the elapse of time. Old bricks and wood boards are kept. Cornices are inclined and the sealing part of the cornice uses the same method with the walls. The end of the cornices can be called "pixel cornice" with a unique visual effect.

榨油厂夹层
车间采用水泥透明自流平地面

松风翠山茶油厂
Song Feng Cui Camellia Oil Plant

江西 婺源　　竣工时间 2014 年

建筑设计
畅想建筑设计事务所
罗四维 卢珊 周伟

摄影 曾江河

Architects
Imagine Architects
LUO Siwei, LU Shan, ZHOU Wei
Location Wuyuan, Jiangxi
Completion 2014
Photo ZENG Jianghe

生产设施

厂房西南

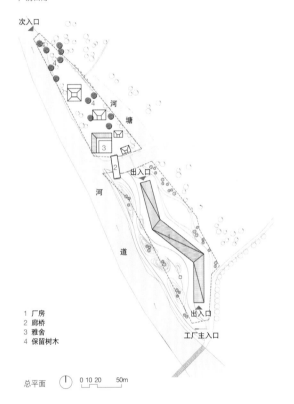

1 厂房
2 廊桥
3 雅舍
4 保留树木

总平面 0 10 20 50m

廊桥及厂房鸟瞰
厂房西侧

1 准备车间
2 榨油车间
3 瓶装车间
4 包装车间
5 成品车间
6 参观走廊
7 辅助用房
8 更衣室
9 消防控制室
10 夹层

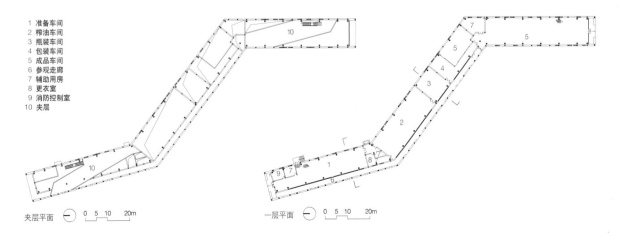

夹层平面 0 5 10 20m

一层平面 0 5 10 20m

地面层公共空间

项目基地位于北京顺义区。业主要求建设一个囊括生产线、物流中心、储藏、科研开发和办公等功能的时尚集合体。设计在布局上采用了外表规整、内部自由的形式。建筑在整体上是规整围合形的，以满足生产要求及应对寒冷气候，同时在建筑的实体内部开出了一条自由曲线形的内街，内街贯穿在整个建筑内，为建筑带来自然的气息和时尚感。在材料和色彩的运用上也内外有别，外表皮为有纵向纹理的铝板幕墙和玻璃幕墙，色彩为工业化的灰色，内廊则采用带有表面质感的挤塑成型水泥板，色彩上沿用内衣常用的粉色系，几种不同深度交错排列。内街上空在13m高度上建有一个空中连廊，形成连接各区域的纽带并增加了趣味性。

The project is sited in Shunyi District, Beijing. The client required that a fashionable complex be built including a production line, distribution center, storage facility, quality control, research and development department, as well as offices. The design is regular on the outside but flexible for the inside. The regular enclosed form responds to production requirements and the cold climate, while an internal curvilinear street is placed throughout the building, bringing natural and fashionable flavor to the building. Application of colors and materials varies from the inside and the outside: aluminum and glass curtain wall with vertical texture on the facades in industrialized grey, while concrete boards molded by extrusion with surface texture in pink normally seen in underwear are used on internal corridors, arranged alternately according to color scale. An airy corridor on the altitude of 13 meters above the internal street is built to connect various parts with added interests.

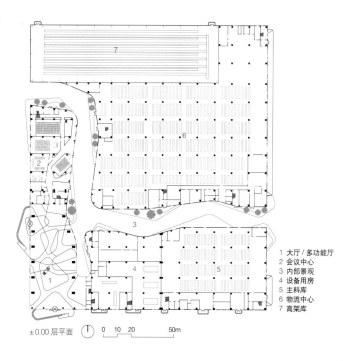

1 大厅/多功能厅
2 会议中心
3 内部景观
4 设备用房
5 主料库
6 物流中心
7 高架库

±0.00 层平面

爱慕时尚工厂
Aimer Lingerie Factory

建筑设计
Crossboundaries
北京市建筑设计研究院有限公司国际工作室
蓝冰可 董灏

Architects
Crossboundaries; BIAD International Studio
Binke Lenhardt, DONG Hao
Location Beijing
Completion 2014
Photo XIA Zhi, YANG Chaoying

北京 | 竣工时间 2014年 | 摄影 夏至 杨超英

生产设施

有机形态的空中连廊

轴测   空中连廊

四层办公室   大堂

从毛竹晾晒场看大车间和办公宿舍楼全景
从东北面的道路上看办公及宿舍楼

办公及宿舍楼西侧外观

武夷山竹筏育制场是武夷山九曲溪旅游漂流用竹排的储存及制作工厂，项目位于武夷山星村镇附近乡野中的一块台地上，由竹子储存仓库（未建成）、竹排制作车间、办公及宿舍楼这三栋建筑及其围合的庭院组成。建筑的布局与朝向结合地形、风向考虑，同时回应气候和项目的功能需求，将当地非常普及的日常性材料与工艺在设计中进行有针对性的运用。项目的工业厂房性质决定了建筑摒弃形式上任何的多余，而在建构上采用最基本的元素，并尽可能呈现其构造逻辑，在营造工业建筑朴素美学的同时获得经济性。

Bamboo rafts of drifting for tourism in Jiuqu Creek in the Wuyi Mountain are produced and stored in this factory, located on a raised wild land close to Xingcun Township. It consists of an unbuilt warehouse, a production factory of rafts, and an office building with a courtyard enclosed by the three buildings. The layout and orientation of the buildings take into considerations the terrain, wind directions and functional requirements in response to the climate, and normal materials widely used in this region are applied in the design with targeted adjustment. The industrial character of the project discourages superfluous design; by using the most basic elements for its construction, the architecture naturally reveals its structural and material logic. The project reconciles aesthetic simplicity with an economy of means, by which the architecture can demonstrate its resolution of form and function.

竹子晾晒场
现状建筑
新建建筑

1 毛竹储存仓库（未建）
2 制作车间
3 办公宿舍楼

入口

总平面 0 10 20 50m

武夷山竹筏育制场
Wuyishan Bamboo Raft Factory

建筑设计
北京迹道建筑设计有限责任公司
华黎

福建 武夷山 　竣工时间 2013 年 　摄影 苏圣亮

Architects Trace Architecture Office HUA Li
Location Wuyishan, Fujian
Completion 2013
Photo SU Shengliang

生产设施

制作车间西面转角外观
大车间室内

1 制作车间（大）
2 制作车间（小）
3 休息间
4 室外庭院
5 工具存放区
6 卫生间

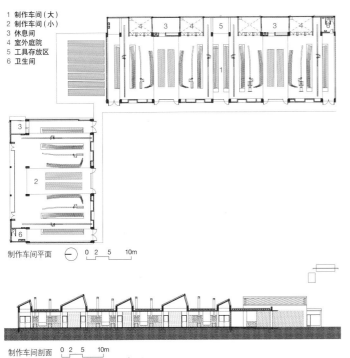

制作车间平面　0 2 5 10m

制作车间剖面　0 2 5 10m

斜屋下面的高空间（小车间）

Urban Design
Others

城市设计
及其他

P384

P417

葫芦口广场鸟瞰

人本性、文学性以及东方式的空间叙述方式，是建筑师长期关注的命题。

通过"情景与共"的东方式空间叙事方式，用多种材料构成设计的语法与词句，将老西门的若干组建筑组成的街区及场所赋予恰当的意义。每栋房子与空间，都可以用诗情画意来加以标识。建筑师仿佛变身为导演，编导出一场接一场蒙太奇一般的城市视觉连续剧。希望通以反观自我，回归历史本原的方式，寻找新的设计依托和起点，避免落入流行时尚审美的巢穴。近30栋建筑的"老西门"区域，从单一的居住商业功能，推演出更多的城市复合功能业态，因其对周边社群居民生活所具有的公共分享意识，地域化的文化传播属性，成为城市新文化的诠释者、朗读者。

Humanism, literariness and oriental-style space narrative method are topics which have drawn the architects attention for a long term.
By utilizing the oriental-style space narrative method harmonizing landscape and emotion, blocks and sites composed of several sets of architectures in Laoximen are endowed with proper connotation. Here, multiple materials are merged in the design. Each building and space is picturesque and poetical. The architects are like directors who write and direct the TV series themed on montage city vision one after another.
The architects are seeking new supports and starting points for design through self-reflection and returning to historical origin as ways of evading the trap of aesthetic fashion.
Apart from the residential & commercial housing, the Laoximen region, with nearly 30 buildings, boast more composite urban functions. The benefits rendered by this region are accessible to citizens in the surrounding communities, and the cultural transmission is unencumbered within the region, which have made it the explainer and representative of the new urban culture.

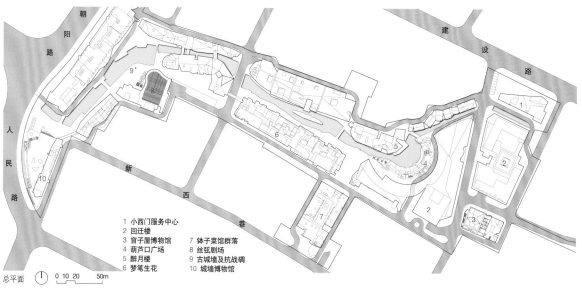

1 小西门服务中心
2 回迁楼
3 窨子屋博物馆
4 葫芦口广场
5 醉月楼
6 梦笔生花
7 钵子菜馆群落
8 丝弦剧场
9 古城墙及抗战碉
10 城墙博物馆

总平面 0 10 20 50m

常德老西门综合片区城市更新
Urban Renewal of Old West City Gate Region

湖南 常德

建筑设计
中旭建筑设计有限公司理想空间工作室
常德市天城规划建筑设计有限公司
曲雷 何勍

竣工时间 在建

摄 影 何勍 张广源

Architects
Ideal Space Studio in Zhongxu
Changde Tiancheng Planning & Architectural Design Co., Ltd.
QU Lei HE Qing
Location Changde Hunan
Completion Under-Construction
Photo HE Qing ZHANG Guangyuan

丝弦剧场及护城河两岸商业街

葫芦口商业街

尼莫桥及护城河倒影

钵子菜博物馆与老药材仓库山墙

窨子屋博物馆鸟瞰

从村外公路上看桥

改造对象为村内一座连接东西两岸的桥。原桥建于2013年，总长度为50m左右，宽度为4m。现状桥面为混凝土铺设、金属圆管栏杆。桥建造的本意是考虑车辆通行，但因为桥两端民房密集、道路狭窄致使无法通车。改造意图是将其转化为村民的公共活动空间，丰富通行的观景感受。设计运用同一种断面的成品木料进行穿插、连接、组合，形成独立的结构体及空间围合。同一种操作方式一以贯之，改造桥面和栏杆。同一种木构件组合成不同的剖面空间，不同的剖面空间再进行组合，把桥划分为步移景异的不同段落，并将桥从车辆的通行尺度转变为人的行为尺度。统一的构件尺度、简化的构造方式，使得施工简单易行，整个桥主要由3名木工完成。

Built in 2013, the bridge to be renovated connects the two sides of the river, 50m long and 4m wide. The bridge was equipped with concrete pavement and metal rails. It was originally designed to open to traffic, but it didn't work out as the houses sited in each end of the bridge are dense and the roads are narrow. The renovation aims to transform the bridge into a public space for residents' activities, enriching the viewing experience. Processed wooden product of the same size are used for penetration, connection and combination to form independent structure and enclosed space. The same method of manipulation is used throughout the construction to renovate the surface and railings of bridge. The same wooden components are assembled into different sections, and the sections are combined as various sectors with varying sceneries. The design transforms the bridge from vehicle scale to human behavior scale. The unified scale of the components and the simplified connection pattern contribute to a simple and easy construction, and the renovation project was completed mainly by three carpenters.

五种剖面空间　0　1　2　　5m

驿道廊桥改造
Lounge Bridge Renovation

建筑设计
合木建筑工作室
张东光　张意姝　刘文娟

Architects Atelier Heimat
ZHANG Dongguang, ZHANG Yishu, LIU Wenjuan
Location Baoding, Hebei
Completion 2017
Photo ZHANG Dongguang, HUO Ying

河北 保定　｜　竣工时间 2017年　｜　摄　影　张东光　霍莹

城
市
设
计
及
其
他

混凝土桥面上的木结构加建
雪后桥内景象

材质与肌理

温暖的色调与光影

穿行中的儿童

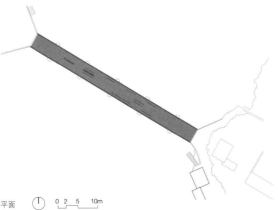

平面　　0 2 5　10m

总平面　　0 8 16　40m

西侧立面
训练场鸟瞰

观礼台圆洞钢梯

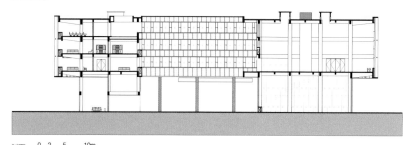
剖面 0 2 5 10m

天府新区公安消防队站
Fire Station of Tianfu New District

四川 成都 | 竣工时间 2016 年

建筑设计
中国建筑西南设计研究院有限公司
刘艺 胡健 王子超

摄影 存在建筑

Architects
China Southwest Architectural Design and Research Institute Crop., Ltd.
LIU Yi, HU Jian, WANG Zichao
Location Chengd, Sichuan
Completion 2016
Photo Arch-Exist Photography

消防站集办公、消防指挥、救援、训练、宣传于一体，成为面向未来的高标准消防站。设计采用了集中布局方式，功能紧凑高效，并留出完整的训练场地。底层采用全架空方式，争取更多的消防停车及活动空间。架空层高8m，上部建筑大尺度悬挑，形成体量的张力。架空层插入4个红色的方盒子体量，形成"消防箱体"的概念。两部造型别致的室外钢楼梯穿插于架空层中间，形成室外流线上的焦点。上部悬挑的体量四周设置通长的机翼形金属遮阳板，有效解决东西晒问题，同时将不同的内部功能统一在完整的表皮肌理之下，凸显整齐划一的部队文化。新区消防站注重公众教育，打造完整的独立参观流线，方便对市民开放。

The fire station integrates office, fire control, rescue, training and propaganda, as a fire station for the future of high standard. The new design adopted centralized layout, so the function becomes compact and efficient, and creates sufficient training ground. In order to create more fire truck parking space and activity space, the entire ground floor is elevated. The overhead space is 8m high, and upper massing has large-scale overhangs that produce tension for the massing. The overhead layer is inserted by four red square boxes, accommodating the functions of "fire box". Two delicate outdoor steel stairs are interspersed between the overhead spaces, creating a focal point on the outdoor circulation. The upper cantilevered part that fits in with long wing-shaped shadings effectively solve the problem of over exposure to sunlight, at the same time unify different internal function under a complete skin texture, highlighting troop's culture of uniformity. The fire station values public education. This convenient strategy for public will become a function highlight for the next generation of fire station.

消防停车区

观礼台

内院

训练馆室内

1 门厅　　5 小餐厅　　9 配电室
2 停车位　6 厨房　　　10 空呼充装站
3 景观庭院 7 烟热训练馆 11 训练器材库
4 大餐厅　8 值班室　　12 战斗服存储架

一层平面　0　5　10　20m

培田村是一个中国南部典型的被大山隔绝的山村。2014年初,一场巨大的洪水将当地很多联通各个分散部分的桥梁和道路冲毁。项目希望可以将培田村和这些孤立的具有历史意义的部分重新联系起来,并恢复原有的历史脉络和道路布局。

风雨桥的建设从福建地区传承已久的木建筑中汲取灵感,采用了相扣的木结构作为主体。在当地仅存的手艺精湛的木匠师傅的指导下,整座桥由265个不同部件拼装而成。这座桥在村子的农田中央创造了一个社群的场所,并在培田村多变的气候中为人们提供了一处可以躲避风雨的小憩之地。项目试图为社区改造提供另一种可能的模式——通过借鉴当地工艺和传统,并利用可再生材料和方法来构建具有社会意义的基础设施。在这一过程中,利用数字技术设计来进行规划和复杂的组装的实验是至关重要的。

Peitian is one of a number of isolated rural villages distributed throughout the mountainous regions of southern China, which, following severe flooding in early 2014, saw much of the infrastructure linking its disparate communities destroyed. This project aims to reconnect Peitian village to that historic network of routes that link these isolated settlements.

"Wind and Rain Bridge" is a reciprocal interlocking timber structure, which draws on the long tradition of wooden buildings native to the region. Each of the bridges' 265 elements is unique and integral, assembled under the supervision of traditional carpenters, who number some of the few remaining exponents of their craft. The bridge creates a community space, located in the heart of the village's fertile farmland, where local people can socialize and exchange. Opening outward towards the village, the bridge negotiates the variable terrain and provides a place of respite from Peitian's changeable climate. This project seeks to offer an alternative mode of community redevelopment that references local crafts and traditions, and utilizes sustainable materials and methods, to create both social and physical infrastructure. Critical to this process is the integration of digital design methodologies, which allow for the planning and testing of complex assemblies. The high level of training and labor associated with these assemblies has been a barrier to the continued viability of complex, long-span, timber structures in China and other developing and transitioning economies.

田间的风雨桥
由桥内向外望

风雨桥
Wind and Rain Bridge

建筑设计
香港大学
Donn Holohan

Architects
The University of Hong Kong
Donn Holohan
Location Longyan, Fujian
Completion 2016
Photo Donn Holohan

福建 龙岩 | 竣工时间 2016年 | 摄 影 Donn Holohan

城市设计及其他

桥身全景
桥内局部

桥内空间

桥身细部

1 2 3 4 5 6 7 8 9 10 11 12

轴测剖面

1 2 3 4 5 6 7 8

1 2 3 4 5 6 7 8 9 10 11 12

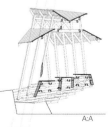

1 2 3

鸟瞰

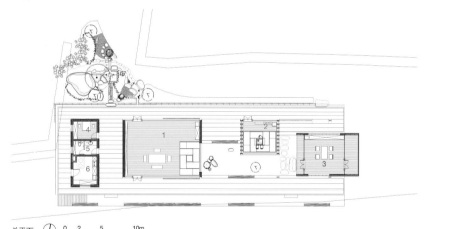

1 见山
2 读水
3 忘愁
4 办公
5 卫生间
6 厨房

总平面 0 2 5 10m

　　项目是建在城市近郊的乡村风景中的一座茶室。方案选择在田地与水体交界处设一座横向展开的一字形建筑，三间茶室，三种意象，一处内院，与山水同坐，观天地人心。一系列竹构设施的建造，赋予农场新的表情。为此莲花荡农场也成为一个周边居民能够自由赏游的场所。建造上，尝试研究宜兴当地特有的建筑材料和建造方式，将竹、木、陶、土等材料悉数表达。在实际建造过程中，由建筑师直接参与从局部施工、景观布置，到室内装修以及家具选择、器物摆设的全部过程。

This project is a teahouse built in rural landscape on the outskirts of the city. The building stretches out horizontally consisting of three tea rooms with three imageries and one courtyard, harmonious with the mountains and water. A series of bamboo facilities gave the farm a new look. For this purpose, Lianhuadang Farm becomes a free entertaining place for people living around. In construction, the architects have studied the local building materials and construction methods in Yixing, using bamboos, woods, potteries, soil, etc. for possible representations. The architects participated in the whole construction processes from partial construction and landscaping to interior decoration, furniture selection and placement.

隐庐莲舍
Lotus Tea House

建筑设计 东南大学建筑学院 唐芃

江苏 宜兴 | 竣工时间 2016 年 | 摄 影 许昊皓 王笑

Architects School of Architecture, Southeast University
TANG Peng
Location Yixing, Jiangsu
Completion 2016
Photo XU Haohao, WANG Xiao

茶室入口　　　　　　　　　　　　　　　　　　大茶室（见山）

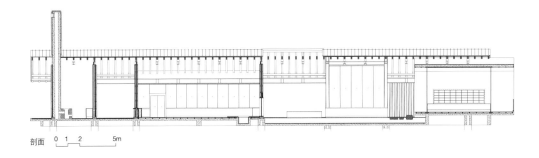

剖面　0　1　2　　5m

内院1　　　　　　　　　　小茶室（读水）
　　　　　　　　　　　　　内院2

HEX-SYS 及其不远处的广州南站

HEX-SYS/ 六边体系是 OPEN 建筑事务所研发的灵活可拆装建筑体系。作为对中国近年来伴随着建造热潮而出现的大量临时建筑的回应,这个可快速建造、可重复使用的建筑体系延长了建筑的生命周期,实现了真正意义上的可持续性。预制化生产和装配式建造,使其像产品一样具备批量生产的可能;而通过模块的不同组合方式,它又会演化出各种各样的版本,灵活适用于不同的场地和功能。它是 OPEN 长期以来对批量定制和建筑可持续性探索的一个方面。基本建筑单元是一个六边形的模块。倒伞状的屋顶钢结构由位于中央的圆柱结构支撑,空心的圆柱兼做雨水管,可将收集到的雨水用于景观灌溉或者注满庭院水池。三种不同的单元——透明的、围合的和室外的,分别适应不同的功能需求。OPEN 建筑事务所拥有此建筑体系的专利。

HEX-SYS is a reconfigurable and reusable building system designed by OPEN. It is the architects' reaction to the unique Chinese phenomenon in the building frenzy of the recent decades: the production of a vast amount of flamboyant but short-lived buildings. This modular building system can easily adapt to many different functions, and can be disassembled and reused, thus extending a building's life cycle and saving a significant amount of resources. Because it is modular and prefabricated, it can be built much faster than traditional buildings. This system is part of OPEN's continuous exploration of mass-customization and the ultimate potential of building sustainably. The basic building cell is a hexagon module with an inverted umbrella roof structure standing atop a single pipe column that double functions as the rain flue. Rainwater is collected and used for landscape irrigation. There are three basic types of cells—indoor-open, indoor-closed, and outdoor-open—to accommodate different functional needs. OPEN holds the design patent for this building system.

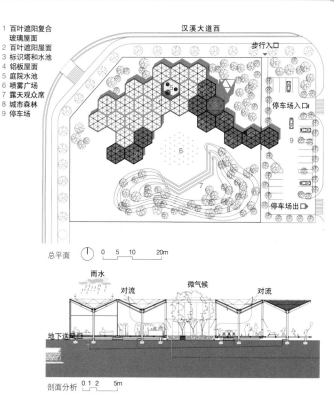

1 百叶遮阳复合玻璃屋面
2 百叶遮阳屋面
3 标识塔和水池
4 铝板屋面
5 庭院水池
6 喷雾广场
7 露天观众席
8 城市森林
9 停车场

总平面

剖面分析

六边体系
HEX-SYS

广东 广州 | 竣工时间 2015 年

建筑设计
OPEN 建筑事务所
建研科技股份有限公司
李虎 黄文菁

摄影 张超

Architects
OPEN Architecture; CABR Technology Co., Ltd.
LI Hu HUANG Wenjing
Location Guangzhou, Guangdong
Completion 2015
Photo ZHANG Chao

从主入口看建筑

室内咖啡区和模型展示区
从入口看室外

建筑屋面细部

公园鸟瞰

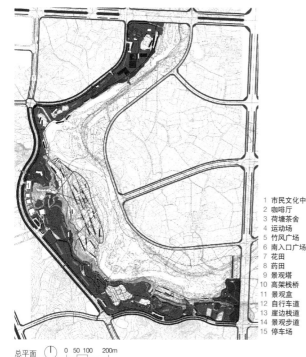

1 市民文化中心
2 咖啡厅
3 荷塘茶舍
4 运动场
5 竹风广场
6 南入口广场
7 花田
8 药田
9 景观塔
10 高架栈桥
11 景观盒
12 自行车道
13 崖边栈道
14 景观步道
15 停车场

总平面 0 50 100 200m

衢州鹿鸣公园位于衢州市西区石梁溪西岸，处于衢州市的新城中心之核心地段，是高密度城市建筑之中的一片"绿洲"。设计将具有生产性的农业景观与低维护的乡土植物融于景观设计之中，创造出一个丰产而美丽的城市公园。一系列漂浮于植被和溪水之上的步行道、栈桥和亭台等构成一个游憩网络，让人悠游于山水自然之中，而又不给自然过程造成过度的干扰。城市遗弃地由此转变成丰产而美丽的景观，同时保留了场地的生态特色与文化遗产。通过探索人工建设与自然元素的平衡，实现人与自然的和谐共生。

Quzhou Luming Park is located in Quzhou, Zhejiang Province. On the site in the central part of the new town surrounded by dense new urban development, the landscape architect created a dynamic urban park by incorporating the agricultural strategy of crop rotation and a low maintenance meadow. An elevated floating network of pedestrian paths, platforms and pavilions create a visual frame for this cultivated swath and the natural features of the terrain and water. Using these strategies, a deserted mismanaged landscape has been dramatically transformed into a productive and beautiful setting for urban living, while preserving the natural and cultural patterns and processes of the site.

衢州鹿鸣公园
Quzhou Luming Park

建筑设计
北京土人城市规划设计有限公司
俞孔坚 刘玉杰 高正敏

浙江 衢州   竣工时间 2015年   摄影 邢春杰 李良 陈鹤

Architects Turenscape Design Institute
YU Kongjian, LIU Yujie, GAO Zhengmin
Location Quzhou, Zhejiang
Completion 2015
Photo XING Chunjie, LI Liang, CHEN He

城市设计及其他

高架凉亭

毓秀阁观景塔

凉亭
观景栈道与步行系统

新河三角洲鸟瞰

长沙滨江文化园包括：博物馆和城市规划展览馆、图书馆、音乐厅。项目位于湘江与浏阳河交汇的新河三角洲，将"顽石"和"沙洲"作为创作的构思原型，引入大地景观元素并将其融入到城市形态之中，重塑了新河三角洲的大地肌理与地表形态。建筑向洲头奔去的动势以及从大地中崛起的形象，象征湖湘文化中勇往直前的性格特点，通过整体建筑形态的隐喻与城市精神形成呼应。设计以"开放共享，交流舞台"为出发点，为市民文化生活打造一个城市客厅。

Riverside cultural park in Changsha includes a museum and city planning exhibition hall, a library, and a concert hall. Located in the new delta of Rivers Xiang and Liuyang, the project uses stone and shoal as the prototype in design, bringing landscape elements in urban form to reconstruct the land texture and form of the terrain in this region. The kinetic trend of the building towards the tip of the shoal and the form emerging from earth symbolizes the brave and unyielding character in Hunan culture, echoing to the spirit of the city through morphological metaphor. Starting with the idea of open and sharing as a stage for exchange, the design aims to create a living room for urban cultural life.

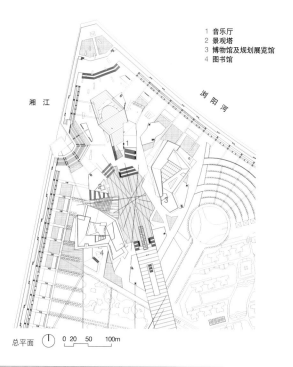

1 音乐厅
2 景观塔
3 博物馆及规划展览馆
4 图书馆

总平面

长沙滨江文化园
Riverside Cultural Park in Changsha

湖南 长沙 | 竣工时间 2015 年

建筑设计
华南理工大学建筑设计研究院
湖南省建筑设计院
陶郅 郭嘉 郭钦恩

摄影 邵峰 涂宇浩

Architects Architectural Design and Research Institute of SCUT; Hunan Architectural Design Institute
TAO Zhi, GUO Jia, GUO Qin'en
Location Changsha, Hunan
Completion 2015
Photo SHAO Feng, TU Yuhao

景观塔
音乐厅室内

音乐厅室外
博物馆室内

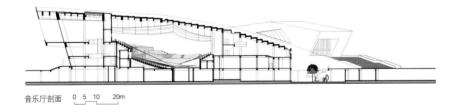

音乐厅剖面　0　5　10　20m

博物馆和城市规划展览馆　　　　　　　　　　　　　　　　　　　图书馆表皮肌理

桥身表面的木纹清水模质感
廊道与环境

天空之桥
Sky Bridge

台湾 云林 | 竣工时间 2015 年

建筑设计
立联合建筑师建筑事务所
(立·建筑工作所)
廖伟立

摄 影 汪德范

Architects AMBi Studio
LIAO Weili
Location Yunlin, Taiwan
Completion 2015
Photo WANG Te-fan

这是一条融入社区生活的景观设施廊道。设计理念是意图制造出由开阔而压缩、由压缩而聚焦的感官体验，仿佛是一种体验时光流动的过程。设计上是通过桥体结构系统上的转换来实现的，桥体在旧铁道护坡尾端以RC深梁板结构的形式出发，于中段时结构转变为"盒"的板墙系统，将原本行走在桥面上的单一路径转向桥体空间内，在桥体透过桥体的框景望见复兴铁路断桥遗址，结构和构造的转换与空间体验的层次感相呼应。桥为钢筋混凝土构造，桥身表面的木纹清水混凝土带出厚实粗犷的体量感，与行人任意穿梭停留的空间体验产生强烈对比。

This is a landscape corridor embedded in community life. The design aims to create sensual experience from broadness to compressedness and then to concentration, just like a process of experiencing the elapse of time. This process is realized through structural transformation in design. The design starts with deep beams and boards of RC structure for the ending part of the revetment of the old railway, and moves the box-like wall and board system for the middle part, which directs the original singular pedestrian route on the bridge into inside the bridge. Structural and tectonic transitions echoes to various layers of spatial experience as the sky bridge frames a view for the remaining site of the broken bridge of Fuxing Railway. Sky Bridge is made of RC structure with massive and rugged exposed concrete, which is a sharp contrast with the freedom experience of people.

桥身局部

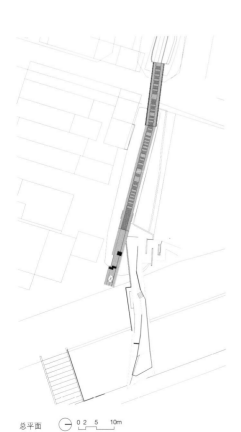

总平面　0 2 5 10m

桥体空间

庭院空间

根据计划,上海西南郊的金山现代农业园正在转变成复合功能的城市郊野公园。核心区域公共空间提升正是这一转变的启动项目。设计构思中游廊成为设计的主题,东西向延伸的游廊将快速路北侧数个孤立的农业旅游项目连接起来,未来还将继续向西延伸,并通过东侧景观高台以步行桥向南整合更多的旅游项目。游廊南侧景观台地隔绝了快速路噪音与视线的干扰。简洁的白色混凝土结构替代了传统游廊的木结构,以与周边现代设施农业环境相匹配,同时有效地控制了造价。整体折线形的布局使得基地东西不同区段节点都可以获得充足的活动空间,特别是中部游廊与景亭一起围合而成的院落已经成为了核心区游客驻足与活动的中心。结合不同区段景观,游廊两侧墙体交替的导向性开口,使"步移景异"的空间体验变为可能。类似于传统高台建筑,底部洗练的台地赋予了游廊与周边环境微妙的高差变化与强有力造型特征。

According to an ambitious plan, the Jinshan Modern Agricultural Park, located on the southwestern outskirts of Shanghai, is being transformed into a suburban park with composite functions, initiated by enhancement of the public space in the core area of the park. In the conception of design, the veranda, an essential element in traditional Chinese gardening, has been selected as the theme of the design. In the park, a recreational corridor extends both to east and west, connecting several isolated agricultural tourism projects on the north side of the freeway, and the corridor will stretch farther westward and will pass through the eastern landscape mound to be integrated with other tourism projects towards the south via a pedestrian bridge. The plateau for sightseeing on the south side of the corridor keeps the park from noise and view of the freeway. Replacing the expensive and complex wooden structure of a traditional corridor, a simple, white concrete structure not only harmonizes the agricultural environment with the surrounding modern facilities, but also effectively lowers cost. The overall folded layout provides sufficient activity space in eastern and western areas of the project, especially for the courtyard enclosed by the central corridor and the sightseeing pavilions as a rest and activity center for visitors. Combined with the landscapes, alternate openings on the walls on each side of the corridor allow the scenery to change along the route. Similar to the traditional high-platform architecture, the bottom platform endows the corridor and surroundings with subtle height changes and powerful stylistic characteristics.

入口空间 1
入口空间 2

上海廊下郊野公园核心区景观
Veranda in Shanghai

建筑设计 珮帕施城市发展咨询 + tf
张宁

Architects PPAS+tf ZHANG Ning
Location Shanghai
Completion 2015
Photo SHEN Qiang

上海　　竣工时间 2015 年　　摄影 申强

城市设计及其他

整体航拍　　　　　　　游廊内部　　　　　　　游廊入口
局部空间

游廊透视

1　游廊
2　景亭
3　绿化斜坡
4　观景平台

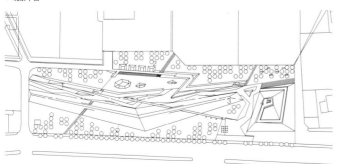

总平面　　0　5　10　20m

线性院落

华山绿工场的设计,使其场所本身成为未来城市生活与绿色地景更加交织混合的范例。另一方面,工场是劳动与生产的场所,这个空间,将不仅成为社区居民种菜的市民菜园,更成为集结更多市民加入成为城市农夫的绿色基地。以线性展开的空间序列,将传统园林形态翻转为多层次的公共院落,涵容了身体劳作、知识传授、经验交换、社群网络连结互动的绿生活场域,创造可游可赏可憩的游园经验,建构了另一种都市风景。建筑主体以模板来构筑,模板是台湾一般建筑建造时常使用的过程材料,建筑师将其转换成了建造材料。

The design of the pavilion for the Green Factory infuses the concept of urban life interacting with green landscapes. The workshop is the place of work and production, where community members can plant produce in a public garden, which acts as a green base that inspires more of the city's residents to become urban farmers. Along the spatial sequence of a linear extension, transformation, and unveiling, the Green Factory converts the form of traditional gardens into a multi-layer public courtyard: a field of green living that incorporates physical labor, the transfer of knowledge, the sharing of experiences, and interactive social network links. Indeed, an alternative type of cityscape is constructed in which people can relax and enjoy a garden environment. The main part of the building is built with molded plates, a widely used temporary material for buildings in Taiwan and the architects recommend that it be used as construction material.

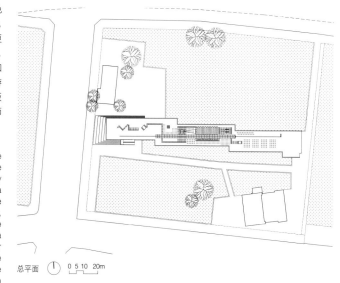
总平面 0 5 10 20m

华山绿工场
Green Factory

台湾 台北 竣工时间 2015 年

建筑设计
禾磊建筑
梁豫漳 蔡大仁 吴明杰

摄 影 李国民 梁豫漳 吴明杰

Architects
Architerior Architects & Associates
LIANG Yuchang, TSAI Daren, WU Mingjie
Location Taipei, Taiwan
Completion 2015
Photo LI Guomin, LIANG Yuchang, WU Mingjie

全区鸟瞰

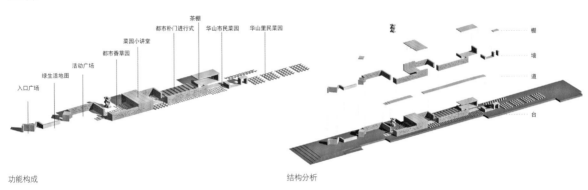

功能构成　　　　　　　　　　　　　　　结构分析

带状延伸的空间层次
都市广场延伸

都市农耕

东侧外观
接待区庭院

　　接待中心位于松鹤墓园的核心位置，东临墓区主干道，北靠墓区停车场，南、西两侧都紧邻墓园。作为墓园内的主要公共建筑，接待中心集业务接待、入葬仪式和办公管理等多种功能于一体，承担了平日的市民购墓、入葬和祭扫事务接待，以及清明、冬至两大高峰时节的大流量接待工作。由于场地有限，同时面对道路和墓园的外向视野或缺乏特色、或不宜引入，这让建筑师在一开始就坚定了"内向性"的策略选择，用江南传统的"宅园"结构中的多重围院模式来组织空间，院落与厅室相依相生，并依托内向的、有差异的庭院景观与使用方式创造出宁静、平和、有冥想性的空间氛围。

The reception center, located in the center of Songhe Cemetery, faces main traffic on the east, the parking lot on the north, and it is adjacent to the cemetery on both south and west sides. The reception center is a public building complex with multiple functions, including a business center, a burial ceremony space and an administrative place. Customers come here for purchases of tomb, burial ceremonies, reception of tomb sweeping, and events during peak seasons, especially for Qingming (Tomb-Sweeping Festival) and Winter Solstice Festival. Due to the lack of space and uninspired surrounding areas, the design focuses passionately on creating a calm, peaceful and meditative inner space at the very beginning. Design methods from traditional "residence garden" in the lower the Yangtze River region are introduced to organize courtyards, rooms and landscapes to be interdependent. Relying on introversive and varying courtyard landscape and relating operative methods, a quiet, peaceful and meditating space is created.

松鹤墓园接待中心
Reception Center for Songhe Cemetery

上海 ｜ 竣工时间 2014 年

建筑设计
致正建筑工作室
上海伊腾建筑设计有限公司

张斌 周蔚

摄　影 陈颢

Architects
Atelier Z+; Shanghai Yiteng Architectural Design Co., Ltd.
ZHANG Bin, ZHOU Wei
Location Shanghai
Completion 2014
Photo CHEN Hao

仪式入口庭院

接待大厅与办公区之间的庭院

业务接待大厅

1 接待出入口　6 配套服务栋
2 仪式出入口　7 办公栋
3 办公出入口　8 庭院
4 业务接待栋　9 接待入口庭院
5 落葬仪式栋　10 仪式入口庭院

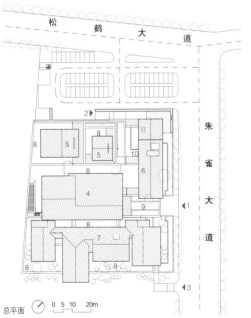

总平面

外立面局部

这是一个地震灾后重建项目，广场中心是格萨尔王雕像，其角度和位置与原址相同。雕像被抬高放在高7m的基座上，基座下是文化展示馆，基座四周借鉴佛教曼陀罗图案——坛城布置，一方一圆。东侧和西侧向河流开放，远方神山、神庙一览无余。南侧布置一组建筑，包括州城市展览馆、档案馆等，长210m。建筑的西南角打开一个小广场，形成了由西南神山到格萨尔王雕像的视觉通廊。建筑大量采用实墙处理，墙面微微倾斜，外饰面采用当地的石材，传递出当地文化特色。

This is a reconstruction project in the aftermath of earthquake, with a statue of Gesaer at center of square aligning with original angle and position. The statue is placed on a 7-meter-high base, beneath which is a cultural exhibition hall. Decorative patterns of Buddhist motifs, i.e. mandala, are used for the base, in both round and square. The east and west sides of the base open to the river, with an amazing view of the sacred mountain and temples. A group of buildings are placed on the south, including Prefecture City Planning Exhibition Hall and archives of 210 meters in length. A small square is arranged at the southwest corner, forming a visual corridor towards the statue of Gesaer. Massive solid walls are used, and the inclined wall surface decorated with local stones convey messages of local cultural characteristics.

鸟瞰
从广场北侧看格萨尔王雕像

格萨尔广场
Gesar Square

建筑设计：天津华汇工程建筑设计有限公司　周恺

Architects Huahui Architectural Design & Engineering Co., Ltd. ZHOU Kai
Location Yushu Tibetan Autonomous Prefecture, Qinghai
Completion 2014
Photo WEI Gang, ZHOU Kai, ZHANG Ning

青海 玉树藏族自治州　竣工时间 2014年　摄影 魏刚 周恺 章宁

城市设计及其他

规划展览馆入口

规划展览馆外广场

1 格萨尔王文化展示馆
2 城市规划展览馆
3 州档案馆
4 向神山和格萨尔王雕像打开的广场
5 室外广场

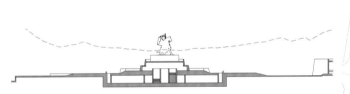

剖面　0 5 10　20m

总平面　0 10 20　50m

规划展览馆入口内看格萨尔王雕像　　　　　　规划展览馆外墙局部

半室外空间
树林中的林建筑

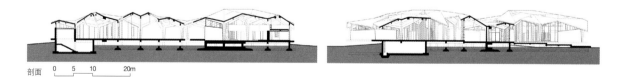

剖面 0 5 10 20m

林建筑
Forest Building

北京

竣工时间 2014 年

建筑设计
北京迹道建筑设计有限责任公司
华黎

摄 影 苏圣亮 夏至

Architects
Trace Architecture Office
HUA Li
Location Beijing
Completion 2014
Photo SU Shengliang, XIA Zhi

林建筑项目位于北京通州运河森林公园。设计受树木形状启发，最初的想法是创造一个树下空间，树的枝干将相互连成一种结构形式并在其遮蔽之下形成空间，以回应场地内树林下的空间。基本单元以柱子为中心并伸出四条悬臂梁，平面基于有些曲折的规律网格，它的边界自由，便于绕开场地现存树木并获得分期施工的灵活性。建筑坐在"飘浮"于地面的混凝土平台上，这样有利于木结构防潮，也将机电设备服务层布置于平台之下，使屋顶下部解放出来，还原为纯粹的结构和空间。建筑主体材料为木和夯土，呼应了场地中的树木和泥土。这些材料能调节室内外的相对湿度温度。立面上木和夯土墙间以玻璃幕墙填充，为室内使用者提供了同时身处人工与自然森林间的感受。

Located in the Grand Canal River Forest Park in Tongzhou District, Beijing, the project is inspired by the shape of trees, and starts with space under trees, using the structural form of connected tree branches and trunks to cover the space beneath in response to the space under trees. The structure is made from the repetition of a basic unit, a tree like column with four cantilevered beams at varied heights. The plan is based on a grid system that is flexible enough to adapt to the site and existing trees. Such flexibility also makes it possible to break down construction into a number of phases. The structure is elevated by a floating concrete platform, which protects the glulam structure from moisture. At the same time, the utilities run underneath the concrete floor to express the structure and space purely. Glue laminated wood and rammed earth are used as main materials, accommodating to the relative humidity and temperature of the site. Clear glass windows fill in the space between wood columns and rammed earth walls, inviting people inside to experience the manmade forest and the natural forest outside at the same time.

木瓦屋顶

就餐空间
"林中"空间

1 门厅
2 吧台
3 厨房
4 问候区
5 办公室
6 休息等候区
7 综合用餐区
8 备餐
9 卫生间
10 音乐露台
11 廊下茶座区
12 临河散座区
13 吧台区
14 休闲娱乐区
15 中庭院广场
16 会议室
17 会议室休息大厅

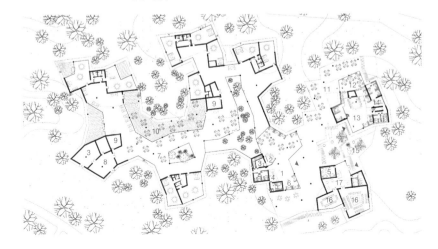

一层平面 0 5 10 20m

该项目是以山区乡村和自然环境为基础的在地社会营造实验。项目基地位于杭州西部临安县山区，场地是140户农户的自然村，是进行中的新兴有机农业种植和休闲旅游试点。建筑材料为原生冬季毛竹及淡竹、小溪挑出的鹅卵石、野生冬季茅草。工匠主要为村中副业为竹工的农户及亲属，形成对在地情况熟悉的复兴乡村营造团队。项目包括在整个山谷乡村范围内进行的系列建造和修建，其中包括动物的养殖圈栏和景观休憩的亭台，以及惠及村人和访客的稻田餐厅、村民戏台及茶室和书院等集体自然营造的活动，正在不断的修建之中。

Located in the mountainous area of Lin'an county, west of Hangzhou city, the project bases itself on social construction experiment in a mountainous villages and the natural environment. The site is composed of 140 household within in a natural villages, as an ongoing pilot for emerging organic farming and leisure tourism. The main part of the building material is bamboo growing in its original site, and pebbles picked from the rivulet and wild winter thatch. The craftsmen are mainly farmers and relatives of the bamboo workers in the village, forming a rejuvenation of the rural construction team, familiar with the situation in the land. The scope of the project includes a series of building and rebuilding activities in the entire valley country, including animal breeding pens and landscape recreation pavilions, and rice restaurant which is in favor of the village and visitors, the villager's stage and teahouse, which are constantly built.

长亭远景
猪舍外景

猪舍剖面 0 0.5 1   2m

鸡舍剖面 0 0.5 1   2m

太阳公社竹构系列
Bamboo Design in Taiyang Farming Commune

建筑设计 山上建筑工作室 陈浩如

浙江 杭州 | 竣工时间 2014年 | 摄影 吕恒中

Architects Citiarc Design Office CHEN Haoru
Location Hangzhou, Zhejiang
Completion 2014
Photo LV Hengzhong

长亭内景

猪舍夜景

猪舍内景

猪舍剖轴测

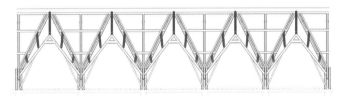
长亭立面

鸡舍剖轴测

鸡舍外景

整体鸟瞰
外景 1

外景 2

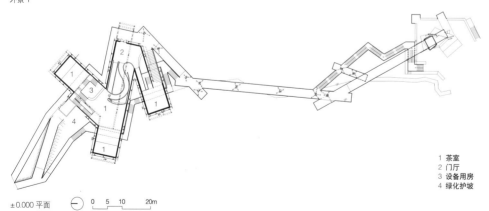

1 茶室
2 门厅
3 设备用房
4 绿化护坡

±0.000 平面

苏仙岭景观瞭望台
Observation Deck in Suxianling

湖南 郴州 | 竣工时间 2014 年

建筑设计
湖南省建筑设计院有限公司
杨瑛

摄影 杨瑛

Architects
Hunan Architectural Design Institute Limited Company
YANG Ying
Location Cheng zhou, Hunan
Completion 2014
Photo YANG Ying

外景

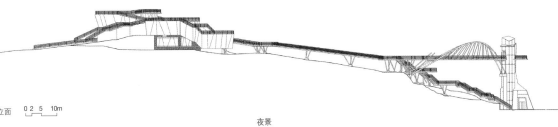

立面　0 2 5　10m

本项目位于苏仙岭主峰之上。设计结合地形地貌，保留原有登山道路和植被，尽量减小对原始场地的破坏。利用现代技术和材料创造出与地域、与时代紧密联结的建筑与空间，形成自然与建筑和谐共存的新景观。设计初始从中国书法抽象的线型艺术中受到启发，从书法的布局和构成中获得创作的源泉。深刻理解场地中佛、道并存的文化特性，将其共同的永恒、长久、延绵的生命述求与意境转化成连续的线性空间，并与3种登顶的路径形成空间上相互链接的回路。

The project sits at the main peak of Suxianling (Suxian Ridge). The design combines the terrain and topographical characteristics to keep original roads leading up to the mountain and vegetation, minimizing damage to the original site. The architects take advantage of modern technologies and materials to create new landscape that harmonizes nature with buildings. The initial design was inspired both by the abstract line art of Chinese calligraphy, and by the layout and composition of calligraphy. With a deeper understanding of cultural characteristics from both Buddhism and Daoism which coexist with each other here and the common appeal on everlasting life into continuous linear space, the architects create a loop in the space with three paths to climb up the mountain.

夜景
室内

Index

项目索引

P418

P440

项目索引

INDEX

公共建筑
Public Architecture

p002
马岔村村民活动中心
Macha Village Center

地点 Location
甘肃省白银市会宁县丁家沟乡马岔村 / Macha Village, Dingjiagou Township, Huining County, Baiyin City, Gansu Province

业主 Client
无止桥慈善基金 /Wu Zhi Qiao (Bridge to China) Charitable Foundation

设计团队 Project Team
[建筑] 蒋蔚 穆钧 李强强 王正阳 黄岩
[结构] 周铁钢 张浩

施工单位 Constructor
马岔村村民及无止桥志愿者

设计时间 Design Period
2013.05 – 2014.04

建造时间 Construction Period
2015.01 – 2016.07

基地面积 Site Area
1860 m²

建筑面积 Floor Area
648 m²

层数 Storey
地上 1层 /1 storey

结构形式 Structure
新型夯土墙承重结构 / New rammed earth wall structure for load bearing

---

p004
张家界国家森林公园游客中心
Zhangjiajie National Forest Park Visitor Center

地点 Location
湖南省武陵源区张家界国家森林公园内 / Zhangjiajie National Forest Park, Wulingyuan District, Hunan Province

业主 Client
张家界国家森林公园管理局 / Administration of Zhangjiajie National Forest Park

设计团队 Project Team
[建筑] 杨志疆 杨程 周妍琳 葛晓峰 叶菁
[结构] 寿刚 王志明
[给排水] 赵元
[电气/智能] 范大勇
[暖通] 吴雁

施工单位 Constructor
张家界市华盛建筑安装有限公司

设计时间 Design Period
2009.03 – 2012.05

建造时间 Construction Period
2013.10 – 2016.02

基地面积 Site Area
8 000 m²

建筑面积 Floor Area
6 300 m²

层数 Storey
地上 3层,地下 1层 /3 storeys overground, 1 storey underground

结构形式 Structure
钢筋混凝土框架结构 / Reinforced concrete frame structure

---

p006
天津泰达居民之家
Residents' Homes of TEDA, Tianjin

地点 Location
天津市西青区迎水道延长线 / Yingshui Street extetion cord Xiqing District, Tianjin City

业主 Client
天津建泰房地产开发有限公司 / Tianjin Jian Tai Real Estate Development Co., Ltd.

设计团队 Project Team
[建筑] 周恺 谢文辉 张莉兰
[结构] 吴峰
[给排水] 牛育辉
[暖通] 郑昕
[电气] 姬宁

施工单位 Constructor
天津建工工程总承包有限公司

设计时间 Design Period
2015.11 – 2016.04

建造时间 Construction Period
2016.04 – 2016.08

基地面积 Site Area
3 000 m²

建筑面积 Floor Area
1 823 m²

层数 Storey
地上 2层 /2 storeys

结构形式 Structure
清水混凝土结构 / Bare concrete structure

---

p008
华夏河图国际艺术家村
River Origins International Artists' Village

地点 Location
宁夏回族自治区银川市鱼悦南街 2号 / No.2, South Yuyue Street, Yinchuan City, Ningxia Hui Autonomous Region

业主 Client
宁夏民生房地产开发有限公司 /Ningxia Minsheng Real Estate Development Co., Ltd.

设计团队 Project Team
[建筑] 王戈 李阳 于宏涛 马笛 李洁苒
[技术] 韩若为 李鹏天
[结构] 张雪驰
[景观] 李雪涛 支筱迪
[室内] 张磊 张镝鸣 王鹏 赵甜甜 刘蕾

施工单位 Constructor
宁夏吴忠市第三建筑工程有限公司

设计时间 Design Period
2013.07 – 2016.04

建造时间 Construction Period
2014.07 – 2016.10

基地面积 Site Area
34 621 m²

建筑面积 Floor Area
18 000 m²

层数 Storey
地上 3层 / 3 storeys

结构形式 Structure
钢筋混凝土结构 / Reinforced concrete structure

---

p010
广州市从化区图书馆
Conghua District Library of Guangzhou

地点 Location
广东省广州市从化区城郊街河滨北路 616号 / No.616, North Hebin Street, Conghua District, Guangzhou City, Guangdong Province

业主 Client
广州市从化区图书馆 / Conghua District Library of Guangzhou

设计团队 Project Team
[建筑] 肖毅强 刘穗杰 齐百慧
[结构] 莫绮琳 (一期) 孙文波 (二期)
[给排水] 关彩玲 (一期) 马双群 (二期)
[电气] 陆晓红 (一期) 莫理莉 (二期)
[暖通] 林小海 (二期) 杜智恒 (二期)

施工单位 Constructor
广东省第四建筑工程公司(一期)、广东省第二建筑工程公司(二期)

设计时间 Design Period
一期 /1st Phase:2005 – 2006；二期 /2nd Phase:2012 – 2014

建造时间 Construction Period
一期 /1st Phase:2006 – 2010；二期 /2nd Phase:2014 – 2016

基地面积 Site Area
7 256 m²

建筑面积 Floor Area
28 323 m² (一期 / 1st Phase:11 546 m², 二期 / 2nd Phase:16 777 m²)

层数 Storey
地上 6层 /6 storeys

结构形式 Structure
钢筋混凝土结构 / Reinforced concrete structure

---

p012
上海棋院
Shanghai Chess Academy

地点 Location
上海市静安区南京西路 595号 /No.595, West Nanjing Road, Jing'an District, Shanghai City

业主 Client
上海棋院 /Shanghai Chess Academy

设计团队 Project Team
[建筑] 曾群 吴敏 汪颖
[结构] 朱圣妤
[暖通] 刘毅
[给排水] 姚思浩
[电气] 蔡玲妹

施工单位 Constructor
上海建工股份有限公司

设计时间 Design Period
2012.09 – 2016.04

建造时间 Construction Period
2013.03 – 2016.12

基地面积 Site Area
6 002 m²

建筑面积 Floor Area
12 424 m²

层数 Storey
地上 5层,地下 1层 / 5 storeys overground, 1 storey underground

结构形式 Structure
混凝土结构 / Concrete structure

---

p014
海边图书馆
Seashore Library

地点 Location
河北省秦皇岛市北戴河新区 / Beidaihe New District, Qinhuangdao City, Hebei Province

业主 Client
天行九州旅游置业开发有限公司 /Beijing Rocfly Investment (Group) Co., Ltd.

合作单位 Cooperative Design
结构、设备:炎黄国际 / Beijing Yanhuang International Architecture

设计团队 Project Team
[主持] 董功
[项目] 梁琛
[驻场] 张艺凡 孙栋平
[建筑] 刘智勇 陈玺兆 谢昕欣
[结构] 纪立新 刘仲瑜

施工单位 Constructor
昌黎县建筑工程有限公司

设计时间 Design Period
2014.02 – 2014.07

建造时间 Construction Period
2014.07 – 2015.04

基地面积 Site Area
310.6 m²

建筑面积 Floor Area
450 m²

层数 Storey
地上 2层 / 2 storeys

结构形式 Structure
混凝土结构 / Concrete structure

---

p016
东湖国家自主创新示范区公共服务中心
Donghu National Innovation District Civic Center

地点 Location
湖北省武汉市光谷核心区高新大道 777号 /No.777, High-Tech Avenue, Guanggu Square Wuhan City, Hubei Province

业主 Client
武汉光谷建设投资有限公司 / Wuhan Optical Valley Construction

设计团队 Project Team
中信:
[建筑] 汤群 杨勇凯 邹淄旻 范志高
[结构] 王新 张浩 李岚 魏丽 王斌
[给排水] 谢丽萍 朱海江
[暖通/空调] 王疆 方云
[电气] 喻辉 刘岗
这方:
[建筑] 赵仲贵 钟明政 刘敏 Luis Balaguer, Ekaggrat Singh Kalsi

施工单位 Constructor
中国建筑第三工程局有限公司

设计时间 Design Period
2010.12 – 2011.12

建造时间 Construction Period
2012.03 – 2015.06

基地面积 Site Area
201 500 m²

建筑面积 Floor Area
147 391 m²

层数 Storey
地上 10层,地下 1层 /10 storeys overground, 1 storey underground

结构形式 Structure
钢筋混凝土框架结构 / Reinforced concrete frame structure

---

p018
重庆桃源居社区中心
Chongqing Taoyuanju Community Center

地点 Location
重庆市渝北区桃源大道 100号桃源公园内 /No.100, Taoyuan Park Taoyuan Avenue, Yubei District, Chongqing City

业主 Client
深圳航空城 (东部) 实业有限公司 / Shenzhen Aviation City (Eastern) Ltd Corporation

合作单位 Cooperative Design
景观 Landscape Design
北京筑境天成建筑设计有限公司 / LAUR Studio
当地设计院 /LDI: 中国机械工业第三设计研究院 / China Ctdi Engineering Corporation

设计团队 Project Team
[主持] 董功
[项目] 王楠
[驻场] 马小凯
[建筑] 陈嘉俊 孔祥栋 常凯申
[结构] 肖从真

施工单位 Constructor
重庆盛源建设 (集团) 公司

设计时间 Design Period
2010.02 – 2011.12

建造时间 Construction Period
2011.12 – 2015.08

基地面积 Site Area
19 900 m²

建筑面积 Floor Area
10 000 m²

层数 Storey
地上 2层 / 2 storeys

结构形式 Structure
混凝土框架结构 +局部钢桁架结构 /Concrete frame structure + Partly steel truss structure

项目索引

p020
罕山生态馆和游客中心
Hanshan Ecological Hall and Tourist Center
地点 Location
内蒙古通辽市扎鲁特旗罕山林场 / Hanshan Forestry Station, Jarud Banner, Tongliao City, Inner Mongolia
业主 Client
通辽市扎鲁特旗罕山林场 / Tongliao Jarud Banner Hanshan Forestry Station
设计团队 Project Team
[建筑] 张鹏举 雷根深
[结构] 孔繁益
[设备] 耿宁波
[电气] 高彦
[室内] 韩超
[景观] 李冰峰
施工单位 Constructor
内蒙古锦基建筑安装工程有限公司
设计时间 Design Period
2011.09 – 2012.05
建造时间 Construction Period
2012.05 – 2015.09
基地面积 Site Area
30 000 m²
建筑面积 Floor Area
9 108 m²
层数 Storey
地上 3层 / 3 storeys
结构形式 Structure
剪力墙 / Shear wall

p022
大厂民族宫
Da Chang Cultural Center
地点 Location
河北省廊坊市大厂回族自治县大安西街68号 / No.68, West Da'an Street, Dachang Hui Autonomous County, Langfang City, Hebei Province
业主 Client
华夏幸福基业有限公司 / China Fortune Land Development Co., Ltd. (CFLD)
设计团队 Project Team
[建筑] 何镜堂 郭卫宏 盘育丹 郑常波 李恺欣
[结构] 郭远翔 孟祥强 林亮洪 徐振楠 廖韶山
[设备] 王淇海 黄璞洁 陈卫彬 耿望阳 何耀炳
施工单位 Constructor
中国新兴建筑工程总公司
设计时间 Design Period
2014.01 – 2014.07
建造时间 Construction Period
2014.07 – 2015.08
基地面积 Site Area
19 042 m²
建筑面积 Floor Area
35 000 m²
层数 Storey
地上 3层,地下 1层 / 3 storeys overground, 1 storey underground
结构形式 Structure
钢筋混凝土框架结构 / Reinforced concrete frame structure

p024
街子古镇梅驿广场
Jiezi Ancient Town Meiyi Square
地点 Location
四川省崇州市街子镇 / Jiezi Town, Chongzhou City, Sichuan Province
业主 Client
成都市琉璃旅游投资开发有限公司 / Chengdu Liuli Tourism Investment Development Co., Ltd.
设计团队 Project Team
[建筑] 蒲еб聿 刘伯英 蒲兵 吴沙沙
[景观] 王琳琳 季烈军 张思思
施工单位 Constructor
四川省轩辕建设实业有限责任公司
设计时间 Design Period
2013 – 2014
建造时间 Construction Period
2014 – 2015
基地面积 Site Area
113 000 m²
建筑面积 Floor Area
12 000 m²
层数 Storey
地上 2层 / 2 storeys
结构形式 Structure
钢结构+钢筋混凝土框架结构 / Steel structure + Reinforced concrete frame structure

p026
天津大学新校区综合体育馆
Gymnasium of New Campus of Tianjin University
地点 Location
天津市津南区百万路 / Baiwan Street, Jinnan District, Tianjin City
业主 Client
天津大学 / Tianjin University
设计团队 Project Team
[建筑] 李兴钢 张音玄 闫昱 易灵洁 梁旭
施工单位 Constructor
中国建筑第二工程局有限公司
设计时间 Design Period
2011.02 – 2013.08
建造时间 Construction Period
2012.06 – 2015.11
基地面积 Site Area
75 000 m²
建筑面积 Floor Area
18 362 m²
层数 Storey
地上 2层,地下 1层 / 2 storeys overground, 1 storey underground
结构形式 Structure
钢筋混凝土框架剪力墙+薄壳屋面 / Reinforced concrete and frame-shear wall structure + Shell roof

p028
盛乐遗址公园游客中心
Tourist Center of Shengle Heritage Park
地点 Location
内蒙古自治区呼和浩特市和林格尔县盛乐镇盛乐经济园区西侧 / The west side of Shengle Economic Park, Shengle Town, Helingeer County, Hohhot City, Inner Mongolia Autonomous Region
业主 Client
内蒙古呼和浩特市和林格尔县人民政府 / People's Government of Helingeer County, Hohhot City, Inner Mongolia
设计团队 Project Team
[建筑] 张鹏举 张恒 郭鹏 李燕 杨耀强
[结构] 刘青华
[电气] 纪华
[设备] 张晶
施工单位 Constructor
呼和浩特市瑞环(集团)有限公司
设计时间 Design Period
2014.05
建造时间 Construction Period
2015.11
基地面积 Site Area
3 526 m²
建筑面积 Floor Area
1 991.99 m²
层数 Storey
地上 2层 / 2 storeys
结构形式 Structure
钢筋混凝土框架结构 / Reinforced concrete frame structure

p030
南矶湿地访客中心
Nanji Wetland Reserve Visitor Center
地点 Location
江西省南昌市鄱阳湖南矶湿地自然保护区 / Panyang Lake Nanji Wetland Reserve, Nanchang City, Jiangxi Province
业主 Client
南矶湿地国家级自然保护区管理局 / Nanji Wetland Natural Reserve Authority
设计团队 Project Team
[建筑] 朱竞翔 吴程辉 韩国日
施工单位 Constructor
南矶乡 10名村民
设计时间 Design Period
2015.01 – 2015.03
建造时间 Construction Period
2015.04 – 2015.07
基地面积 Site Area
375 m²
建筑面积 Floor Area
375 m²
层数 Storey
地上 1层 / 1 storey
结构形式 Structure
模块楼板与承重家具柜结构+装配式底层架空柱 / Modular floor panels and load-bearing cabinets + Prefab stilts system

p032
南京牛首山景区游客中心
Tourist Center in Niushou Scenic, Nanjing
地点 Location
江苏省南京市江宁区牛首山景区 / Niushou Scenic, Jiangning District, Nanjing City, Jiangsu Province
业主 Client
南京牛首山文化旅游集团有限公司 / Nanjing Niushou Culture Tourism Group Co., Ltd.
设计团队 Project Team
[建筑] 王建国 朱渊 吴云鹏 姚昕悦 张航
[结构] 孙逊 梁沙河 蒋剑峰 胥建华 夏仕洋
[设备] 孙毅 龚德建 史海山 张磊
[景观] 王晓俊 钱筠
施工单位 Constructor
中国建筑第八工程局有限公司
设计时间 Design Period
2013.02 – 2014.02
建造时间 Construction Period
2014.02 – 2015.10
基地面积 Site Area
69 881 m²
建筑面积 Floor Area
91 670 m²
层数 Storey
地上 2层,地下 1层 / 2 storeys overground, 1 storey underground
结构形式 Structure
混凝土框架结构+钢框架-支撑结构(屋盖) / Concrete frame structure + Steel frame and supporting structure (roof)

p034
武汉光谷国际网球中心网球馆
Wuhan Optics Valley International Tennis Center, The Tennis Court
地点 Location
湖北省武汉市东湖新技术开发区高新二路与佛祖岭一路交汇处 / Intersection with 1st Fozuling Road and 2nd Hi-Tech road, East Lake New Technology Development Zone, Wuhan City, Hubei Province
业主 Client
武汉光谷建设投资有限公司 / Wuhan Optics Valley Construction Investment Co., Ltd.
设计团队 Project Team
[建筑] 陆晓明 叶炜 姜瀚 郭雷 李鸣宇
[结构] 刘文잖 温四清 董卫国 赵文争
[设备] 谢丽萍 王疆 喻辉 蔡雄飞
施工单位 Constructor
浙江省一建建设集团有限公司
设计时间 Design Period
2013.05 – 2013.12
建造时间 Construction Period
2013.12 – 2015.09
基地面积 Site Area
137 263 m²
建筑面积 Floor Area
54 340 m²
层数 Storey
地上 5层 / 5 storeys
结构形式 Structure
钢筋混凝土框架+空间网格钢结构 / Reinforced concrete frame structure + Spatial grid steel structure

p036
码头书屋
Library on the Quay
地点 Location
安徽省铜陵市滨江生态公园 / Riverside Ecopark, Tongling City, Anhui Province
业主 Client
铜陵市住房和城乡建设委员会 / The Housing and Urban Construction Commission of Tongling City
设计团队 Project Team
[建筑] 李竹 王嘉峻 滕衍泽
[结构] 孙逯 张翀 方立新
[给排水] 鲍迎春 葛启龙 贺海涛
[暖通] 许东晟 杨媛茹 陈俊
[电气] 袁星 赵鸿鑫 范大勇
施工单位 Constructor
合肥环业建工程有限公司
设计时间 Design Period
2015.02 – 2015.05
建造时间 Construction Period
2015.06 – 2015.12
基地面积 Site Area
853 m²
建筑面积 Floor Area
517 m²
层数 Storey
地上 1层 / 1 storey
结构形式 Structure
钢结构 / Steel structure

p038
2014青岛世界园艺博览会天水、地池综合服务中心
Tianshui/Dichi Service Center of International Horticultural Exposition 2014
地点 Location
山东省青岛市李沧区百果山 / Baiguo Mountain, Licang District, Qingdao City, Shandong Province
业主 Client
2014青岛世界园艺博览会 / 2014 Qingdao International Horticultural Exposition
设计团队 Project Team
[天水] 王振飞 王鹿鸣 李宏宇 王凌柱 季方
[地池] 王振飞 王鹿鸣 李宏宇 汪琪 潘昊
设计时间 Design Period
2011.08 – 2012.05
建造时间 Construction Period

2012.06 – 2014.04
基地面积 Site Area
天水 / Tianshui; 23 000 m²;
地池 / Dichi; 32 000 m²
建筑面积 Floor Area
天水 / Tianshui; 6 200 m²;
地池 / Dichi; 9 300 m²
层数 Storey
地上 2层 / 2 storeys
结构形式 Structure
混凝土框架结构 / Concrete frame structure

p040
南京万景园小教堂
Nanjing Wanjing Garden Chapel
地点 Location
江苏省南京市建邺区扬子江大道228号万景园内 / No.228, Yangzi River Avenue, Jianye District, Nanjing City, Jiangsu Province
业主 Client
南京河西新城建设发展有限公司 / Nanjing Hexi Xincheng Construction Development Co., Ltd.
设计团队 Project Team
[建筑] 张雷 戚威 王莹
施工单位 Constructor
苏州奥纳木结构设计工程有限公司
设计时间 Design Period
2014.06
建造时间 Construction Period
2014.06 – 2014.07
基地面积 Site Area
800 m²
建筑面积 Floor Area
200 m²
层数 Storey
地上 1层 / 1 storey
结构形式 Structure
钢木结构 / Steel and wood structure

p042
无锡阳山田园综合体 I 期田园生活馆
The 1st Phase of Yangshan Rural Life Complex in Wuxi, Rural Life Pavilion
地点 Location
江苏省无锡市惠山区阳山镇 / Yangshan Town, Huishan District, Wuxi City, Jiangsu Province
业主 Client
东方园林产业集团,无锡阳山田园东方投资有限公司 / Orient Landscape Industry Group; Pastoral Oriental Investment Co., Ltd.
设计团队 Project Team
[建筑] 钱强 张丹 李晨成 汪淑靓 辜克威
设计时间 Design Period
2013.02 – 2013.12
建造时间 Construction Period
2013.12 – 2014.03
建筑面积 Floor Area
1 337 m²
层数 Storey
地上 1层,局部 2层 / 1 storey, partly 2 storeys
结构形式 Structure
钢结构 / Steel structure

p044
康巴艺术中心
Kangba Art Center
地点 Location
青海省玉树自治州结古镇胜利路 / Shengli Street, Jiegu Town Yushu Tibetan Autonomous Prefecture, QingHai Province
业主 Client
玉树州三江源投资建设有限公司 / Yushu Sanjiangyuan investment and Construction Corp. Ltd.
设计团队 Project Team
[建筑] 崔愷 关飞 曾瑞 高凡
[结构] 王文宇
[给排水] 董超
[暖通] 尹奎超
[电气] 陈佩仁
施工单位 Constructor
中国建筑第八工程局有限公司青岛公司
设计时间 Design Period
2010.10 – 2012.02
建造时间 Construction Period
2011.10 – 2014.07
基地面积 Site Area
24 563 m²
建筑面积 Floor Area
20 610 m²
层数 Storey
地上 6层,地下 1层 / 6 storeys overground, 1 storey underground
结构形式 Structure
钢筋混凝土框架剪力墙结构 + 钢筋混凝土框架结构 / Reinforced concrete frame-shear wall structure + Reinforced concrete frame structure

p046
盛泽文化中心
Shengze Cultural Center
地点 Location
江苏省苏州市吴江区和田路 88号 / No.88, Hetian Street, Wujiang District, Suzhou City, Jiangsu Province
业主 Client
江苏盛泽投资有限公司 / Jiangsu Shengze Investment Co., Ltd.
设计团队 Project Team
[建筑] 李立 谭亮 郑聪 许宁 许鑫
施工单位 Constructor
吴江市舜新建筑工程有限公司
设计时间 Design Period
2012.06 – 2012.12
建造时间 Construction Period
2013.01 – 2014.11
基地面积 Site Area
30 800 m²
建筑面积 Floor Area
28 076 m²
层数 Storey
地上 4层,地下 1层 / 4 storeys overground, 1 storey underground
结构形式 Structure
钢筋混凝土框架结构 / Reinforced concrete frame structure

p048
第十三届全国冬季运动会冰上运动中心
Ice Sports Center of the 13th National Winter Games
地点 Location
新疆维吾尔自治区乌鲁木齐市西沟乡 / Xigou Town, Urumqi City, Xinjiang Uygur Autonomous Region
业主 Client
新疆维吾尔自治区体育局 / Sport Administration of Xinjiang Uygur Autonomous Region
设计团队 Project Team
[建筑] 梅洪亮 陆诗亮 初晓 梁斌 魏治平 等
[结构] 戴大志
[给排水] 冷润海
[暖通] 卢艳秋
[电气] 史建雷
施工单位 Constructor
新疆城建(集团)股份有限公司
设计时间 Design Period
2012.03 – 2012.12
建造时间 Construction Period
2013.03 – 2014.12
基地面积 Site Area
300 600 m²
建筑面积 Floor Area
78 334 m²
层数 Storey
速滑馆 / Speed skating hall; 3层 / 3 storeys; 冰球馆 / Ice Hockey Hall; 3层 / 3 storeys; 冰壶馆 / Curling hall; 3层 / 3 storeys; 媒体中心 / Media center; 3层 / 3 storeys; 餐厅及公寓 / Restaurant and apartment; 5层 / 5 storeys
结构形式 Structure
钢筋混凝土框架结构 + 局部屋面张弦杆系结构 / Reinforced concrete frame structure + truss string structure (partial roof)

p050
黄河口生态旅游区游客服务中心
Visitor Center of Ecological Tourism Area at Delta of Yellow River
地点 Location
山东省东营市黄河口生态旅游区 / Ecological Tourism Area at Delta of Yellow River, Dongying City, Shandong Province
业主 Client
东营市旅游开发有限公司 / Tourism Development Corporation of Dongying
设计团队 Project Team
[建筑] 李麟学 刘旸 周凯锋 李欢欢 王瑾瑾 等
施工单位 Constructor
江苏天虹建设集团东营分公司
设计时间 Design Period
2011.09
建造时间 Construction Period
2011 – 2014
基地面积 Site Area
49 000 m²
建筑面积 Floor Area
9 900 m²
层数 Storey
地上 3层 / 3 storeys
结构形式 Structure
钢筋混凝土结构 / Reinforced concrete structure

p052
淮安市体育中心
Huai'an City Sports Center
地点 Location
江苏省淮安市清江浦区宁连路以南,枚皋路以北 / South of Ninglian Street, North of Meigao Street, Qingjiangpu District, Huai'an City, Jiangsu Province
业主 Client
淮安市体育局,淮安市体育中心发展有限公司 / Huai'an City Sports Bureau, Huai'an City Sports Center Development Co., Ltd.
设计团队 Project Team
[建筑] 孙一民 陶亮 叶伟康 申永刚 邓芳
[规划] 黄烨勍 梁艳艳 李敏稚 冷天翔 杨定
[景观] 徐莹 任振华
[结构] 孙文波
[给排水] 王琪海
[空调] 林伟强
[电气] 高飞
[节能] 胡文斌
施工单位 Constructor
南通建工集团总承包有限公司
设计时间 Design Period
2009.11 – 2010.05
建造时间 Construction Period
2010.06 – 2014.05
基地面积 Site Area
428 949 m²
建筑面积 Floor Area
95 483 m²
层数 Storey
地上 3层,地下 1层 / 3 storeys overground, 1 storey underground
结构形式 Structure
钢筋混凝土结构 + 钢结构 / Reinforced concrete structure + Steel structure

p054
天津武清文化中心
Wuqing Cultural Center, Tianjin
地点 Location
天津市武清区泉旺路和振华西道交口 / The Intersection of Quanwang Road and Zhenhua West Road, Wuqing District, Tianjin City
业主 Client
天津市武清区政府 / Government of the Wuqing District, Tianjin City
设计团队 Project Team
[建筑] 周恺 唐敏 王建平 黄菲
[结构] 毛文俊
[给排水] 陈太洲
[暖通] 邵海
[电气] 王裕华
施工单位 Constructor
天津奥林鑫建设投资集团有限公司
设计时间 Design Period
2011.12 – 2012.06
建造时间 Construction Period
2012.06 – 2014.10
基地面积 Site Area
29 000 m²
建筑面积 Floor Area
34 300 m²
层数 Storey
地上 5层,地下 1层 / 5 storeys overground, 1 storey underground
结构形式 Structure
图书馆 / Library; 钢筋混凝土框架结构 / Reinforced concrete frame structure; 博物馆 / Museum; 钢梁 — 钢管砼柱混合框架结构 / Steel beam - mixed concrete filled steel tube column frame structure

p056
遵义市娄山关红军战斗遗址陈列馆
The Site Museum of Loushanguan Battle, Zunyi
地点 Location
贵州省遵义市汇川区娄山关景区 / Loushanguan Scenic, Huichuan District, Zunyi City, Guizhou Province
业主 Client
遵义市娄山关管理处 / Loushanguan Administration
设计团队 Project Team
[建筑] 任九之 李楚婧 廖凯 邹昊阳 王金蕾
[室内] 任亚慧 吴杰 邰燕荣
[景观] 宋利骏 滕华伟
[结构] 李学平 高之楠
[给排水] 杨玲 陈文祥
[暖通] 谭立民
[强弱电] 彭岩
施工单位 Constructor
贵州东恒(集团)建筑工程有限公司
设计时间 Design Period
2015.06 – 2016.04
建造时间 Construction Period
2016.02 – 2017.05
基地面积 Site Area
9 000 m²
建筑面积 Floor Area
6 056 m²
层数 Storey
地上 1层,地下 1层 / 1 storey overground, 1 storey underground
结构形式 Structure
钢筋混凝土框架结构 / Reinforced concrete frame structure

p058
上海世博会博物馆
Shanghai Expo Museum

地点 Location
上海市黄浦区蒙自路818号 / No.818, Mengzi Road Huangpu District, Shanghai City

业主 Client
上海世博会博物馆 / Shanghai Expo Museum

设计团队 Project Team
[建筑] 汪孝安 杨刚 刘海洋 俞楠 等
[结构] 姜文伟 包联进 黄永强 赵雪莲 闫琪 等
[机电] 田建强 左鑫 胡明 韩风明 吕宁 等

施工单位 Constructor
上海建工四建集团有限公司

设计时间 Design Period
2012.03 – 2014.08

建造时间 Construction Period
2013.12 – 2017.05

基地面积 Site Area
40 000 m²

建筑面积 Floor Area
46 550 m²

层数 Storey
地上5层,地下1层 / 5 storeys overground, 1 storey underground

结构形式 Structure
钢+混凝土结构 / Steel + Concrete structure

p060
苏州非物质文化遗产博物馆
Suzhou Intangible Cultural Heritage Museum

地点 Location
江苏省苏州市吴中区园艺博览会会址东侧 / Estern Part of Suzhou Horticultural Expo, Wuzhong District, Suzhou City, Jiangsu Province

业主 Client
苏州太湖园博实业发展有限公司 / Suzhou Taihu Lake Horticultural Exposition Industrial Development Co., Ltd.

设计团队 Project Team
直向建筑:
[主持] 董功
[项目] 刘晨
[驻场] 周飓
[建筑] 王艺祺 孙栋平 赵丹 李柏 侯瑞瑶 叶品晨 王依伦 张恺
启迪设计集团股份有限公司:
[项目] 蔡爽
[建筑] 王颖 张晓峰 王威
[结构] 叶永毅 卞克俭 谭骞
[机电] 张广仁 陈凯旋 王海港 季健 祝合虎 等

施工单位 Constructor
苏州第一建筑集团有限公司

设计时间 Design Period
2014.10 – 2015.02

建造时间 Construction Period
2015.02 – 2016.04

基地面积 Site Area
26 000 m²

建筑面积 Floor Area
24 520.31 m²

层数 Storey
地上4层,地下2层 / 4 storeys overground, 2 storeys underground

结构形式 Structure
钢筋混凝土结构 / Reinforced concrete structure

p062
延安大剧院
Yan'an Grand Theatre

地点 Location
陕西省延安市新区剧院路 / Theatre Street, Yan'an New District, Yan'an City Shaanxi Province

业主 Client
延安新区管委会 / Yan'an New District Management Committee

设计团队 Project Team
[建筑/景观] 赵元超 李强 李彬 高令奇 张晶
[结构] 王洪臣 郭东
[机电] 张军

施工单位 Constructor
陕西建工集团有限公司

设计时间 Design Period
2014.08 – 2014.11

建造时间 Construction Period
2014.11 – 2016.07

基地面积 Site Area
43 330 m²

建筑面积 Floor Area
33 134 m²

层数 Storey
地上3层,地下1层 / 3 storeys overground, 1 storey underground

结构形式 Structure
主体:框架剪力墙结构+屋顶:网架结构 / Main part: Frame-shear wall structure+Roof: Grid structure

p064
宜昌规划展览馆
Yichang Planning Exhibition Hall

地点 Location
湖北省宜昌市伍家岗区柏临河路北侧 / Bolinhe Street, Wujiagang District, Yichang City, Hubei Province

业主 Client
宜昌市规划局 / Yichang City Planning Bureau

设计团队 Project Team
[建筑] 孙晓恒 丁蓉 张冉 李斌 夏嘉蓉
[结构] 陈思力 罗清华
[电气] 刘小丽
[给排水] 周雪松
[暖通] 郑兵

施工单位 Constructor
宜昌市城市建设投资开发有限公司

设计时间 Design Period
2013.05

建造时间 Construction Period
2013.05 – 2016.02

基地面积 Site Area
10 670.1 m²

建筑面积 Floor Area
20 960 m²

层数 Storey
地上3层,地下1层 / 3 storeys overground, 1 storey underground

结构形式 Structure
钢筋混凝土框架结构 / Reinforced concrete frame structure

p066
2015米兰世博会中国馆
China Pavilion for Expo Milano, 2015

地点 Location
意大利米兰 / Milano, Italy

业主 Client
中国国际贸易促进会 / China Council for the Promotion of International Trade

设计团队 Project Team
[建筑] 陆轶辰 Kenneth Namkung 蔡沁雯
[结构] Simpson Gumpertz & Heger, F&M Ingegneria
[机电] 北京清尚, F&M Ingegneria

合作设计 Cooperative Design
[项目总负责] 苏丹
[项目总监] 张月 杜异
[技术顾问] 王长钢 涂山
[室内设计] 汪建松
[景观设计] 崔笑声
[展陈设计] 周艳阳 赵华森 李彩丽
[装置设计] 师丹青 洗枫
[灯光设计] 杜异 刘晓希
[VI系统] 管云嘉
[视觉传达] 顾欣 王之纲

施工单位 Constructor
中艺建筑装饰, 永一阁, Bodino Engineering

设计时间 Design Period
2013.10 – 2014.09

建造时间 Construction Period
2014.10 – 2015.04

基地面积 Site Area
4 590 m²

建筑面积 Floor Area
3 500 m²

层数 Storey
地上2层 / 2 storeys

结构形式 Structure
胶合木结构 / Glued-laminated Timber structure

p068
银川当代美术馆
Museum of Contemporary Art Yinchuan

地点 Location
宁夏回族自治区银川市永宁县禾乐路12号 / No.12, Hele Street, Yongning County, Yinchuan City, Ningxia Hui Autonomous Region

业主 Client
宁夏民生房地产开发有限公司 / Ningxia Minsheng Real Estate Development Co., Ltd.

设计团队 Project Team
[建筑] 张迪 Jack Young
[结构] 蔡卫宁
[给排水] 邹艳容
[暖通] 吴刚
[电气] 李永宁

施工单位 Constructor
中国中铁建工集团有限公司

设计时间 Design Period
2011 – 2015

建造时间 Construction Period
2012 – 2015

基地面积 Site Area
13 200 m²

建筑面积 Floor Area
15 000 m²

层数 Storey
地上3层,地下1层 / 3 storeys overground, 1 storey underground

结构形式 Structure
钢筋混凝土框架结构 / Reinforced concrete frame structure

p070
张家界博物馆
Zhangjiajie Museum

地点 Location
湖南省张家界市永定区大庸路 / Dayong Street, Yongding District, Zhangjiajie City, Hunan Province

业主 Client
张家界市文化局 / Zhangjiajie Municipal Bureau of Press and Publication

设计团队 Project Team
[建筑] 魏春雨 齐靖 刘海力 宋明星 李煦
[结构] 朱建华
[给排水] 郑少平
[暖通] 黄文胜
[电气] 彭成生

施工单位 Constructor
湖南顺天建设集团有限公司

设计时间 Design Period
2009.11 – 2010.05

建造时间 Construction Period
2010.08 – 2015.01

基地面积 Site Area
14 180 m²

建筑面积 Floor Area
15 395.73 m²

层数 Storey
地上4层,地下1层 / 4 storeys overground, 1 storey underground

结构形式 Structure
混凝土框架结构 / Concrete frame structure

p072
商丘博物馆
Shangqiu Museum

地点 Location
河南省商丘市睢阳区华商大道西段 / West Section of Huashang Avenue, Suiyang District, Shangqiu City, Henan Province

业主 Client
商丘市文化广电新闻出版局 / Bureau of Culture, Radio, Film, TV, Press and Publication of Shangqiu

设计团队 Project Team
[建筑] 李兴钢 谭泽阳 付邦保 张哲 郭佳
[景观] 李力

施工单位 Constructor
中国建筑第七工程局有限公司

设计时间 Design Period
2008.12 – 2010.09

建造时间 Construction Period
2011.03 – 2015.06

基地面积 Site Area
73 613 m²

建筑面积 Floor Area
29 672 m²

层数 Storey
地上3层,地下1层 / 3 storeys overground, 1 storey underground

结构形式 Structure
钢筋混凝土框架剪力墙结构 + 钢结构 / Reinforced concrete frame-shear wall structure + Steel structure

p074
侵华日军南京大屠杀遇难同胞纪念馆三期扩容工程
The Memorial Hall of the Victims in Nanjing Massacre by Japanese Invaders, Phase III

地点 Location
江苏省南京市江东中路 / Jiangdong Middle Road, Nanjing City, Jiangsu Province

业主 Client
南京市建邺区河西建设指挥部办公室 / Hexi Construction Headquarters Office, Jianye District of Nanjing

设计团队 Project Team
[建筑] 何镜堂 倪阳 刘宇波 包莹 李恺欣
[结构] 方小丹 赖洪涛 孙传伟
[景观] 晏忠 张莎玮 廖俊峰
[室内] 郑炎 梁景韶 李靖灵
[电气] 俞洋 过仕佳
[给排水] 陈欣焘 李家泰
[智能化] 陈铢虹 黄光伟
[空调] 黄璞洁 何耀炳 李艳霞

施工单位 Constructor
中国建筑第八工程局有限公司

设计时间 Design Period
2013 – 2014

建造时间 Construction Period
2014 – 2015

基地面积 Site Area
28 307.39 m²

建筑面积 Floor Area
54 636.3 m²

层数 Storey
地上1层,地下2层 / 1 storey overground, 2 storeys underground

结构形式 Structure
混凝土框架结构 / Concrete frame structure

p076
哈尔滨大剧院
Harbin Opera House
地点 Location
黑龙江省哈尔滨市松北区文化中心岛内 / Cultural Center Island, Songbei District, Harbin City, Heilongjiang Province
业主 Client
哈尔滨松北投资发展集团有限公司 / Harbin Songbei Investment and Development Group Co., Ltd.
合作单位 Cooperative Design
[景观] 北京土人景观与规划设计研究院 / Beijing Turenscape Institute
设计团队 Project Team
马岩松 党群 早野洋介 Jordan Kanter Daniel Gillen 等
施工单位 Constructor
北京市第三建筑工程有限公司
设计时间 Design Period
2010 – 2015
建造时间 Construction Period
2010 – 2015
基地面积 Site Area
18 000 m²
建筑面积 Floor Area
79 000 m²
层数 Storey
地上 8层, 地下 1层 /8 storeys overground, 1 storey underground
结构形式 Structure
钢筋混凝土框架 +钢结构 / Reinforced concrete frame structure + Steel structure

p078
云南省博物馆新馆
New Yunnan Province Museum
地点 Location
云南省昆明市官渡区广福路 6393号 /No.6393, Guangfu Road, Guandu District, Kunming City, Yunnan Province
业主 Client
云南省博物馆新馆建设指挥部 / The Construction Command of Yunnan Museum of Yunnan Province
设计团队 Project Team
[建筑] 严迅奇 陈邦贤 谭伟霖 王权 唐筱
[结构] 廖述江 吴宏雄 刘涛 黄汝强
[给排水] 许学华 李扬
[电气] 蔡月确 王健
[暖通] 潘京平 吴江
施工单位 Constructor
云南省建设总承包公司
设计时间 Design Period
2009.03 – 2010.06
建造时间 Construction Period
2010.07 – 2015.05
基地面积 Site Area
91 006.7 m²
建筑面积 Floor Area
57 787.4 m²
层数 Storey
地上 5层, 地下 2层 /5 storeys overground, 2 storeys underground
结构形式 Structure
型钢混凝土框架 — 钢支撑结构 / Steel Reinforced concrete frame and steel support structure

p080
木心美术馆
Muxin Art Museum
地点 Location
浙江省桐乡市乌镇西栅大街 1508号 / No.1508, West Gate Street, Wuzhen Town, Tongxiang City, Zhejiang Province
业主 Client
乌镇旅游股份有限公司 / Wuzhen Tourism Co., Ltd.
设计团队 Project Team
冈本博 林兵 法比安 Steve Hopkins 陈柏全 等
施工单位 Constructor
浙江巨匠控股集团有限公司
设计时间 Design Period
2011.09 – 2015.11
建造时间 Construction Period
2012.06 – 2015.11
基地面积 Site Area
8 770 m²
建筑面积 Floor Area
6 770 m²
层数 Storey
地上 2层, 地下 1层 / 2 storeys overground, 1 storey underground
结构形式 Structure
框架剪力墙结构 / Frame-shear wall structure

p082
金陵大报恩寺遗址博物馆
Site Museum of Jinling Grand Bao'en Temple
地点 Location
江苏省南京市秦淮区雨花路 1号 / No.1 Yuhua Road, Qinhuai District, Nanjing City, Jiangsu Province
业主 Client
南京大明文化实业有限责任公司 / Nanjing Daming Culture Industry Co., Ltd.
设计团队 Project Team
[建筑] 韩冬青 陈薇 王建国 马晓东 孟媛 胡明皓 吴国栋 马俊华
[结构] 孙逊 张翀 黄凯
[景观] 杨冬辉 伍清辉
[给排水] 鲍迎春
[暖通] 陈俊
[电气] 屈建球
[智能] 张程
施工单位 Constructor
中国建筑第八工程局有限公司
设计时间 Design Period
2011.09 – 2013.04
建造时间 Construction Period
2013.04 – 2015.12
基地面积 Site Area
75 300 m²
建筑面积 Floor Area
60 800 m²
层数 Storey
博物馆 / Museum: 地上 1-2层, 地下 1层 / 1-2 storeys overground, 1 storey underground;
报恩新塔 /New Bao'en Pagoda: 地上 9层 / 9 storeys
结构形式 Structure
博物馆 / Museum: 单层钢管柱平面桁架结构 / Plane truss structure composed of steel pipe;
报恩新塔 / New Bao'en Pagoda: 钢框架结构 / Steel frame structure

p084
临安市体育文化会展中心
Lin'an Sports and Culture Center
地点 Location
浙江省临安市锦南新城九州路 / Jiuzhou Street, Jinnan New Town, Lin'an City, Zhejiang Province
业主 Client
临安市新锦投资开发有限公司 / Lin'an Xinjin Investment Development Co., Ltd.
设计团队 Project Team
[建筑] 董丹申 陈建 倪剑 蔡弋 雷持平
施工单位 Constructor
浙江大华建设集团有限公司, 杭州临安荣大建设工程有限公司
设计时间 Design Period
2010.03 – 2012.02
建造时间 Construction Period
2012.02 – 2015.05
基地面积 Site Area
51 520 m²
建筑面积 Floor Area
74 986 m²
层数 Storey
地上 3层, 地下 1层 / 3 storeys overground, 1 storey underground
结构形式 Structure
钢筋混凝土框架结构 +钢结构 / Reinforced concrete frame structure + Steel structure

p086
毓绣美术馆
Yu-Hsiu Museum of Art
地点 Location
台湾南投县草屯镇平林村健行路 150巷 26号 / No.26, Lane 150, Jianxing Street, Pinglin Village, Nantou County, Taiwan
业主 Client
叶毓绣 / YE Yuxiu
设计团队 Project Team
[建筑] 廖伟立
[机电] 冠升工程设计事务所
[结构] 鼎匠工程顾问有限公司
施工单位 Constructor
清水建筑工坊
设计时间 Design Period
2011.04 – 2012.11
建造时间 Construction Period
2013.04 – 2015.01
基地面积 Site Area
4 680.8 m²
建筑面积 Floor Area
主馆 / Exhibition Hall: 323.01 m²; 餐厅 / Restaurant: 147.07 m²; 假日学校 / Holiday school: 276.1 m²
层数 Storey
主馆 / Exhibition Hall: 地上 3层, 地下 1层 / 3 storeys overground, 1 storey underground; 餐厅 / Restaurant: 地上 1层 / 1 storey; 假日学校 / Holiday school: 地上 2层 / 2 storeys
结构形式 Structure
清水混凝土结构 / Bare concrete structure

p088
银川韩美林艺术馆
Yinchuan Han Meilin Art Museum
地点 Location
宁夏回族自治区银川市贺兰山岩画遗址公园内 / Helan Mountain Rock Painting Site Park, Yinchuan City, Ningxia Hui Autonomous Region
业主 Client
银川市贺兰山岩画管理处 / Helan Mountain Yinchuan City Rock Painting Management Office
设计团队 Project Team
[建筑] 张华 范黎 孙睿 李江源 谭庆君
[结构] 崔世敏 刘龙飞 魏利会 史炎生 雷阳
[给排水] 赵小新 伍胜春 罗小辉
[电气] 尤世刚 谢鑫 石玉
[暖通] 郑珊珊 曾志华 吴介轩 吴鹏飞
[总图] 雒展 王璞实
设计时间 Design Period
2013
建造时间 Construction Period
2015
基地面积 Site Area
15 868 m²
建筑面积 Floor Area
6 694 m²
层数 Storey
地上 3层 /3 storeys
结构形式 Structure
钢筋混凝土结构 /Reinforced concrete structure

p090
蚌埠博物馆及规划档案馆
Bengbu Museum, Urban Planning Exhibition and Archive Centre
地点 Location
安徽省蚌埠市蚌山区东海大道 / Donghai Avenue, Bengshan District, Bengbu City, Anhui Province
业主 Client
蚌埠城建投资发展有限公司 / Bengbu Urban Construction Investment Co., Ltd.
设计团队 Project Team
[建筑] 孟建民 邢立华 徐昀超 周富 曾智
施工单位 Constructor
中国建筑第五工程局有限公司
设计时间 Design Period
2011.06 – 2015.01
建造时间 Construction Period
2012.08 – 2015.11
基地面积 Site Area
98 649 m²
建筑面积 Floor Area
68 833 m²
层数 Storey
地上 5层, 地下 1层 /5 storeys overground, 1 storey underground
结构形式 Structure
混凝土框架结构 +钢悬索结构 / Concrete frame structure + Steel cable structure

p092
刘海粟美术馆
Liu Haisu Art Museum
地点 Location
上海市长宁区延安西路 1609号 /No.1609, West Yan'an Road, Changning District, Shanghai City
业主 Client
刘海粟美术馆 /Liu Haisu Art Museum
设计团队 Project Team
[建筑] 陈剑秋 吴靖杰 郭辛怡
[室内] 颜敏
施工单位 Constructor
上海建工四建集团有限公司
设计时间 Design Period
2012.03 – 2012.09
建造时间 Construction Period
2012.10 – 2015.02
基地面积 Site Area
6 000 m²
建筑面积 Floor Area
12 540 m²
层数 Storey
地上 3层, 地下 2层 / 3 storeys overground, 2 storeys underground
结构形式 Structure
钢结构 /Steel structure

p094
龙美术馆（西岸馆）
Long Museum (West Bund)
地点 Location
上海市徐汇区龙腾大道 3398号 / No.3398, Longteng Avenue, Xuhui District, Shanghai City
业主 Client
上海徐汇滨江开发投资建设有限公司 / Shanghai Xuhui Waterfront Development, Investment & Construction Co., Ltd.
合作单位 Cooperative Design
[结构与机电] 同济大学建筑设计研究院（集团）有限公司
设计团队 Project Team
[建筑] 柳亦春 陈屹峰 王龙海 王伟宏 伍正辉
[结构/机电] 巢斯 张准 邵晓健 邵喆 张颖
施工单位 Constructor
上海汇成建设（集团）有限公司

项目索引

设计时间 Design Period
2011.11 - 2012.07
建造时间 Construction Period
2012.03 - 2014.03
基地面积 Site Area
19 337 m²
建筑面积 Floor Area
33 007 m²
层数 Storey
地上 2层,地下 2层 /
2 storeys overground, 2 storeys underground
结构形式 Structure
钢筋混凝土结构 / Reinforced concrete structure

p096
无锡阖闾城遗址博物馆
Helv City Historic Site Museum
地点 Location
江苏省无锡市滨湖区环太湖公路马山闾江 2号 / No.2, Mashanlv River Huanhu Street, Binhu District, Wuxi City, Jiangsu Province
业主 Client
无锡吴都阖闾古城发展有限公司 / Wuxi Wu Ancient City Development Co., Ltd.
设计团队 Project Team
[建筑] 李立 王文胜 叶雯 高山 郭婵姣
施工单位 Constructor
江苏汉中建设集团有限公司
设计时间 Design Period
2009.10 - 2010.06
建造时间 Construction Period
2010.12 - 2014.06
基地面积 Site Area
47 000 m²
建筑面积 Floor Area
26 500 m²
层数 Storey
地上 3层,地下 1层 /3 storeys overground, 1 storey underground
结构形式 Structure
钢筋混凝土框架结构 / Reinforced concrete frame structure

p098
柯力博物馆
Keli Museum
地点 Location
江苏省宁波市江北区长兴路 199号 /No.199, Changxing Street, Jiangbei District, Ningbo City, Jiangsu Province
业主 Client
宁波柯力电气制造有限公司 /Keli Sensing Technology (Ningbo) Co., Ltd.
设计团队 Project Team
[建筑] 王籍 马学鑫 徐丹 田园 国山
[结构] 周俊才
施工单位 Constructor
王志平施工队
设计时间 Design Period
2010 - 2011
建造时间 Construction Period
2011 - 2014
基地面积 Site Area
1 800 m²
建筑面积 Floor Area
4 500 m²
层数 Storey
地上 3层,地下 1层 /3 storeys overground, 1 storey underground
结构形式 Structure
型钢混凝土 /Steel reinforced concrete

p100
范曾艺术馆
Fan Zeng Art Gallery
地点 Location
江苏省南通市啬园路 9号南通大学校区内 /Nantong University, No.9, Seyuan Street, Nantong City, Jiangsu Province
业主 Client
南通大学 /Nantong University
设计团队 Project Team
[建筑] 章明 张姿 李雪峰 孙嘉龙
[结构] 洪文明
[给排水] 施锦岳
[电气] 张逸峰
[暖通] 朱伟昌
施工单位 Constructor
南通市中房建设工程有限公司
设计时间 Design Period
2010.11
建造时间 Construction Period
2014.01
基地面积 Site Area
20 529 m²
建筑面积 Floor Area
7 028 m²
层数 Storey
地上 4层,地下 1层 /4 storeys overground, 1 storey underground
结构形式 Structure
混凝土框架结构 / Concrete frame structure

p102
乌海市黄河渔类增殖站及展示中心
Wuhai Yellow River Fishing Station and Exhibition Center
地点 Location
内蒙古自治区乌海市黄河左岸水利枢纽工程下游 /Downstream of Yellow River Left Bank Water Control Project, Wuhai City, Inner Mongolia Autonomous Region
业主 Client
黄河海勃湾水利枢纽工程建设管理局 /Yellow River Haibowan Water Control Project Construction Administration Bureau
设计团队 Project Team
[建筑] 张鹏举 郭嵩
[结构] 丁旭栋
[设备] 钱才贵
[电气] 钱春节
施工单位 Constructor
江苏大汉建设实业集团有限责任公司
设计时间 Design Period
2013.01
建造时间 Construction Period
2014.05
基地面积 Site Area
4 525 m²
建筑面积 Floor Area
1 872 m²
层数 Storey
地上 1层 /1 storey
结构形式 Structure
砖混结构 /Brick-concrete structure

p104
浙江湖州梁希纪念馆
Huzhou Liangxi Memorial Hall
地点 Location
浙江省湖州市吴兴区 G104梁希森林公园 / G104 Liangxi Forest Park, Wuxing District, Huzhou City, Zhejiang Province
业主 Client
湖州园林绿化管理有限公司 / Huzhou Landscaping Co., Ltd.
设计团队 Project Team
[建筑] 张应鹏 王凡 肖蓉婷 钱舟 谢磊
[结构] 龚明华 石立彬
[暖通] 梁羽晴
[给排水] 张琦
[电气] 薛青
施工单位 Constructor
浙江大东吴集团建筑有限公司
设计时间 Design Period
2010.07 - 2011.10
建造时间 Construction Period
2011.08 - 2014.10
基地面积 Site Area
3 955 m²
建筑面积 Floor Area
3 972.9 m²
层数 Storey
地上 2层,地下 1层 /2 storeys overground, 1 storey underground
结构形式 Structure
混凝土框架结构 /Concrete Frame structure

p106
北京菜市口输变电站综合体（电力科技馆）
Beijing Caishikou Power Transformer Substation Complex (Electric Power Science and Technology Museum)
地点 Location
北京市西城区菜市口大街 /Caishikou Avenue, Xicheng District, Beijing City
业主 Client
国家电网公司 / State Grid Corporation of China
设计团队 Project Team
[建筑] 庄惟敏 张维 杜爽 任飞 梁思思
[结构] 陈宏 李征宇 王岚
[给排水] 邵强
[暖通] 李晖
[电气] 刘力红
施工单位 Constructor
中国建筑第六工程局有限公司
设计时间 Design Period
2009.03 - 2013.06
建造时间 Construction Period
2013.01 - 2014.05
基地面积 Site Area
7 502.9 m²
建筑面积 Floor Area
47 767.75 m²
层数 Storey
地上 12层,地下 6层 /12 storeys overground, 6 storeys underground
结构形式 Structure
混凝土框架剪力墙结构 /Concrete frame-shear wall structure

p108
中国版画艺术博物馆
China Scratchboard Art Museum
地点 Location
广东省深圳市观澜湖 / Guanlan Lake, Shenzhen City, Guangdong Province
业主 Client
中国版画博物馆 / China Scratchboard Art Museum
设计团队 Project Team
朱雄毅 凌鹏志 罗俊松 黄虹 程昀 等
设计时间 Design Period
2009 - 2011
建造时间 Construction Period
2011 - 2014
基地面积 Site Area
16 800 m²
建筑面积 Floor Area
18 600 m²
层数 Storey
地上 4层,地下 1层 / 4 storeys overground, 1 storey underground
结构形式 Structure
钢筋混凝土结构 +局部钢结构 / Reinforced concrete structure + Partly steel structure

p110
玉树州博物馆
Yushu Museum
地点 Location
青海省玉树藏族自治州结古镇民主路 /Minzhu Street, Jiegu Town, Yushu Tibetan Autonomous Prefecture, Qinghai Province
业主 Client
玉树州三江源投资建设有限公司 / Yushu Sanjiangyuan Investment and Construction Co., Ltd.
设计团队 Project Team
[建筑] 何镜堂 郭卫宏 丘建发 包莹 盘育丹 等
[结构] 郭远翔 越国明 孟祥强 林焦乐
[给排水] 岑洪金
[电气] 俞洋 过仕佳
[空调] 黄璞洁 陈卓伦 何耀炳
施工单位 Constructor
中铁二十一局集团第二有限责任公司
设计时间 Design Period
2010 - 2011
建造时间 Construction Period
2011 - 2014
基地面积 Site Area
9 427 m²
建筑面积 Floor Area
11 500 m²
层数 Storey
地上 2层 /2 storeys
结构形式 Structure
混凝土框架结构 / Concrete frame structure

p112
国家会展中心（上海）
Shanghai National Exhibition and Convention Center
地点 Location
上海市青浦区崧泽大道 333号 / No.333, Songze Street, Qingpu District, Shanghai City
业主 Client
上海博览会有限公司 / Shanghai Exhibition Co., Ltd.
设计团队 Project Team
[建筑] 庄惟敏 张俊杰 单军 傅海聪 刘念雄
[结构] 周建龙 包联进 刘彦生 黄永强 李果
[暖通] 马伟骏 贾昭凯 魏炜 万嘉凤 韩家宝
[电气] 邹民杰 王雷 王晔 徐华 王磊
[给排水] 徐青 陈立宏 王婷 张威 吉兴亮
施工单位 Constructor
上海建工集团
设计时间 Design Period
2012.01 - 2013.12
建造时间 Construction Period
2013.12 - 2014.12
基地面积 Site Area
855 946.8 m²
建筑面积 Floor Area
1 470 000 m²
层数 Storey
地上 2-10层,地下 1层 /2-10 storeys overground, 1 storey underground
结构形式 Structure
钢结构 /Steel structure

p114
又见五台山剧场
Encore Wutai Mountain Theater
地点 Location
山西省五台山第三区马圈沟片区 / No.3 Maquangou Area Wutai Mountain, Shanxi Province
业主 Client
山西五台山文化旅游集团有限公司 / Mount Wutai Culture and Tourism Development Co., Ltd.
设计团队 Project Team
[建筑] 朱小地 高博 朱颖 孔繁锦 罗文 等
[结构] 田立宗 田玉香 王越
[设备] 赵伟 江雅卉
[电气] 赵阳
[灯光] 郑见伟
施工单位 Constructor
中冶天工集团有限公司

设计时间 Design Period
2011.10 – 2014.05
建造时间 Construction Period
2013.09 – 2014.09
基地面积 Site Area
152 909.03 m²
建筑面积 Floor Area
27 836.72 m²
层数 Storey
地上 1层，局部 2层 / 1 storey overground, partly 2 storeys
结构形式 Structure
钢筋混凝土结构 +钢结构 / Reinforced concrete structure + Steel structure

p116
盘锦城市文化展示馆
Panjin City Cultural Exhibition Center
地点 Location
辽宁省盘锦市辽东湾新区湘江街 / Xiangjiang Street, Liaodong Bay New District, Panjin City, Liaoning Province
业主 Client
盘锦市重点工程建设管理办公室 / Panjin Key Project Construction Management Office
设计团队 Project Team
[建筑] 张伶伶 赵伟峰 王靖 刘万里 武威 等
[室内] 王哲民
[景观] 陈宇
施工单位 Constructor
中国建筑第八工程局有限公司
设计时间 Design Period
2010.02 – 2011.12
建造时间 Construction Period
2010.10 – 2013.08
基地面积 Site Area
85 600 m²
建筑面积 Floor Area
25 619 m²
层数 Storey
地上 4层 / 4 storeys
结构形式 Structure
钢筋混凝土框架结构 / Reinforced concrete frame structure

p118
上海嘉定保利大剧院
Jiading New Town Poly Grand Theater
地点 Location
上海市嘉定区白银路 159号 / No.159, Baiyin Road, Jiading District, Shanghai City
业主 Client
上海保利茂佳房地产开发有限公司 / Maojia Shanghai Poly Real Estate Development Co., Ltd.
设计团队 Project Team
[建筑] 安藤忠雄 陈剑秋 戚鑫 张瑞 汤艳丽
[景观] 陆伟宏
[结构] 林建萍
[暖通] 谭立民
[电气] 蔡英琪 严志峰
施工单位 Constructor
中国建筑股份有限公司(上海)
设计时间 Design Period
2009.09 – 2013.05
建造时间 Construction Period
2010.12 – 2014.12
基地面积 Site Area
30 235 m²
建筑面积 Floor Area
55 904 m²
层数 Storey
地上 6层，地下 1层 / 6 storeys overground, 1 storey underground
结构形式 Structure
框架剪力墙结构 / Frame-shear wall structure

p120
玉树文成公主纪念馆
Memorial of Princess Wencheng, Yushu
地点 Location
青海省玉树藏族自治州结古镇文成公主庙东 / The East of The Princess Wencheng Temple, Jiegu Town, Yushu Tibetan Autonomous Prefecture, Qinghai Province
业主 Client
玉树三江源投资建设有限公司 / Yushu Sanjiangyuan Investment Ltd.
设计团队 Project Team
[建筑] 钱方 黄怀海 李峰 王笑南 冯泰林 等
[结构] 冯远 陈文明 付刊林 雷云 王飞 等
[给排水] 孙钢 杜毅 谭古今
[电气] 银雷 郑祖雷 温蕊芳
[暖通] 戎向阳 司鹏飞
[幕墙] 董彪 陈恩莉
[建筑经济] 张瑜梅 雷波
施工单位 Constructor
中国中铁二局集团有限公司
设计时间 Design Period
2010.12 – 2012.10
建造时间 Construction Period
2013.03 – 2014.11
基地面积 Site Area
6 000 m²
建筑面积 Floor Area
2 307 m²
层数 Storey
地上 1层 / 1 storey
结构形式 Structure
框架剪力墙结构 / Frame-shear wall structure

p122
绩溪博物馆
Jixi Museum
地点 Location
安徽省宣城市绩溪县华阳镇良安路 100号 / No.100, Liang'an Street, Huayang Town, Jixi County, Xuancheng City, Anhui Province
业主 Client
绩溪县文化广电新闻出版局 / Bureau of Culture, Radio, Film, TV, Press and Publication of Jixi
设计团队 Project Team
[建筑] 李兴钢 张音玄 张哲 邢迪 张一婷
[结构] 王力波 杨威 梁伟
[景观] 李力 于超
施工单位 Constructor
绩溪县良安建筑工程有限责任公司
设计时间 Design Period
2009.11 – 2010.12
建造时间 Construction Period
2010.12 – 2013.11
基地面积 Site Area
9 500 m²
建筑面积 Floor Area
10 003 m²
层数 Storey
地上 2层，地下 1层 / 2 storeys overground, 1 storey underground
结构形式 Structure
钢筋混凝土框架结构 +钢屋架结构 / Reinforced concrete frame structure + Steel roof truss structure

p124
贾平凹文化艺术馆
Jia Pingwa Culture & Art Gallery
地点 Location
陕西省西安市临潼区芷阳三路 / Lintong District, Xi'an City, Shaanxi Province
业主 Client
西安曲江临潼旅游商业发展有限公司 / Xi'an Qujiang-Lintong Tourism & Business Development Ltd.
设计团队 Project Team
[建筑] 屈培青 阎飞 李大为 张恒岩
施工单位 Constructor
陕西建工第六建设集团有限公司
设计时间 Design Period
2011.03 – 2012.04
建造时间 Construction Period
2012.04 – 2013.06
基地面积 Site Area
6 667 m²
建筑面积 Floor Area
4 621.5 m²
层数 Storey
地上 2层 / 2 storeys
结构形式 Structure
钢筋混凝土框架结构 / Reinforced concrete frame structure

p126
南京博物院改扩建工程
Renovation and Expansion Project of Nanjing Museum
地点 Location
江苏省南京市玄武区中山东路 321号 / No.321, East Zhongshan Street, Xuanwu District, Nanjing City, Jiangsu Province
业主 Client
南京博物院 / Nanjing Museum
设计团队 Project Team
[建筑] 程泰宁 王幼芬 王大鹏 周红雷 柴敬
施工单位 Constructor
南通四建工程公司
设计时间 Design Period
2008.05 – 2009.10
建造时间 Construction Period
2009.12 – 2013.10
基地面积 Site Area
20 300 m²
建筑面积 Floor Area
84 500 m²
层数 Storey
地上 2-5层，地下 1层 / 2-5 storeys overground, 1 storey underground
结构形式 Structure
钢筋混凝土框架结构 / Reinforced concrete frame structure

p128
乌镇剧院
Wuzhen Theater
地点 Location
浙江省桐乡市乌镇西栅景区 / West Gate Scenic Area, Wuzhen Town,Tongxiang City, Zhejiang Province
业主 Client
乌镇旅游股份有限公司 / Wuzhen Tourism Co., Ltd.
设计团队 Project Team
[建筑] 姚仁喜 袁建平 朱文弘 沈国健 苏昶
[结构] 施从伟 唐小辉 张朕磊
[暖通] 万阳
[给排水] 赵俊
[电气] 胡戎
施工单位 Constructor
巨匠建设集团有限公司
设计时间 Design Period
2010.05 – 2010.12
建造时间 Construction Period
2011.01 – 2013.04
基地面积 Site Area
54 980 m²
建筑面积 Floor Area
6 920 m²
层数 Storey
地上 2层，地下 1层 / 2 storeys overground, 1 storey underground
结构形式 Structure
钢筋混凝土结构 +钢骨结构 / Reinforced concrete structure + Steel framing structure

p130
金陵美术馆
Jinling Art Museum
地点 Location
江苏省南京市秦淮区剪子巷 50号 / No.50, Jianzi Alley, Qinhuai District, Nanjing City, Jiangsu Province
业主 Client
南京市文化广电新闻出版局 / Nanjing City Cultural Broadcasting Press and Publication Bureau
设计团队 Project Team
[建筑] 刘克成 肖莉 吴超 装钊 同庆楠
[室内] 陈卫新
施工单位 Constructor
南京金中建幕墙装饰有限公司
设计时间 Design Period
2011.10 – 2012.10
建造时间 Construction Period
2012.10 – 2013.10
基地面积 Site Area
4 424 m²
建筑面积 Floor Area
12 974 m²
层数 Storey
地上 4层 / 4 storeys
结构形式 Structure
混凝土框架结构 / Concrete frame structure

p132
南开大学新校区核心教学区
Core Teaching Quarter of the New Campus of Nankai University
地点 Location
天津市津南区海河教育园区南侧 / Southern Part of Haihe Educational Area, Jinnan District, Tianjin City
业主 Client
南开大学 / Nankai University
设计团队 Project Team
[建筑] 章明 张姿 肖镭 冯珊珊
[结构] 吕军
[给排水] 冯玮
[电气] 王坚
[暖通] 曾刚
施工单位 Constructor
天津市建工工程总承包有限公司、天津天一建设集团有限公司
设计时间 Design Period
2011.08 – 2013.06
建造时间 Construction Period
2012.10 – 2016.07
基地面积 Site Area
90 940 m²
建筑面积 Floor Area
112 030 m²
层数 Storey
地上 5层 / 5 storeys
结构形式 Structure
钢筋混凝土框架剪力墙结构 / Reinforced concrete frame-shear wall structure

p134
上海德富路初中
De Fu Junior High School
地点 Location
上海市嘉定区洪德路 618号 / No.618, Hongde Street, Jiading District, Shanghai City
业主 Client
上海嘉定新城发展有限公司 / Shanghai Jiading New Town Development Co., Ltd.
设计团队 Project Team
张佳晶 徐文斌 黄巍 徐聪 易博文 等
施工单位 Constructor
浙江万汇建设集团有限公司
设计时间 Design Period
2010 – 2014

项目索引

建造时间 Construction Period
2014 – 2016
基地面积 Site Area
27 816 m²
建筑面积 Floor Area
12 783 m²
层数 Storey
地上 3层,地下 1层 /3 storeys overground, 1 storey underground
结构形式 Structure
框架结构 /Frame structure

p136
苏州实验中学原址重建项目
Reconstruction Project of Suzhou Experimental Middle School

地点 Location
江苏省苏州市高新区金山路76号 / No.76, Jinshan Street, Hi-tech District, Suzhou City, Jiangsu Province
业主 Client
江苏省苏州实验中学（建设单位）; 苏州永新置地有限公司（代建单位）/ Suzhou Experimental Middle School (Construction Unit); Suzhou Yongxin Land Co., Ltd. (Agent Construction Unit)
设计团队 Project Team
[建筑] 曾群 文小琴 汪颖 张艳 李荣荣
[结构] 吴树勋 余思谨 陈凯
[给排水] 施锦岳 肖蓝
[暖通] 邵华厦 叶耀蔚
[电气] 徐建栋 施国平
施工单位 Constructor
苏州建鑫建设集团有限公司
设计时间 Design Period
2013.11 – 2014.06
建造时间 Construction Period
2014.11 – 2016.05
基地面积 Site Area
62 487 m²
建筑面积 Floor Area
75 510 m²（地上 / Overground：56 484 m²,地下 / Underground：19 026 m²）
层数 Storey
地上 6层,地下 1层 / 6 storeys overground, 1 storey underground
结构形式 Structure
混凝土框架结构 / Concrete frame structure

p138
山西兴县 120师学校
Instruction Building of the 120th Division School in Xing County, Shanxi Province

地点 Location
山西省吕梁市兴县新区蔡家崖 / Caijiaya, New District, Xing County, Lvliang City, Shanxi Province
业主 Client
兴县教育体育局 /Bureau of Education & Sports in Xing Country
设计团队 Project Team
[建筑] 吴林寿 赵向莹 黄瑞言 傅湘嫒 何少俊
[结构] 吴晖 王冬松
[暖通] 潘北川 詹展谋 郑德金
[电气] 胡明红 刘旭
[给排水] 左正渊 杨华
施工单位 Constructor
北京港源建筑装饰工程有限公司
设计时间 Design Period
2013 – 2014
建造时间 Construction Period
2014 – 2016
基地面积 Site Area
80 000 m²
建筑面积 Floor Area
36 000 m²
层数 Storey
地上 4层,地下 1层 /4 storeys overground, 1 storey underground
结构形式 Structure
混凝土框架结构 /Concrete frame structure

p140
清华大学海洋中心
Tsinghua Ocean Center

地点 Location
广东省深圳市南山区西丽大学城 /Xili University Town, Nanshan District, Shenzhen City, Guangdong Province
业主 Client
深圳市建筑工务署 /Bureau of Public Works of Shenzhen Municipality
设计团队 Project Team
李虎 黄文菁 Victor Quiros 赵耀 张汉仰 等
施工单位 Constructor
四川华西集团有限公司
设计时间 Design Period
2011.11 – 2014.08
建造时间 Construction Period
2013.06 – 2016.12
基地面积 Site Area
2 439 m²
建筑面积 Floor Area
15 884 m²
层数 Storey
地上 14层,地下 2层 / 14 storeys overground, 2 storeys underground
结构形式 Structure
框架剪力墙结构 /Frame-shear wall structure

p142
天颐湖儿童体验馆
Tianyi Lake Children's Edutainment Mall

地点 Location
山东省泰安市岱岳区滨湖路 / Binhu Street, Daiyue District, Tai'an City, Shandong Province
业主 Client
山东泰安市泰山大汶口旅游置业有限公司 / Tai'an Tai Mountain Dawenkou Tourism Development Ltd.
设计团队 Project Team
[方案] iDEA 建筑设计事务所（高岩 郭馨）
[建筑] 郭馨 吴彬
[运营/策划] 幺强
[景观] 张海霞
[设计] 黄文颖 李静仪 许宏杰 杨丹
[施工图设计] 上海同键强华建筑设计有限公司
施工单位 Constructor
山东泰山普惠建工有限公司
设计时间 Design Period
2015.09 – 2016.01
建造时间 Construction Period
2016.01 – 2016.09
基地面积 Site Area
20 000 m²
建筑面积 Floor Area
9 500 m²
层数 Storey
地上 2层 /2 storeys
结构形式 Structure
钢结构 /Steel structure

p144
苏州湾实验小学
Suzhou Bay Experimental Primary School

地点 Location
江苏省苏州市吴江区开平路以北、夏蓉街以西、春兰街以东 / North of Kaiping Street, West of Xiarong Street, East of Chunlan Street, Wujiang District, Suzhou City, Jiangsu Province
业主 Client
江苏省吴江实验小学 /Wujiang Experimental Primary School
设计团队 Project Team
[建筑] 张应鹏 黄志强 唐超乐 董霄霜 王濛杨
[结构] 李红星 吴玉英 屈磊 杨威
[给排水] 仲文彬
[电气] 薛青 胡鑫
[暖通] 梁羽晴
施工单位 Constructor
苏州市庙港建筑有限公司
设计时间 Design Period
2014.10 – 2015.06
建造时间 Construction Period
2015.02 – 2016.09
基地面积 Site Area
76 282.8 m²
建筑面积 Floor Area
69 974.65 m²
层数 Storey
地上 4层,地下 1层 / 4 storeys overground, 1 storey underground
结构形式 Structure
混凝土框架结构 /Concrete frame structure

p146
北京华为环保园 J地块数据通信研发中心
Beijing Huawei R&D Center

地点 Location
北京市海淀区北清路 156号环保园 / No.156, Beiqing Street, Haidian District, Beijing City
业主 Client
华为投资控股有限公司 /Huawei Investment & Holding Co., Ltd.
设计团队 Project Team
[建筑] 陆静 林雷 郑飞 余洁
[总图] 刘文
[结构] 曹清 董洲
[给排水] 黎松 王存凤
[暖通] 梁琳 王春雷 高丽颖
[电气] 胡桃 崔振辉
[智能] 张雅 陈玲玲
[景观] 雷洪强 方威
[室内] 邓雪映
施工单位 Constructor
中建三局集团有限公司
设计时间 Design Period
2008.01 – 2015.12
建造时间 Construction Period
2011.06 – 2016.02
基地面积 Site Area
96 300 m²
建筑面积 Floor Area
147 800.32 m²
层数 Storey
地上 4层,地下 2层 /4 storeys overground, 2 storeys underground
结构形式 Structure
框架剪力墙结构 /Frame-shear wall structure

p148
FAST工程观测基地综合楼
Comprehensive Building of FAST Observation

地点 Location
贵州省黔南布依族苗族自治州平塘县大洼凼洼地 / Taipa Depressions, Dawa Pingtang County, Qiannan Buyi and Miao Autonomous Prefecture, Guizhou Province
业主 Client
中国科学院国家天文台 /National Astronomical Observatories
设计团队 Project Team
[建筑] 于一平 李凯 马建 王瀚辰 徐银军
[结构] 王成虎 王伟 钟艺
[给排水] 张旭 周佐辉
[暖通] 刘星
[电气] 陶戍驹 石咸胜
[总图] 陈景来 郭伟华
[精装修设计] 李长胜 陈绍华 黄欣
施工单位 Constructor
中铁十一局集团有限公司
设计时间 Design Period
2012.03 – 2015.05
建造时间 Construction Period
2015.06 – 2016.10
基地面积 Site Area
240.00 m²
建筑面积 Floor Area
7 065 m²
层数 Storey
地上 3层 /3 storeys
结构形式 Structure
钢筋混凝土框架结构 /Reinforced concrete frame structure

p150
岱山小学 岱山幼儿园
Daishan Primary School and Kindergarten

地点 Location
江苏省南京市雨花台区西善桥街道 /Xishan Bridge, Yuhua District, Nanjing City, Jiangsu Province
业主 Client
南京商贸港房地产开发有限责任公司 /Nanjing Commerce Real Estate Development Co., Ltd.
设计团队 Project Team
[建筑] 周凌 杨海 汪愫憬 邹丰 朱莹辉
[结构] 刘子洁
[给排水] 田小晶
[电气] 李锦喜
[暖通] 周美象
施工单位 Constructor
南京天泉建筑安装工程有限公司
设计时间 Design Period
2011 – 2012
建造时间 Construction Period
2012 – 2015
基地面积 Site Area
小学 / Primary school: 18 912 m²; 幼儿园 / Kindergarden: 6 500 m²
建筑面积 Floor Area
小学 / Primary school: 11 910 m²; 幼儿园 / Kindergarden: 4 279.6 m²
层数 Storey
地上 4层 / 4 storeys
结构形式 Structure
钢筋混凝土框架结构 /Reinforced concrete frame structure

p152
中衡设计集团研发中心
The Design and Research Building of ARTS Group Co., Ltd.

地点 Location
江苏省苏州市工业园区八达街 111号 /No.111, Bada Street, Industrial Park, Suzhou City, Jiangsu Province
业主 Client
中衡设计集团股份有限公司 /Arts Group Co., Ltd.
设计团队 Project Team
[建筑] 冯正功 高霖 平家华 黄琳
[室内] 宋洋
[景观] 王翔 陈天花 黄蓓丽
施工单位 Constructor
中亿丰建设集团股份有限公司
设计时间 Design Period
2010.12 – 2011.06
建造时间 Construction Period
2011.07 – 2015.10
基地面积 Site Area
14 100 m²
建筑面积 Floor Area
77 000 m²
层数 Storey
地上 22层,地下 3层 /22 storeys overground, 3 storeys underground
结构形式 Structure
框架剪力墙结构 /Frame-shear wall structure

p154
华东师范大学附属双语幼儿园
East China Normal University Affiliated Bilingual Kindergarten
地点 Location
上海市嘉定区安亭镇 /Anting Town, Jiading District, Shanghai City
业主 Client
安亭国际汽车城 /Anting International Automobile City
设计团队 Project Team
[建筑] 祝晓峰 李启同 丁鹏华 杜济 石延安
施工单位 Constructor
甘肃第五建设集团公司
设计时间 Design Period
2012 – 2015
建造时间 Construction Period
2015
基地面积 Site Area
7 400 m²
建筑面积 Floor Area
6 196 m²
层数 Storey
地上 3层 / 3 storeys
结构形式 Structure
钢筋混凝土结构 / Reinforced concrete structure

p156
中福会浦江幼儿园
Pujiang China Welfare Institute Kindergarten
地点 Location
上海市闵行区浦江镇江柳路、浦秀路 /Puxiu Road and & Jiangliu Road, Pujiang New Town, Minhang District, Shanghai City
业主 Client
上海浦江镇投资发展有限公司 / Shanghai Pujiang Town Investment Development Co., Ltd.
设计团队 Project Team
[建筑] 周蔚 张斌 袁怡 王佳绮 李姿娜 等
施工单位 Constructor
上海广厦建筑工程有限公司
设计时间 Design Period
2011.12 – 2014.03
建造时间 Construction Period
2013.06 – 2015.05
基地面积 Site Area
5 092 m²
建筑面积 Floor Area
15 329 m²
层数 Storey
地上 3层,地下 1层 /3 storeys overground, 1 storey underground
结构形式 Structure
钢筋混凝土框架结构 /Reinforced concrete frame structure

p158
上海国际汽车城研发港D地块
Plot D, The R&D and Innovative Port of Anting International Automobile City
地点 Location
上海市嘉定区安虹路 / Anhong Road, Jiading District, Shanghai City
业主 Client
上海国际汽车城发展有限公司 /Anting Shanghai International Automobile City
设计团队 Project Team
[建筑] 陈屹峰 柳亦春 宋崇芳 李珺
施工单位 Constructor
嘉定区建设工程(集团)有限公司
设计时间 Design Period
2009.12 – 2010.12
建造时间 Construction Period
2013.07 – 2015.06
基地面积 Site Area
8 707 m²
建筑面积 Floor Area
36 600 m²
层数 Storey
地上 4层,地下 1层 / 4 storeys overground, 1 storey underground
结构形式 Structure
钢筋混凝土框架结构 / Reinforced concrete frame structure

p160
芭莎·阳光童趣园
BAZAAR · Sunshine Playhouse
地点 Location
甘肃省白银市会宁县大沟乡厍去小学 /Shequ Village Primary School, Huining County, Baiyin City, Gansu Province
业主 Client
北京市西部阳光农村发展基金会 / Beijing Western Sunshine Rural Development Foundation
设计团队 Project Team
[建筑] 朱竞翔 韩国日 吴程辉 Tibor Franek
施工单位 Constructor
厍去村村民与小学老师、志愿义工
设计时间 Design Period
2014.12 – 2015.04
建造时间 Construction Period
2015.04
基地面积 Site Area
50 m²
建筑面积 Floor Area
50 m²
层数 Storey
地上 1层 /1 storey
结构形式 Structure
空间板式系统 (桉木胶合板) / Spatial panel system (eucalyptus plywood)

p162
天津大学新校区图书馆
Library on the New Campus of Tianjin University
地点 Location
天津市津南区海河教育园区雅观路 135号 /No.135, Yaguan Street, Haihe Education Park Jinnan District, Tianjin City
业主 Client
天津大学 /Tianjin University
设计团队 Project Team
[建筑] 周恺 张莉兰 章宁 王力新
[结构] 郭恩建 左克伟
[给排水] 田书韦
[暖通] 杨琳
[电气] 张月洁
施工单位 Constructor
中国建筑第三工程局有限公司
设计时间 Design Period
2010.10 – 2013.05
建造时间 Construction Period
2013.05 – 2015.10
基地面积 Site Area
47 000 m²
建筑面积 Floor Area
53 000 m²
层数 Storey
地上 4层,局部地下 1层 /4 storeys overground, partly 1 storey underground
结构形式 Structure
钢筋混凝土框架结构 /Reinforced concrete frame structure

p164
浙江音乐学院
Zhejiang Conservatory of Music
地点 Location
浙江省杭州市西湖区转塘镇浙音路 1号 / No.1, Zheyin Street, Zhuantang Town, Xihu District, Hangzhou City, Zhejiang Province
业主 Client
浙江省文化厅 /Zhejiang Provincial Department of Culture
设计团队 Project Team
[建筑] 王宇虹 张微 朱培栋 宋萍 王静 徐亮 程越 宋一村 李明 钱明一 朱峰
[结构] 任光勇 徐凌峰 李建军
[机电设备] 崔大梁 何剑 陆柏庆 吴文坚 黄国华 郭丽梅
[景观] 张铁群 荣耀 李婷云 周润
[室内] 陈耀光 沈雷 孙云 胡昕
施工单位 Constructor
浙江省建工集团有限责任公司
设计时间 Design Period
2012.09 – 2013.10
建造时间 Construction Period
2013.10 – 2015.09
基地面积 Site Area
401 333 m²
建筑面积 Floor Area
359 648 m²
层数 Storey
地上 13层,地下 1层 /13 storeys overground, 1 storey underground
结构形式 Structure
钢筋混凝土结构 /Reinforced concrete structure

p166
清华大学南区学生食堂
Central Canteen of Tsinghua University
地点 Location
北京市海淀区双清路 30号 / No. 30, Shuangqing Street, Haidian District, Beijing City
业主 Client
清华大学 /Tsinghua University
设计团队 Project Team
[建筑] 宋晔皓 孙菁芬 解丹 陈晓娟 王丽娜
[结构] 蒋炳丽
[暖通] 吴晓燕
[电气] 费洪凤
施工单位 Constructor
北京市第三建筑工程有限公司
设计时间 Design Period
2011.11 – 2013.11
建造时间 Construction Period
2013.08. – 2015.08
基地面积 Site Area
6 380 m²
建筑面积 Floor Area
21 000 m²
层数 Storey
地上 3层,地下 3层 / 3 storeys overground, 3 storeys underground
结构形式 Structure
钢筋混凝土框架结构 / Reinforced concrete structure

p168
杭州师范大学仓前校区
Hangzhou Normal University Cangqian Campus
地点 Location
浙江省杭州市余杭区余杭塘路 2318号 /No.2318, Yuhangtang Street, Yuhang District, Hangzhou City, Zhejiang Province
业主 Client
杭州师范大学 / Hangzhou Normal University
设计团队 Project Team
[建筑] 吴钢 陈凌 曲克明 杨易栋
[幕墙] 陶善钧
施工单位 Constructor
浙江省三建建筑集团有限公司、深圳市奇信建设集团股份有限公司
设计时间 Design Period
2010.07 – 2013.10
建造时间 Construction Period
2013.07 – 2015.09
基地面积 Site Area
61 648 m²
建筑面积 Floor Area
161 426 m²
层数 Storey
地上 13层,地下 1层 /13 storeys overground, 1 storey underground
结构形式 Structure
框架剪力墙结构 / Frame-shear wall structure

p170
苏州科技城实验小学
Experimental Primary School of Suzhou Science and Technology Town
地点 Location
苏州高新技术产业开发区科技城科业路 88号 /No.88, Keye Street, Science and Technology Town, National New and Hi-tech Industrial Development Zone, Suzhou City, Jiangsu Province
业主 Client
苏州科技城社会事业服务中心 / Social Programs Service Center of Suzhou Science and Technology Town
设计团队 Project Team
张斌 李硕 陈颢 丁心慧 吴人洁 等
施工单位 Constructor
苏州建鑫建设集团有限公司
设计时间 Design Period
2013.06 – 2014.02
建造时间 Construction Period
2014.03 – 2015.09
基地面积 Site Area
14 478 m²
建筑面积 Floor Area
53 422 m²
层数 Storey
地上 4层 /4 storeys
结构形式 Structure
钢筋混凝土框架+局部钢结构 / Reinforced concrete frame+Partly steel frame

p172
清控人居科技示范楼
THE-Studio
地点 Location
贵州省贵安新区 / Gui'an New District, Guizhou Province
业主 Client
清控人居控股集团有限公司 / Tsinghua Holdings Habitat Development Group
设计团队 Project Team
[建筑/室内] 宋晔皓 孙菁芬 陈晓娟 林正豪
[结构] 孙晓彦
[暖通/给排水] 张玥
[电气] 李高楼
施工单位 Constructor
北京清控水木建筑工程有限公司第四分公司
设计时间 Design Period
2015.02 – 2015.03
建造时间 Construction Period
2015.03 – 2015.06
基地面积 Site Area
1 826 m²
建筑面积 Floor Area
701 m²
层数 Storey
地上 2层,地下 1层 / 2 storeys overground, 1 storey underground
结构形式 Structure
木结构 +钢结构 /Timber Structure+ Steel structure

p174
武汉理工大学南湖校区图书馆
Wuhan University of Technology Nanhu Campus Library

地点 Location
湖北省武汉市洪山区骆狮路122号 / No. 122, Luoshi Street, Hongshan District, Wuhan City, Hubei Province

业主 Client
武汉理工大学 /Wuhan University of Technology

设计团队 Project Team
[建筑] 陶郅 郭钦恩 陈健生 陈子坚 陈向荣
[结构] 柳一心 舒宣武
[给排水] 王学峰
[电气] 黄晓峰 伍尚仁
[空调] 王钊 陈卓伦 邹玉进
[景观] 陈天宁 刘伟庆 邓寿朋

施工单位 Constructor
中建三局集团有限公司

设计时间 Design Period
2015.02 – 2015.03

建造时间 Construction Period
2012.01 – 2015.01

基地面积 Site Area
18 914 m²

建筑面积 Floor Area
47 557.6 m²

层数 Storey
地上 11层,地下 1层 /11 storeys overground, 1 storey underground

结构形式 Structure
钢筋混凝土框架结构 /Concrete frame structure

p176
东北大学浑南新校园风雨操场
Hunnan Campus of Northeastern University Gymnasium

地点 Location
辽宁省沈阳市浑南区创新路195号 /No.195, Chuangxin Street, Hunnan District, Shenyang City, Liaoning Province

业主 Client
东北大学 /Northeastern university

设计团队 Project Team
[建筑] 梅洪元 苑雪飞 张晓航 张刘阳 胡建丽
[结构] 刘志伟 张英华 刘海峰
[设备] 孙振宇 李莹莹 常忠海

施工单位 Constructor
上海宝冶集团有限公司

设计时间 Design Period
2012.02

建造时间 Construction Period
2012.02 – 2015.06

基地面积 Site Area
20 577.06 m²

建筑面积 Floor Area
16 599.68 m²

层数 Storey
地上 2层 /2 storeys

结构形式 Structure
钢筋混凝土框架+钢结构 / Reinforced concrete frame structure + Steel structure

p178
北京四中房山校区
Beijing No.4 High School Fangshan Campus

地点 Location
北京市房山区长阳镇怡和北路10号 / No.10, North Yihe Street, Changyang Town, Fangshan District, Beijing City

业主 Client
北京中粮万科房地产开发有限公司 / Cofco Vanke Property Group Co., Ltd.

设计团队 Project Team
李虎 黄文菁 Daijiro Nakayama 叶青 张浩 等

施工单位 Constructor
中兴建设有限公司

设计时间 Design Period
2010.04 – 2014.08

建造时间 Construction Period
2012.02 – 2014.08

基地面积 Site Area
45 000 m²

建筑面积 Floor Area
57 000 m²

层数 Storey
地上 5层,地下 1层 /5 storeys overground, 1 storey underground

结构形式 Structure
钢筋混凝土框架结构 /Reinforced concrete frame structure

p180
松江名企艺术产业园区
Songjiang Art Campus

地点 Location
上海市松江区泗砖南路 / South Sizhuan Street, Songjiang District, Shanghai City

业主 Client
上海万居德实业有限公司 / Shanghai Wanjude Group

设计团队 Project Team
[建筑] 袁烽 韩力 孟浩 顾红兵 孔祥亭 王欧 等
[结构] 李俊民 刘宇宏 周军
[给排水] 时荣伟
[暖通] 陆仁瑞
[机电] 张新华

设计时间 Design Period
2006 – 2009

建造时间 Construction Period
2013 – 2014

基地面积 Site Area
72 000 m²

建筑面积 Floor Area
150 000 m²

层数 Storey
地上 4-6层 /4-6 storeys

结构形式 Structure
现浇混凝土框架结构砖填充墙体系 / Cast-in-place concrete frame structure with brick infilled wall system

p182
中新生态城滨海小外中学部
Binhai Xiaowai High School, Sino-Singapore Tianjin Eco-City

地点 Location
天津市中新生态城 /Sino-Singapore Eco-City, Tianjin City

业主 Client
天津中新生态城 /Sino-Singapore Tianjin Eco-City

设计团队 Project Team
[方案] 王振飞 王鹿鸣 李宏宇 范靓 Thomas Clifford Bennett
[建筑] 范靓 贾隽 黄苇 姚红湖
[结构] 巨江涛 郭恩建
[给排水] 安君 边疆 牛育辉
[暖通] 郑昕 邵海
[电气] 王裕华 张峻

设计时间 Design Period
2008.10 – 2012.05

建造时间 Construction Period
2012.06 – 2014.09

基地面积 Site Area
20 000 m²

建筑面积 Floor Area
53 000 m²

层数 Storey
地上 5层,地下 1层 /5 storeys overground, 1 storey underground

结构形式 Structure
钢结构 /Steel structure

p184
寒地建筑研究中心
Cold Region Architecture Research Center

地点 Location
黑龙江省哈尔滨市南岗区黄河路73号 /No.73, Huanghe Street, Nangang District, Harbin City, Heilongjiang Province

业主 Client
哈尔滨工业大学 /Harbin Institute of Technology

设计团队 Project Team
[建筑] 梅洪元 王飞 张伟玲 梁海岚 富永亮
[结构] 冯阿巧 张煜
[给排水] 彭晶 刁克炜

施工单位 Constructor
中十冶建筑工程有限公司

设计时间 Design Period
2012.02 – 2013.03

建造时间 Construction Period
2013.06 – 2014.12

基地面积 Site Area
4 500 m²

建筑面积 Floor Area
10 400 m²

层数 Storey
地上 5层,地下 1层 / 5 storeys overground, 1 storey underground

结构形式 Structure
钢筋混凝土框架结构 /Reinforced concrete frame structure

p186
北京工业大学第四教学楼组团
A Complex of the Teaching Facilities at Beijing University of Technology

地点 Location
北京市朝阳区平乐园100号 / No.100, Pingleyuan, Chaoyang District, Beijing City

业主 Client
北京工业大学 /Beijing University of Technology

设计团队 Project Team
[设计指导] 崔愷
[设计主持人] 柴培根 于海为 谢悦 张东
[建筑] 田海鸥 潘天佑 李楠
[结构] 任庆英 曾金盛 刘新国
[给排水] 黎松
[暖通] 韦航
[电气] 胡桃
[室内] 韩文文
[景观] 刘环
[总图] 王玮

施工单位 Constructor
河北建设集团有限公司, 中铁建设集团

设计时间 Design Period
2009.07 – 2012.08

建造时间 Construction Period
2011.09 – 2014.12

基地面积 Site Area
33 760 m²

建筑面积 Floor Area
154 292 m²

层数 Storey
地上 9层,地下 2层 / 9 storeys overground, 2 storeys underground

结构形式 Structure
钢筋混凝土框架减力墙+框架结构 / Reinforced concrete Frame-shear Wall structure+ Frame structure

p188
同济大学浙江学院图书馆
Zhejiang Campus Library, Tongji University

地点 Location
浙江省嘉兴市商务大道168号 / No.168, Commercial Avenue, Jiaxing City, Zhejiang Province

业主 Client
同济大学浙江学院 /Zhejiang College, Tongji University

设计团队 Project Team
[建筑] 张斌 周蔚
[方案] 陆均
[初步/施工图] 袁怡
[室内] 王佳绮 李姿娜
[景观] 何茜

施工单位 Constructor
浙江嘉兴福达建设股份有限公司

设计时间 Design Period
2008.12 – 2013.12

建造时间 Construction Period
2010.08 – 2014.10

基地面积 Site Area
4 590 m²

建筑面积 Floor Area
30 840 m²

层数 Storey
地上 10层,地下 1层 /10 storeys overground, 1 storey underground

结构形式 Structure
钢筋混凝土框架剪力墙结构 / Reinforced concrete frame-shear wall structure

p190
浙江科技学院安吉新校区
New Campus of Zhejiang University of Science and Technology in Anji City

地点 Location
浙江省湖州市安吉县中德路1号 / No.1, Zhongde Street, Anji County, Huzhou City, Zhejiang Province

业主 Client
浙江科技学院 /Zhejiang University of Science and Technology

设计团队 Project Team
[建筑] 秦洛峰 魏薇 沈晓鸣 威圣飞 俞淳流
[结构] 顾正维
[电气] 李平 冯百乐
[给排水] 易家松
[暖通] 潘大红

施工单位 Constructor
浙江引拓建设有限公司

设计时间 Design Period
2013

建造时间 Construction Period
2014

基地面积 Site Area
333 600 m²

建筑面积 Floor Area
180 000 m²

层数 Storey
地上 2-7层,地下 1层 /2-7 storeys overground, 1 storey underground

结构形式 Structure
混凝土框架结构 /Concrete frame structure

p192
北京育翔小学回龙观学校
Beijing Yuxiang Primary School Huilongguan School

地点 Location
北京市昌平区回龙观镇龙域中路 / Middle Longyu Street, Huilongguan Town, Changping District, Beijing City

业主 Client
北京西城区教育委员会 / Beijing West District Commission of Education

设计团队 Project Team
[建筑] 石华 周娅妮 王小工 褚奕爽 王英童 等
[室内] 王童
[景观] 郭雪

施工单位 Constructor

恒万实业有限公司
设计时间 Design Period
2010.03 – 2012.12
建造时间 Construction Period
2012.12 – 2014.07
基地面积 Site Area
26 800 m²
建筑面积 Floor Area
43 606 m²
层数 Storey
地上 5层，地下 2层 / 5 storeys overground, 2 storeys underground
结构形式 Structure
框架剪力墙结构 / Frame-shear wall structure

p194
上海嘉定桃李园实验学校
Shanghai Jiading Tao Li Yuan Experimental School
地点 Location
上海市嘉定区树屏路 2065号 / No.2065, Shuping Street, Jiading District, Shanghai City
业主 Client
上海市嘉定区国有资产经营有限公司 / Shanghai Jiading National Assets Management Limited Company
设计团队 Project Team
[建筑] 柳亦春 陈屹峰 高林 王龙海 宋崇芳
施工单位 Constructor
浙江凯德建设有限公司
设计时间 Design Period
2009.03 – 2013.03
建造时间 Construction Period
2013.06 – 2015.12
基地面积 Site Area
24 223 m²
建筑面积 Floor Area
35 688 m²
层数 Storey
地上 3层 / 3 storeys
结构形式 Structure
钢筋混凝土结构 / Reinforced concrete structure

p196
北京大学光华管理学院西安分院
Peking University Xi'an Branch of Guanghua School of Management
地点 Location
陕西省西安市临潼区凤凰大道 6号 / No.6, Fenghuang Avenue, Lintong District, Xi'an City, Shaanxi Province
业主 Client
西安曲江临潼旅游商业发展有限公司 / Xi'an Qujiang-lintong Tourism & Business Development Ltd.
设计团队 Project Team
[建筑] 屈培青 高伟 张超文 崔丹 闫文秀
[结构] 王世斌
[电气] 任万娣
施工单位 Constructor
中铁十二局集团建筑安装工程有限公司
设计时间 Design Period
2012.04 – 2012.10
建造时间 Construction Period
2012.10 – 2014.08
基地面积 Site Area
14 915 m²
建筑面积 Floor Area
54 994 m²
层数 Storey
地上 2-7层，地下 1层 / 2-7 storeys overground, 1 storey underground
结构形式 Structure
框架剪力墙结构 / Frame-shear wall structure

p198
大连华信（国际）软件园
Dalian Hi-Think (International) Software Park
地点 Location
辽宁省大连市高新园区 / Hi-Tech Zone, Dalian City, Liaoning Province
业主 Client
大连华信计算机技术股份有限公司 / Dalian Hi-Think Computer Technology, Corp.
设计团队 Project Team
[建筑] 黄星元 李瑞林 陈凡
[结构] 李娜 王愉
[公共专业] 彭玉翠 廖卓卫 蔡广会
施工单位 Constructor
大连阿尔滨集团有限公司
设计时间 Design Period
2009 – 2011
建造时间 Construction Period
2011 – 2014
基地面积 Site Area
87 500 m²
建筑面积 Floor Area
161 200 m²
层数 Storey
地上 11层，地下 2层 / 11 storeys overground, 2 storeys underground
结构形式 Structure
框架剪力墙 + 钢桁架结构 / Frame-shear wall structure + Frame construction

p200
同济大学嘉定校区留学生宿舍及专家公寓
The Foreign Students' Dormitory and Experts' Apartment in Jiading Campus of Tongji University
地点 Location
上海市嘉定区曹安公路 4800号 / No.4800, Cao'an Road, Jiading District, Shanghai City
业主 Client
同济大学 / Tongji University
设计团队 Project Team
[建筑] 李振宇 刘红 卢斌 李颂锋 李小群
[结构] 陆平
[电气] 朱亚君
[暖通] 季金星
[给排水] 张勇杰
施工单位 Constructor
上海润玛建设工程有限公司
设计时间 Design Period
2010.11 – 2012.06
建造时间 Construction Period
2012.07 – 2014.01
基地面积 Site Area
22 143 m²
建筑面积 Floor Area
36 150 m²
层数 Storey
地上 5-9层，地下 1层 / 5-9 storeys overground, 1 storey underground
结构形式 Structure
钢筋混凝土框架结构 / Reinforced concrete structure

p202
瓦山——中国美术学院象山校区专家接待中心
Tiles Hill : New Reception Center for the Xiangshan Campus, China Academy of Art
地点 Location
浙江省杭州市西湖区转塘镇象山路 352号 / No.352, Xiangshan Street, Zhuantang Town, Xihu District, Hangzhou City, Zhejiang Province
业主 Client
中国美术学院 / China Academy of Art
设计团队 Project Team
[建筑] 王澍 陆文宇 陈立超
[结构] 申屠团兵
[水电] 孙明亮
[夯土技术] 马克 朱迪 魏超超 张雯
[设计范围] 建筑设计、景观设计、室内设计、家具设计
设计时间 Design Period
2010.06 – 2013.05
建造时间 Construction Period
2011.08 – 2013.08
基地面积 Site Area
7 500 m²
建筑面积 Floor Area
6 200 m²
层数 Storey
结构形式 Structure
钢筋混凝土框架与局部钢结构 + 夯土围护墙体 + 木结构 / Reinforced concrete frame structure (partly steel) + Rammed earth retaining wall + Timber structure

p204
南京三宝科技集团物联网工程中心
Networking Engineering Center, Nanjing Sample Sci-Tech Park
地点 Location
江苏省南京市马群金马路 / Jingma Road, Maqun, Nanjing City, Jiangsu Province
业主 Client
南京三宝科技股份有限公司 / Nanjing Sample Technology Co., Ltd.
设计团队 Project Team
[建筑] 张彤 殷伟韬
[室内] 耿涛
[景观] 杨冬辉
施工单位 Constructor
南京东坝建设安装工程有限公司
设计时间 Design Period
2010.07 – 2012.12
建造时间 Construction Period
2012.02 – 2013.10
基地面积 Site Area
24 600 m²
建筑面积 Floor Area
21 272 m²
层数 Storey
地上 6层 / 6 storeys
结构形式 Structure
钢筋混凝土框架结构 / Reinforced concrete frame structure

p206
大连理工大学辽东湾校区
The Liaodong Bay Campus of Dalian University of Technology
地点 Location
辽宁省盘锦市辽东湾新区大工路 2号 / No. 2, Dagong Street, Liaodong Bay Area, Panjin City, Liaoning Province
业主 Client
盘锦地方大学和中心医院工程建设办公室 / Panjin Local University and Central Hospital Engineering Construction Office
设计团队 Project Team
[建筑] 张伶伶 赵伟峰 刘万里 黄勇 高雪松
[规划] 袁敬诚
[景观] 夏柏树
[室内] 钟兆康
施工单位 Constructor
北京建工集团有限公司、中国建筑第六工程局有限公司、中国建筑第八工程局有限公司
设计时间 Design Period
2011.05 – 2011.12
建造时间 Construction Period
2011.10 – 2014.10
基地面积 Site Area
563 100 m²
建筑面积 Floor Area
383 298 m²
层数 Storey
国际交流中心 / International exchange center: 地上 24层，地下 1层 / 24 storeys overground, 1 storey underground; 其余教学区、图书信息区、行政区、生活区、科技研发区建筑 / The buildings of teaching areas, book information areas, administrative areas, living areas, science and technology research and development zones: 地上 4-6层 / 4-6 storeys
结构形式 Structure
钢筋混凝土框架结构 / Reinforced concrete frame structure

p208
国电新能源技术研究院
Guodian New Energy Technology Research Institute
地点 Location
北京市昌平区未来科学城北二街 / North 2 Street, Future Science Park City, Changping District, Beijing City
业主 Client
中国国电集团公司 / China Guodian Corporation
设计团队 Project Team
[建筑] 叶依谦 刘卫纲 薛军 段伟 从振
[结构] 周箏 王雪生 石光磊 郭丽萍 王洋
[设备] 徐宏庆 陈莉 富晖 赵墨 张春平
[电气] 骆平 刘洁 李超 杨帆 杨奕
施工单位 Constructor
中建一局集团建设发展有限公司、中国建筑第七工程局有限公司
设计时间 Design Period
2010.09 – 2012.07
建造时间 Construction Period
2011.03 – 2013.12
基地面积 Site Area
141 922.486 m²
建筑面积 Floor Area
243 100 m²
层数 Storey
地上 1-17层，地下 1-2层 / 1-17 storeys overground, 1-2 storeys underground
结构形式 Structure
钢筋混凝土框架剪力墙结构 / Reinforced concrete frame-shear wall structure

p210
大乐之野庾村民宿
Lostvilla Boutique Hotel in Yucun
地点 Location
浙江省湖州市德清县莫干山镇黄郛西路 85号 / No.85, West Huangfu Street, Moganshan Town, Deqing County, Huzhou City, Zhejiang Province
业主 Client
上海野舍酒店管理有限公司 / Shanghai Yeshe Hotel Management Co., Ltd.
设计团队 Project Team
[建筑] 水雁飞 苏亦奇 马圆融 邓丹 徐翰骅
[室内] 李格格 柴燕妮 周晓燕 王珂一 朱颖
[景观] 孙晶
[结构] 张维
[机电] 陈哲 陆鹏飞 陈强
施工单位 Constructor
汪洪永、上海野舍酒店管理有限公司

项目索引

设计时间 Design Period
2015.03 – 2017.06
建造时间 Construction Period
2015.10 – 2017.06
基地面积 Site Area
1 735 m²
建筑面积 Floor Area
1 491 m²
层数 Storey
地上三层 / 3 storeys
结构形式 Structure
砖混钢木结构 / Mixed masonry structure, wood and steel truss

p212
湖南城陵矶综合保税区通关服务中心
Hunan Chenglingji Free Trade Zone Customs Clearance Service Center

地点 Location
湖南省岳阳市云溪区云港路 / Yungang Road, Yunxi District, Yueyang City, Hunan Province
业主 Client
湖南城陵矶临港产业新区开发投资有限公司 / Chenglingji Hunan New Port Area Invests & Development Co., Ltd.
设计团队 Project Team
[建筑] 苏昶 谭春晖 金欢 陈迪 沈逸斐
[结构] 刘宏欣 马泽峰
[机电] 干红 蒋明 徐雪芳
施工单位 Constructor
五矿二十三冶建设集团有限公司
设计时间 Design Period
2015.01 – 2015.12
建造时间 Construction Period
2016.01 – 2017.02
基地面积 Site Area
36 935 m²
建筑面积 Floor Area
49 250 m²
层数 Storey
地上6层, 地下1层 / 6 storeys overground, 1 storey underground
结构形式 Structure
框架结构 / Frame structure

p214
湖上村舍
The House by the Lake

地点 Location
江苏省苏州市澄林路88号 / No.88, Chenglin Street, Suzhou City, Jiangsu Province
业主 Client
苏州春上文旅发展有限公司 / Chun Shang Culture & Tourism Development Ltd.
设计团队 Project Team
[建筑] 王斌 谢选集 刘凯强
[结构] 李锋清
[设备] 顾晓峰
施工单位 Constructor
苏州七鑫工程有限公司
设计时间 Design Period
2015.05 – 2015.09
建造时间 Construction Period
2015.11 – 2016.10
基地面积 Site Area
2 000 m²
建筑面积 Floor Area
1 680 m²
层数 Storey
地上2层 / 2 storeys
结构形式 Structure
钢筋混凝土结构 / Reinforced concrete structure

p216
石塘互联网会议中心
Shitang Village Internet Conference Center

地点 Location
江苏省南京市江宁区石塘村 / Shitang Village, Jiangning District, Nanjing City, Jiangsu Province
业主 Client
南京市江宁区人民政府 / The Government of Jiangning District, Nanjing
合作单位 Cooperative Design
结构 /Structure: 上海同基钢结构技术有限公司 / Shanghai Tongji Steel Structure Technology Ltd.
设计团队 Project Team
[建筑] 钟华颖 张雷 戚威 席弘
[结构] 袁鑫
施工单位 Constructor
上海康业建筑装饰工程有限公司
设计时间 Design Period
2016.07 – 2016.08
建造时间 Construction Period
2016.07 – 2016.10
基地面积 Site Area
2 376 m²
建筑面积 Floor Area
3 000 m²
层数 Storey
地上2层 / 2 storeys
结构形式 Structure
钢木混合结构 / Steel and wood structure

p218
上海中心大厦
Shanghai Tower

地点 Location
上海市浦东新区银城中路501号 / No.501, Middle Yincheng Street, Pudong New District, Shanghai City
业主 Client
上海中心大厦建设发展有限公司 / Shanghai Tower Construction and Development Co., Ltd.
设计团队 Project Team
[工程总负责人] 丁洁民 任力之
[项目经理] LI Xiaomei 陈继良
[项目协调] 周瑛 胡宇滨 周鹏
[建筑] 张洛先 孙晔 张鸿武 张丽萍 高一鹏 谢春 陈奕
[结构] 巢斯 贾坚 万月荣 虞终军 何志军 金炜 赵昕 阮永辉 毛华 张峥 吴宏磊 邹智兵
[给排水] 归谈纯 杨民 龚海宁 张晓燕
[暖通] 王健 刘毅 周谨 张心刚 王希星
[电气] 夏林 包顺强 钱大勋 罗武 廖述龙 施国平
[BIM专业] 张东升 刘建
施工单位 Constructor
上海建工集团
设计时间 Design Period
2008 – 2016
建造时间 Construction Period
2008 – 2016
基地面积 Site Area
30 368 m²
建筑面积 Floor Area
573 223 m²
层数 Storey
地上127层, 地下5层 /127 storeys overground, 5 storeys underground
结构形式 Structure
巨型框架结构 / Mega-frame structure

p220
天赐新能源企业总部
Headquarters of Zhejiang TCI Ecology & New Energy Technology

地点 Location
浙江省嘉善县世纪大道归谷科技园 / Guigu Science Park, Jiashan Country, Jiaxing City, Zhejiang Province
业主 Client
浙江天赐新能源科技有限公司 / Zhejiang TCI Ecology & New Energy Tech Co., Ltd.
设计团队 Project Team
[建筑] 冯路 唐加超 赵青 张雨 杨锦雄
施工单位 Constructor
浙江广地建设工程有限公司
设计时间 Design Period
2011 – 2013
建造时间 Construction Period
2012 – 2016
基地面积 Site Area
6 970 m²
建筑面积 Floor Area
13 991 m²
层数 Storey
地上5层 /5 storeys
结构形式 Structure
钢筋混凝土框架结构 /Reinforced concrete frame structure

p222
南昌绿地紫峰大厦
Jiangxi Nanchang Greenland Zifeng Tower

地点 Location
江西省南昌市 /Nanchang City, Jiangxi Province
业主 Client
南昌绿地申新置业有限公司 / Greenland Group in Nanchang
设计团队 Project Team
华东院总:
[建筑] 亢智毅 董丽丽 刘瑾 叶琪瑚
[结构] 陆道渊 宫伟智
[给排水] 李鸿奎
[暖通] 吴国华
[电气] 钱蓉
SOM:
Michael Pfeffer, Mark Nagis, Yue Zhu, Gregory Smith, Henry Chan 等
施工单位 Constructor
上海建工集团有限公司
设计时间 Design Period
2010.11 – 2012.03
建造时间 Construction Period
2012.03 – 2014.12
基地面积 Site Area
40 022 m²
建筑面积 Floor Area
209 158 m²
层数 Storey
地上56层, 地下2层 / 56 storeys overground, 2 storeys underground
结构形式 Structure
框架 +核心筒结构 / Frame+Core tube structure

p224
大舍西岸工作室
Atelier Deshaus Westbund

地点 Location
上海市徐汇区龙腾大道 /Longteng Avenue, Xuhui District, Shanghai City
业主 Client
大舍建筑设计事务所 / Atelier Deshaus
设计团队 Project Team
[建筑] 柳亦春 陈屹峰 王伟实 王龙海
施工单位 Constructor
同济室内装饰工程有限公司
设计时间 Design Period
2015.01 – 2015.03
建造时间 Construction Period
2015.03 – 2015.09
基地面积 Site Area
550 m²
建筑面积 Floor Area
430 m²
层数 Storey
地上2层 / 2 storeys
结构形式 Structure
砖混 + 轻钢结构 / Brick-concrete structure + Light steel structure

p226
滨江休闲广场商业用房
Binjiang Leisure Plaza and Commercial Housing

地点 Location
江苏省常熟市滨江体育公园东侧 /East of Binjiang Sports Park, Changshu City, Suzhou Province
业主 Client
常熟市滨江城市建设经营投资有限责任公司 / Changshou City Binjiang City Construction and Operation Investment Co., Ltd.
设计团队 Project Team
[建筑] 查金荣 程伟 陈苏琳 金帆 张筠之
[结构] 张志
[机电] 周晓东
施工单位 Constructor
江苏金土木建设集团华星工程有限公司
设计时间 Design Period
2013.11 – 2014.06
建造时间 Construction Period
2014.04 – 2015.04
基地面积 Site Area
7 579m²
建筑面积 Floor Area
11 400m²
层数 Storey
地上3层 /3 storeys
结构形式 Structure
钢筋混凝土框架结构 /Reinforced concrete frame structure

p228
外滩 SOHO
Bund SOHO

地点 Location
上海市黄浦区中山东二路近新开河路 / Zhongshandong Road, Near Xinkaihe Road, Huangpu District, Shanghai City
业主 Client
Soho中国 / SOHO China
设计团队 Project Team
[建筑] 施特凡·胥茨 施特凡·瑞沃勒 乔伟·叶琪卿 鲍威
[结构] 岑伟 陆道渊 黄良 周婷婷 瞿鹏
[给排水] 王珏 江涛 任国栋 王华星 叶俊
[暖通] 魏烨 曹斌 李扬 张琳
[强电] 包昀毅 许士杰 朱安辉 顾佳
[弱电] 瞿二澜
[动力] 吕宁 崔岚
施工单位 Constructor
上海建工一建集团有限公司
设计时间 Design Period
2010.05 – 2013.10
建造时间 Construction Period
2011.01 – 2015.01
基地面积 Site Area
22 462 m²
建筑面积 Floor Area
189 909 m²
层数 Storey
地上31层 /31 storeys
结构形式 Structure
框架 +核心筒混合结构 / Frame+Core tube mixed structure

p230
水西工作室
Shui Xi Studio

地点 Location
天津市南开区宾水西道146号 / No.146, West Binshui Street, Nankai District, Tianjin City
业主 Client
周恺 /ZHOU Kai
设计团队 Project Team
[建筑] 周恺 黄菲
[室内] 张莉兰
[暖通] 朱元

[电气] 曾永捷
[给排水] 田书韦
施工单位 Constructor
天津利丰装饰工程有限公司
设计时间 Design Period
2015.01 – 2015.08
建造时间 Construction Period
2015.07 – 2015.12
基地面积 Site Area
457.5 m²
建筑面积 Floor Area
775 m²
层数 Storey
地上 2层 / 2 storeys
结构形式 Structure
混凝土框架结构 / Concrete frame structure

p232
北京绿地中心
Beijing Greenland Center
地点 Location
北京市朝阳区大望京 / Great Wangjing, Chaoyang District, Beijing City
业主 Client
北京绿地京华置业有限公司 / Greenland Group Beijing
设计团队 Project Team
[建筑] 汪恒 安澎 孟海港
[结构] 范重
[给排水] 王耀堂
[暖通] 刘玉春
[电气] 曹磊
施工单位 Constructor
北京六建集团有限责任公司
设计时间 Design Period
2011.07 – 2013.09
建造时间 Construction Period
2013.07 – 2015.12
基地面积 Site Area
14 670 m²
建筑面积 Floor Area
179 079 m²
层数 Storey
地上 55层,地下 5层 / 55 storeys overground, 5 storeys underground
结构形式 Structure
框架核心筒混合结构 / Frame+Core tube mixed structure

p234
宝龙城市广场集装箱售楼处
Baolong Qingpu Plaza Container Sales Office
地点 Location
上海市青浦区汇金路秀源路 / Huijin Street and Xiuyuan Street, Qingpu District, Shanghai City
业主 Client
宝龙地产控股有限公司 / Powerlong Real Estate Holdings Limited
设计团队 Project Team
[建筑] 刘可南 张旭 孙闻岚 陈骁 李哲元 等
施工单位 Constructor
江苏省第一建筑安装股份有限公司上海分公司
设计时间 Design Period
2014.06 – 2014.07
建造时间 Construction Period
2014.08 – 2014.10
基地面积 Site Area
2 000 m²
建筑面积 Floor Area
270 m²
层数 Storey
地上 1层 / 1 storey
结构形式 Structure
钢结构 / Steel structure

p236
中国商务部驻印尼商务馆舍
Office Building of Chinese Ministry of Commerce in Indonesia
地点 Location
印度尼西亚雅加达特区
南雅加达区 / Jalan Mega Kuningan Barat 7号街 / Jalan Mega Kuningan Barat VII Street, South Jakarta City, Special Capital Region of Jakarta, Indonesia
业主 Client
中华人民共和国商务部 /Ministry of commerce of P.R.C
设计团队 Project Team
单军 刘玉龙 铁雷 赵瞳 孙显 等
施工单位 Constructor
北京城建集团有限责任公司
设计时间 Design Period
2007.05 – 2012.10
建造时间 Construction Period
2012.10 – 2014.08
基地面积 Site Area
3 185 m²
建筑面积 Floor Area
3 160 m²
层数 Storey
地上 7层 / 7 storeys
结构形式 Structure
钢筋混凝土框架结构 /Reinforced concrete frame structure

p238
苏州科技城国家知识产权局苏州中心办公楼
Patent Examination Cooperation Jiangsu Center of the Patent Office, SIPO
地点 Location
江苏省苏州市高新区光启路 88号 / No.88, Guangqi Street, Suzhou New District, Suzhou City, Jiangsu Province
业主 Client
苏州高新软件园有限公司 /Suzhou Software Technology Park Co., Ltd.
设计团队 Project Team
[建筑] 张应鹏 王凡 许宁 陆泓成 王苏嘉
[结构] 苗平洲 王永杰 吴玉英
[暖通] 陈云高
[给排水] 张晓明
[电气] 薛青
施工单位 Constructor
苏州建鑫建设集团有限公司
设计时间 Design Period
2012.06 – 2012.12
建造时间 Construction Period
2012.10 – 2014.01
基地面积 Site Area
28 778 m²
建筑面积 Floor Area
59 536 m²
层数 Storey
地上 6层,地下 1层 / 6 storeys overground, 1 storey underground
结构形式 Structure
混凝土框架结构 /Concrete Frame structure

p240
北方长城宾馆三号楼
NO.3 Building of Northen Great Wall Hotel
地点 Location
北京市昌平区中国兵器工业第二〇八研究所区 / No.208, Research Institute of China Ordnance Industries district, Changping District, Beijing City
业主 Client
中国兵器工业第二〇八研究所 / No.208 Research Institute of China Ordnance Industries

合作单位 Cooperative Design
北京都林国际工程设计咨询有限公司 / Beijing Dulin International Engineering Consulting Corporation
设计团队 Project Team
[建筑] 陈一峰 杨光 尚佳 郭琪伟 何继舜
[室内] 刘烨 饶劢 张晔
[施工图] 日新 郑亚娟
施工单位 Constructor
中国核工业二四建设有限公司
设计时间 Design Period
2011.10 – 2012.04
建造时间 Construction Period
2012.06 – 2014.10
基地面积 Site Area
9 090 m²
建筑面积 Floor Area
5 000 m²
层数 Storey
地上 2层 / 2 storeys
结构形式 Structure
混凝土框架结构 / Concrete frame structure

p242
玉树藏族自治州行政中心
Yushu Tibetan Autonomous Prefecture Administrative Center
地点 Location
青海省玉树藏族自治州结古镇民主路 / Minzhu Road, Jiegu Town, Yushu Tibetan Autonomous Prefecture, Qinghai Province
业主 Client
玉树州三江源投资建设有限公司 / Yushu Sanjiangyuan Construction Investment Co., Ltd.
设计团队 Project Team
[建筑] 庄惟敏 张维 姜魁元 龚佳振 屈张
[结构] 汤鹿 张涛 刘忆川
[电气] 徐华 杨莉
[暖通] 张菁华 米忠
[给排水] 罗新宇
施工单位 Constructor
中国建筑第八工程局有限公司
设计时间 Design Period
2010.08 – 2012.05
建造时间 Construction Period
2012.07 – 2014.11
基地面积 Site Area
63 278 m²
建筑面积 Floor Area
72 638 m²
层数 Storey
地上 12层,地下 1层 / 12 storeys overground, 1 storey underground
结构形式 Structure
框架剪力墙结构 / Frame-shear wall structure

p244
西村大院
West Village
地点 Location
四川省成都市青羊区贝森北路 1号 / No.1, North Basis Street, Qingyang District, Chengdu City, Sichuan Province
业主 Client
四川迈伦实业有限责任公司 / Sichuan Myron Industrial Co., Ltd.
设计团队 Project Team
[主持] 刘家琨
[设计] 杨磊 靳洪铎 刘速 杨鹰 蔡克非 等
施工单位 Constructor
中国华西企业股份有限公司
设计时间 Design Period
2009.12 – 2013.03
建造时间 Construction Period
2013.11 – 2014.09
基地面积 Site Area
41 863 m²

建筑面积 Floor Area
135 552 m²
层数 Storey
地上 5层,地下 2层 / 5 storeys overground, 2 storeys underground
结构形式 Structure
钢筋混凝土框架结构 /Reinforced concrete frame structure

p246
成都远洋太古里
Sino-Ocean Taikoo Li, Chengdu
地点 Location
四川省成都市锦江区中纱帽街 8号 / No.8, Middle Shamao Street, Jinjiang District, Chengdu City, Sichuan Province
业主 Client
太古地产有限公司、远洋地产控股有限公司 /Swire Properties Limited, Sino-Ocean Land Holdings Limited
合作单位 Cooperative Design
[施工图,结构,机电] 中国建筑西南设计研究院有限公司 / China Southwest Architecture Design and Research Institute Corp. Ltd.
[景观] 雅邦规划设计有限公司 / Urbis Limited
[商场地下一层室内设计] Spawton Architecture Ltd. & Elena Gali Giallini Ltd.
设计团队 Project Team
[建筑] 郝琳 区美德 郭兆荣 黄颂威 李荣昊 等
施工单位 Constructor
中国建筑第三工程局有限公司
设计时间 Design Period
2008 – 2014
建造时间 Construction Period
2011– 2014
基地面积 Site Area
74 000 m²
建筑面积 Floor Area
114 000 m²
层数 Storey
地上 2-3层,地下 3层 / 2-3 storeys overground, 3 storeys underground
结构形式 Structure
钢结构 / Steel structure

p248
凌空 SOHO
Sky SOHO
地点 Location
上海市虹桥临空经济园区 / Linkong Economic Park of Hongqiao, Shanghai City
业主 Client
搜候（上海）投资有限公司 /SOHO China
设计团队 Project Team
[建筑] 刘恩芳 Satoshi Ohashi Patrik Schumacher 姜世峰 周燕
[结构] 李亚明 石硕
[给排水] 包虹 陆文慷
[电气] 万洪 胡戎
[暖通] 于红
施工单位 Constructor
中国建筑第八工程局有限公司
设计时间 Design Period
2011.02 – 2012.05
建造时间 Construction Period
2012.05 – 2014.10
基地面积 Site Area
86 164 m²
建筑面积 Floor Area
343 447 m²
层数 Storey
地上 11层,地下 2层 /11 storeys overground, 2 storeys underground
结构形式 Structure
钢筋混凝土框架 – 抗震墙结构 / Reinforced concrete frame and seismic wall structure

p250
尤努斯中国中心陆口格莱珉乡村银行
Yunus China Center Lukou Grameen Village Bank

地点 Location
江苏徐州窑湾镇陆口村尹庄 / Yinzhuang Lukou Village, Yaowan Town, Xuzhou City, Jiangsu Province
业主 Client
格莱珉中国银行 / Yunus China Center Lukou Grameen Village Bank
设计团队 Project Team
[建筑] 朱竞翔 夏珩 韩国日
[景观] 赵妍
施工单位 Constructor
沂尹庄 20位村民
设计时间 Design Period
2014.10 – 2014.11
建造时间 Construction Period
2014.11 – 2014.12
基地面积 Site Area
95 m²
建筑面积 Floor Area
220 m²
层数 Storey
地上 2层 / 2 storeys
结构形式 Structure
C型轻钢骨架与填充板材形成的复合结构 / LGS skeleton strengthened by Rigid Board

p252
凤凰中心
Phoenix Center

地点 Location
北京市朝阳区朝阳公园南路 3号 / No.3 South Chaoyang Park Street, Chaoyang District, Beijing City
业主 Client
凤凰卫视 / Phoenix Satellite Television(East) Beijing
设计团队 Project Team
[建筑] 邵韦平 刘宇光 陈颖 周凯湦 吴锡
[结构] 束伟农 朱忠义 周思红 张世忠 王毅
[设备] 张铁辉 杨扬 钱强 刘均
[电气] 孙成群 金红 郑波
[总图] 吕娟
施工单位 Constructor
北京天润建设有限公司
设计时间 Design Period
2007.09 – 2013.07
建造时间 Construction Period
2008.09 – 2013.07
基地面积 Site Area
18 821.83 m²
建筑面积 Floor Area
72 478 m²
层数 Storey
地上 11层,地下 3层 / 11 storeys overground, 3 storeys underground
结构形式 Structure
外壳 / Shell:钢结构 / Steel structure;
内部 / Interior:框架剪力墙 / Frame shear wall structure

p254
上海虹桥国际机场T1航站楼改造及交通中心工程
Shanghai Hongqiao International Airport T1 Renovation & Traffic Center

地点 Location
上海市长宁区迎宾一路 699号 / No.699, Yingbin First Street, Changning District, Shanghai City
业主 Client
上海机场(集团)有限公司 / Shanghai Airport (Group) Company, Ltd.
设计团队 Project Team
[建筑] 郭建祥 张宏波 吕程 孙芸
[结构] 周健 张耀康
[机电] 陆燕 马海渊
施工单位 Constructor
上海建工集团
设计时间 Design Period
2012.02 – 2015.12
建造时间 Construction Period
2014.11 – 2016.10
基地面积 Site Area
103 277 m²
建筑面积 Floor Area
203 746 m²
层数 Storey
地上 4层,地下 2层 / 4 storeys overground, 2 storeys underground
结构形式 Structure
钢筋混凝土框架结构 / Reinforced concrete frame structure

p256
上海东方肝胆医院
Shanghai Oriental Hepatic Hospital

地点 Location
上海市嘉定区墨玉北路 700号 / No. 700, North Moyu Street, Jiading District, Shanghai City
业主 Client
中国人民解放军第二军医大学东方肝胆外科医院 / Dongfang Department of Hepatobiliary Surgery Hospital of The Second Military Medical University of PLA
设计团队 Project Team
[建筑] 陈国亮 邵宇卓 唐茜嵘 华君良
[结构] 周雪雁 周宇庆 李敏华 杨洋
[给排水] 朱建荣 张隽
[暖通] 朱学锦 朱喆
[电气] 朱文 刘兰 康辉
施工单位 Constructor
上海建工一建集团有限公司
设计时间 Design Period
2008.12 – 2013.07
建造时间 Construction Period
2013.08 – 2015.10
基地面积 Site Area
94 825 m²
建筑面积 Floor Area
180 576 m²
层数 Storey
地上 13层,地下 2层 / 13 storeys overground, 2 storeys underground
结构形式 Structure
框架剪力墙结构 / Frame shear-wall structure

p258
昂洞卫生院
Angdong Hospital

地点 Location
湖南省保靖县千陵镇昂洞村 / Angdong Village, Qianling Town, Baojing County, Hunan Province
业主 Client
昂洞卫生院 / Angdong Health Bureau
设计团队 Project Team
[主持] Joshua Bolchover 林君翰
[项目] 马洁怡
[建筑] Mark Kingsley Jeffery Huang 关帼盈 黄稚沄 梁卓嘉
施工单位 Constructor
保靖县第一工程公司
设计时间 Design Period
2011.05 – 2012.06
建造时间 Construction Period
2013.02 – 2014.08
基地面积 Site Area
756 m²
建筑面积 Floor Area
1 450 m²
层数 Storey
地上 3层,地下 1层 / 3 storeys overground, 1 storey underground
结构形式 Structure
混凝土结构 / Concrete structure

p260
太原南站
Taiyuan South Railway Station

地点 Location
山西省太原市小店区农科北路 / North Nongke Street, Xiaodian District, Taiyuan City, Shanxi Province
业主 Client
太原铁路局 / Taiyuan Railway Administration
设计团队 Project Team
[建筑] 李春舫 王力 张继 孙行 陈勇
[结构] 周德良 曹登武 张卫 李功标
[机电] 秦晓梅 冯星明 马友才 刘华斌
施工单位 Constructor
中铁北京工程局集团有限公司
设计时间 Design Period
2006.01 – 2010.06
建造时间 Construction Period
2009.05 – 2014.06
基地面积 Site Area
166 500 m²
建筑面积 Floor Area
183 952 m²
层数 Storey
地上 2层,地下 1层 / 2 storeys overground, 1 storey underground
结构形式 Structure
钢结构 / Steel structure

p262
都江堰大熊猫救护与疾病防控中心
Dujiangyan Giant Panda Conservation and Disease Control Center

地点 Location
四川省都江堰市青城山镇石桥村 / Shiqiao Village, Qingchenshan Town, Dujiangyan City, Sichuan Province
业主 Client
四川卧龙国家级自然保护区管理局 / Wolong Sichuan National Nature Reserve Administration
设计团队 Project Team
[建筑] 钱方 茅锋 刘磊 胡佳 刘晓伟 等
[绿建] 戎向阳 高庆龙
[结构] 吴小宾 熊耀清
[电气] 李先进 何海波
[暖通] 杨玲 陈英杰
[给排水] 李波 杜欣
施工单位 Constructor
华北建设集团有限公司
设计时间 Design Period
2009.06 – 2010.07
建造时间 Construction Period
2010.09 – 2013.12
基地面积 Site Area
505 123.67 m²
建筑面积 Floor Area
12 398.83 m²
层数 Storey
地上 1层,局部 2-3层 / 1 storey overground, partly 2-3 storeys
结构形式 Structure
钢框架 + 加气混凝土砌体结构 + 砖混结构 / Steel frame + Aerated concrete masonry structure + Brick-concrete structure

改造及修复
Renovation Heritage Preservation

p266
七园居
The Hotel of Septuor

地点 Location
浙江省湖州市德清县莫干山镇对河口村西岑坞石门坑 / Shimen Pit, Xicen Cove, Duihekou Village, Moganshan Town, Deqing County, HuzhouCity, Zhejiang Province
业主 Client
私人 / Private
设计团队 Project Team
王方戟 董晓 肖潇 张婷 陈长山 等
施工单位 Constructor
曹建珍施工队
设计时间 Design Period
2015.05 – 2017.02
建造时间 Construction Period
2015.10 – 2017.02
基地面积 Site Area
950 m²
建筑面积 Floor Area
645 m²
层数 Storey
地上 2层 / 2 storeys
结构形式 Structure
木结构(保留)+钢筋混凝土(新建) / Timber Structure (reserved) + Reinforced concrete structure (new)

p268
五龙庙环境整治工程
The Environmental Upgrade of the Five Dragons Temple

地点 Location
山西省运城市芮城县中龙泉村 / Middle Dragon Spring Village, Ruicheng County, Yuncheng City, Shanxi Province
业主 Client
万科企业股份有限公司,芮城县旅游文物局 / China Vanke Co.,Ltd.; The Tourism Cultural Relics Bureau of Ruicheng City
合作单位 Cooperative Design
[施工图] 清华大学建筑设计研究院有限公司 / Architectural Design and Research Institute of Tsinghua University Co., Ltd.
loma 陆玛景观规划设计有限公司 / Loma Landscape Co., Ltd.
设计团队 Project Team
[建筑] 王辉 邹德华 杜爱宏
施工单位 Constructor
太原播翠园林绿化工程有限公司
设计时间 Design Period
2013 – 2015
建造时间 Construction Period
2015 – 2016
基地面积 Site Area
5 838 m²
建筑面积 Floor Area
267 m²
层数 Storey
地上 1层 / 1 storey
结构形式 Structure
框架结构 / Frame construction

p270
四叶草之家
Clover House

地点 Location
日本爱知县冈崎市 / Okazaki, Aichi, Japan
业主 Client
奈良健太郎,奈良珠纪 / Kentaro Nara, Tamaki Nara

设计团队 Project Team
马岩松 早野洋介 党群 米津孝祐
李悠焕 等
施工单位 Constructor
Kira Construction INC
设计时间 Design Period
2012 – 2016
建造时间 Construction Period
2012 – 2016
基地面积 Site Area
283 m²
建筑面积 Floor Area
300 m²
层数 Storey
地上 2层 / 2 storeys
结构形式 Structure
木结构 / Timber structure

p272
乌镇北栅丝厂改造
Renovation of Beizha Silk Factory in Wuzhen
地点 Location
浙江省桐乡市乌镇环河路 / Huanhe Street, Wuzhen Town, Tongxiang City, Zhejiang Province
业主 Client
乌镇旅游股份有限公司 / Wuzhen Tourism Co., Ltd.
设计团队 Project Team
[建筑] 陈强 付娜 陈剑如 郑英玉
施工单位 Constructor
桐乡市乌镇内利仿古园林建设有限公司
设计时间 Design Period
2015.06 – 2015.11
建造时间 Construction Period
2015.11 – 2016.03
基地面积 Site Area
16 900 m²
建筑面积 Floor Area
8 502 m²
层数 Storey
地上 2层 / 2 storeys
结构形式 Structure
砖混结构 + 钢结构 + 钢混屋架 + 木屋架 / Brick-concrete structure + Steel structure + Reinforced concrete roof + Wooden roof

p274
五原路工作室
Wuyuan Rd. Studio
地点 Location
上海市徐汇区五原路 281弄 9号 / No.9, Lane 281, Wuyuan Road, Shanghai City
业主 Client
刘宇扬建筑事务所 / Atelier Liu Yuyang Architects
设计团队 Project Team
刘宇扬 曹飞乐 王珂一 陈芸菲
施工单位 Constructor
上海卓浩装饰工程有限公司
设计时间 Design Period
2014 – 2016
建造时间 Construction Period
2015 – 2016
基地面积 Site Area
355 m²
建筑面积 Floor Area
400 m²
层数 Storey
地上 2层 /2 storeys
结构形式 Structure
砖混 + 钢结构加固 / Partial concrete frames + Load-bearing brick walls + Steel beam reinforcement

p276
西浜村昆曲学社
The Society of Kun Opera at Xibang Village
地点 Location
江苏省昆山市巴城镇西浜村 / Xibang Village, Bacheng Town, Kunshan City, Jiangsu Province
业主 Client
昆山城市建设投资发展集团有限公司 / Kunshan City Construction Investment Development Group Co., Ltd.
设计团队 Project Team
[建筑] 崔愷 郭海鞍 沈一婷
[景观] 冯磊
[结构] 何相宇
[暖通] 王加
[给排水] 安明阳
[电气] 胡思宇
施工单位 Constructor
昆山经济技术开发区联合建筑工程有限公司, 苏州苏园古建园林有限公司
设计时间 Design Period
2014.09 – 2015.05
建造时间 Construction Period
2015.12 – 2016.09
基地面积 Site Area
2 775 m²
建筑面积 Floor Area
1 405.81 m²
层数 Storey
地上 2层 / 2 storeys
结构形式 Structure
钢结构 / Steel structure

p278
乡宿上泗安
Shangsi'an Cottage
地点 Location
浙江省湖州市长兴县上泗安村 / Shangsi'an Village, Changxing County, Huzhou City, Zhengjiang Province
业主 Client
杭州隐居乡创投资管理有限公司 / Hangzhou Seclusive Genealand Investment Mangement Co., Ltd.
设计团队 Project Team
[建筑] 范蓓蕾 孔锐 薛喆 章迅 陈晓艺
[结构] 张准
[机电] 李亮 吴强 马祥胜 卫曦超 陈俊涛
施工单位 Constructor
上海中圣建设工程有限公司
设计时间 Design Period
2015.05 – 2015.12
建造时间 Construction Period
2015.12 – 2016.02
基地面积 Site Area
1 608 m²
建筑面积 Floor Area
1 097 m²
层数 Storey
地上 2层 / 2 storeys
结构形式 Structure
混合结构 / Mixed structure

p280
池社
Chi She
地点 Location
上海市徐汇滨江龙腾大道 2555号 4栋 / Building D, No.2555, Longteng Avenue, Xuhui District, Shanghai City
业主 Client
池社 / Chi She
设计团队 Project Team
[建筑] 袁烽 韩力 孔祥平 朱天睿 刘秦榕
[室内] 王徐伟
[结构] 沈骏超
施工单位 Constructor
上海一造建筑智能工程有限公司
设计时间 Design Period
2016.02 – 2016.03
建造时间 Construction Period
2016.04 – 2016.09
基地面积 Site Area
164 m²
建筑面积 Floor Area
199 m²
层数 Storey
地上 2层 /2 storeys
结构形式 Structure
砖混结构 / Brick-concrete structure

p282
四行仓库修缮工程
Protection and Restoration of the Joint Trust Warehouse
地点 Location
上海市闸北区光复路 1号、21号 / No.1 & No.21, Guangfu Street, Zhabei District, Shanghai City
业主 Client
百联集团置业有限公司 / Shanghai Bailian Real Estate Co., Ltd.
设计团队 Project Team
[建筑] 唐玉恩 邹勋 刘寄珂 邱致远 吴胄婧
[结构] 张坚 英明 刘桂 苏朝阳 范沈龙
[暖通] 干红 陈叶青
[电气] 汪海良 马立果 高笠源
[给排水] 徐雪芳 周海山 赵旻
[绿建] 孙斌
施工单位 Constructor
上海建工五建集团有限公司
设计时间 Design Period
2014.06 – 2015.11
建造时间 Construction Period
2014.10 – 2016.06
基地面积 Site Area
4 558 m²
建筑面积 Floor Area
25 550 m²
层数 Storey
地上 6层 /6 storeys
结构形式 Structure
主体: 钢筋混凝土无梁楼盖结构 / 局部: 钢筋混凝土框架结构 / Main part:Reinforced concrete slab-column frame structure + Partial: Reinforced concrete frame structure

p284
谢店村传统村落保护与再生规划设计
Planning of Xiedian Traditional Village Protection and Regeneration
地点 Location
湖北省麻城市谢店村 / Xiedian Village, MaCheng City, Hubei Province
业主 Client
麻城市宋埠镇谢店古村 / Xiedian Village, Songbu Town, Macheng City
设计团队 Project Team
[建筑] 肖伟 何欣然 岑伟强
[传统村落保护] 杨青松 吴航
施工单位 Constructor
湖北钰丰苑园林景观有限公司
设计时间 Design Period
2015.07 – 2016.01
建造时间 Construction Period
2016.01 – 2016.09
基地面积 Site Area
73 260 m²
建筑面积 Floor Area
9 400 m²
层数 Storey
地上 2层 /2 storeys
结构形式 Structure
砖混结构 / Brick-concrete structure

p286
天目湖微酒店
Tianmu Lake VIP Club
地点 Location
江苏省溧阳市天目湖旅游度假区环湖西路 8号 / No.8, West Lake Street, Tianmu Lake Resort, Liyang City, Jiangsu Province
业主 Client
西本新干线溧阳有限公司 / Xiben New Line Stock Co., Ltd.
设计团队 Project Team
[建筑] 程明性 张萌 王维
[结构] 包佐
[机电] 吴泉
施工单位 Constructor
江苏溧阳城建集团有限公司
设计时间 Design Period
2013.09 – 2014.07
建造时间 Construction Period
2014.01 – 2016.01
基地面积 Site Area
7 385.2 m²
建筑面积 Floor Area
8 998.13 m²
层数 Storey
地上 2层, 地下 1层 /2 storeys overground, 1 storey underground
结构形式 Structure
钢筋混凝土结构 / Reinforced concrete structure

p288
隐居江南精品酒店
Seclusive Jiangnan Boutique Hotel
地点 Location
浙江省杭州市大兜路历史街区 188,190号 / No.188 & 190, Historic District, Dadou Street, Hangzhou City, Zhejiang Province
业主 Client
浙江隐居集团有限公司 / Zhejiang Seclusion Group Co., Ltd.
合作单位 Cooperative Design
GFD广飞室内设计 / GFD Interior Designs
设计团队 Project Team
[建筑] 孟凡浩 李昕光
[结构] 任光勇 汪小娣
[机电] 崔大梁 吴文坚 陆柏庆
[室内] 叶飞 赵春光
施工单位 Constructor
浙江东邦装饰设计工程有限公司
设计时间 Design Period
2016.01 – 2016.04
建造时间 Construction Period
2016.05 – 2016.12
基地面积 Site Area
1 237 m²
建筑面积 Floor Area
2 816 m²
层数 Storey
地上 4层 /4 storeys
结构形式 Structure
混凝土框架结构 / Concrete frame structure

p290
北京民生现代美术馆
Minsheng Museum of Modern Art
地点 Location
北京市朝阳区酒仙桥北路 9号 / No. 9, North Jiuxiangjiao Street, Chaoyang District, Beijing City
业主 Client
中国民生银行 / China Minsheng Banking Corp., Ltd.
设计团队 Project Team
[建筑] 朱锫 Edwin Lam 何帆 Virginia Melnyk 柯军 等
施工单位 Constructor
上海建工七建集团有限公司
设计时间 Design Period
2011.05 – 2012.04
建造时间 Construction Period
2014.04 – 2015.06
基地面积 Site Area
20 336 m²
建筑面积 Floor Area
32 910 m²
层数 Storey
地上 3层 /3 storeys
结构形式 Structure
混凝土 + 钢结构 / Steel-concrete structure

p292
首都电影院装修改造
Preservation and Reparation of the Capital Cinema
地点 Location
北京市西城区天桥南大街 3号楼 / No.3 building, South Tianqiao Street, Xicheng District, Beijing City
业主 Client
北京天桥盛世投资集团有限责任公司 / Beijing Tianqiao Zenith Investment Group Co., Ltd.
设计团队 Project Team
[建筑] 郭骏 刘鼎纳 刘洪涛 韩莉
[结构] 柴万先 桂平 赵兴忠
[给排水] 赵薇
[暖通] 杨红
[电气] 王琛
施工单位 Constructor
中国建筑第二工程局有限公司
设计时间 Design Period
2014.09 – 2015.06
建造时间 Construction Period
2015.07 – 2016.02
基地面积 Site Area
743 m²
建筑面积 Floor Area
2 232 m²
层数 Storey
地上 2层, 地下1层 /2 storeys overground, 1 storey underground
结构形式 Structure
钢筋混凝土框架结构 /Reinforced concrete frame structure

p294
居住集合体 L
Housing L
地点 Location
山东省莱阳市马山路 /Mashan Street, Laiyang City, Shandong Province
业主 Client
山东成功路投资担保有限公司 / Shandong Chenggong Road Investment Guarantee Co., Ltd.
合作单位 Cooperative Design
莱阳设计院 (水暖电) / Laiyang Design Institute
设计团队 Project Team
徐千禾 梁幸
施工单位 Constructor
艺铭工程公司
设计时间 Design Period
2013.09 – 2013.11
建造时间 Construction Period
2013.11 – 2015.12
基地面积 Site Area
2 704 m²
建筑面积 Floor Area
15 000 m²
层数 Storey
地上 10层 /10 storeys
结构形式 Structure
钢筋混凝土结构 + 钢结构 / Reinforced concrete structure +Steel structure

p296
薇园
Wei Yuan Garden
地点 Location
江苏省南京市江宁区将军路 6号 /No.6, Jiangjun Street, Jiangning District, Nanjing City, Jiangsu Province
业主 Client
南京筑微堂装饰工程有限公司 / Nanjing Zhuweitang Decoration Engineering Co., Ltd.
设计团队 Project Team
[建筑] 葛明 陈洁萍 孔德钟 唐静寅 姜文林 等
[结构] 淳庆
[景观] 陈洁萍
施工单位 Constructor
南京筑微堂装饰工程有限公司
设计时间 Design Period
2012.01 – 2014.01
建造时间 Construction Period
2013.01 – 2015.05
基地面积 Site Area
2 162 m²
建筑面积 Floor Area
2 002 m²
层数 Storey
地上 2层 /2 storeys
结构形式 Structure
钢筋混凝土框架剪力墙结构 +钢筋混凝土排架结构 (经加固与扩建) / Reinforced concrete frame-shear wall structure + Reinforced concrete bracket structure (Strengthened and expanded)

p298
陈化成纪念馆移建改造
Removal Renovation of Chen Huacheng Memorial
地点 Location
上海市宝山区临江公园内 / Riverside Park, Baoshan District, Shanghai City
业主 Client
上海宝山区文物保护管理局 / Baoshan District Shanghai Cultural Relics Protection Administration
合作单位 Cooperative Design
上海源规建筑结构设计事务所 (普通合伙) /Shanghai Wildness Structural Des. Firm Inc. (General Partnership)
设计团队 Project Team
[建筑] 庄慎 任皓 唐煜 朱捷 方夏 等
[结构] 张业巍 李明蔚
施工单位 Constructor
南通华胜建设工程有限公司
设计时间 Design Period
2014
建造时间 Construction Period
2015
基地面积 Site Area
812 m²
建筑面积 Floor Area
198 m²
层数 Storey
地上 1层 /1 storey
结构形式 Structure
混凝土、木、钢混合 /Concrete, wood, steel

p300
竞园 22号楼改造
Jingyuan No.22 Transformation
地点 Location
北京市朝阳区广渠路 3号 /No.3, Guanggu Street, Chaoyang District, Beijing City
业主 Client
北京捷越联合信息咨询有限公司 /Jieyue United
设计团队 Project Team
[建筑] 程艳春 王亚坤 谢秉佑 郑一春 李琳杰
施工单位 Constructor
北京南通启益建设集团有限公司
设计时间 Design Period
2015.01 – 2015.04
建造时间 Construction Period
2015.04 – 2015.07
基地面积 Site Area
300 m²
建筑面积 Floor Area
600 m²
层数 Storey
地上 2层 /2 storeys
结构形式 Structure
钢结构、砖混结构 /Steel structure, brick-concrete structure

p302
富丽服装厂改造
Renovation of Fuli Clothing Factory
地点 Location
上海市杨浦区凤城路 /Fengcheng Street, Yangpu District, Shanghai City
业主 Client
上海同和文化创意产业投资有限公司 /Shanghai Tonghe Cultural Creative Industry Investment Co., Ltd.
设计团队 Project Team
[建筑] 范蓓蕾 孔锐 陈晓艺 薛喆 刘洋
[结构] 张准
[机电] 张正传 文秀琼 李亮 吴强 卫曦超
施工单位 Constructor
上海恒想建设工程有限公司
设计时间 Design Period
2014.09 – 2015.01
建造时间 Construction Period
2015.01 – 2015.12
基地面积 Site Area
3 275 m²
建筑面积 Floor Area
8 600 m²
层数 Storey
地上 5层 /5 Storeys
结构形式 Structure
预装配式混凝土结构 /Pre-assembled concrete structure

p304
上海联创国际设计谷
Shanghai UDG International Design Valley
地点 Location
上海市杨浦区控江路 1500弄 / 1500th Alley, Kongjiang Street, Yangpu District, Shanghai City
业主 Client
UDG联创国际设计集团 / United Design Group Co., Ltd.
设计团队 Project Team
[建筑] 钱强 迟晓昱 常视 陈宁 蔡哲理
施工单位 Constructor
深圳市洪涛装饰股份有限公司
设计时间 Design Period
2014.04 – 2014.08
建造时间 Construction Period
2014.08 – 2015.06
基地面积 Site Area
19 122 m²
建筑面积 Floor Area
16 987 m²
层数 Storey
主楼 1栋 3层, 小楼 8栋 2层, 地下 1层 / One three-storied main building, eight two-storied assistant building, one level beneath the ground
结构形式 Structure
原有建筑 /Original building: 钢筋混凝土结构 / Reinforced concrete structure
加建改建 /Reconstruction: 钢结构 /Steel structure

p306
上海延安中路 816号修缮项目——解放日报社
Renovation Project of 816# Middle Yan'an Road: the Jiefang Daily Office
地点 Location
上海市静安区延安中路 816号 /No. 816, Middle Yan'an Road, Jingan District, Shanghai City
业主 Client
上海文新经济发展有限公司 /
设计团队 Project Team
[建筑] 章明 张姿 肖镭 席伟东 冯珊珊 等
[结构] 徐伟栋
施工单位 Constructor
天津市建工工程总承包有限公司、天津天一建设集团有限公司
设计时间 Design Period
2014.11 – 2015.08
建造时间 Construction Period
2015.01 – 2015.11
基地面积 Site Area
3 816 m²
建筑面积 Floor Area
5 370 m²
层数 Storey
地上 5层 /5 storeys
结构形式 Structure
框架结构 +砖混结构 /Frame structure + Brick-concrete structure

p308
大理慢屋·揽清度假酒店
Dali Munwood Lakeside Resort Hotel
地点 Location
云南省大理市环海西路葭蓬村 / Jiapeng Village, West Street around, Erhai Lake, Dali City, Yunnan Province
业主 Client
重庆慢屋酒店管理有限公司 / Chongqing Munwood Hotel Management Co., Ltd.
设计团队 Project Team
[建筑] 苏云锋 陈俊 宗德新 李舸
[室内] 邓陈
[结构] 李超 陈功 李元初
施工单位 Constructor
当地农民施工队
设计时间 Design Period
2013.03 – 2014.09
建造时间 Construction Period
2013.08 – 2015.06
基地面积 Site Area
820 m²
建筑面积 Floor Area
1 000 m²
层数 Storey
地上 3层 /3 storeys
结构形式 Structure
钢筋混凝土框架结构 +钢结构 / Reinforced concrete frame structure+Steel structure

p310
徐家汇观象台修缮工程
Refurbishment of L'Observatoire de ZI-KAWEI, Xuhui, Shanghai
地点 Location
上海徐汇区蒲西路 166号 /No.166, Puxi Street, Xuhui District, Shanghai City
业主 Client
上海市气象局 /Shanghai Meteorological Bureau
设计团队 Project Team
[建筑] 张斌 周蔚 金燕琳 刘昱 胡丽瑶 等
施工单位 Constructor
上海建筑装饰 (集团) 有限公司
设计时间 Design Period
2013.10 – 2014.09
建造时间 Construction Period
2014.12 – 2015.10
基地面积 Site Area
833 m²
建筑面积 Floor Area
2 986 m²
层数 Storey
地上 3层, 局部 4层 /3 storeys, partly 4 storeys
结构形式 Structure
砖木结构 +局部混凝土框架和钢结构 /Masonry-timber structure + Partly reinforced concrete frame and steel construction

p312
上海油雕院美术馆及咖啡厅
SPSI Art Museum & Chimney Cafe

地点 Location
上海市长宁区金珠路111号 / No.111, Jinzhu Street, Changning District, Shanghai City

业主 Client
上海油画雕塑院·大烟囱(上海)文化发展有限公司 /Shanghai Oil-painting & Sculpture Institute, Chimney (Shanghai) Culture Development Co., Ltd.

设计团队 Project Team
[建筑] 王彦 王一博
[结构] 邵宏政
[设备] 章琳杰
[室内] 王庐浩

施工单位 Constructor
上海普宏建设工程有限公司

设计时间 Design Period
美术馆 / Museum :2007 - 2008;
咖啡厅 / café: 2015

建造时间 Construction Period
美术馆 / Museum:2008 - 2010;
咖啡厅 / café:2015

基地面积 Site Area
美术馆 / Museum:1 000 m²;
咖啡厅 / café: 60 m²

建筑面积 Floor Area
美术馆 / Museum:2 400 m²;
咖啡厅 / café: 60 m²

层数 Storey
美术馆 / Museum:地上3层 / 3 storeys;咖啡厅 / café:地上1层 / 1 storey

结构形式 Structure
美术馆 / Museum:混凝土框架结构 / Concrete frame structure;咖啡厅 / café:钢结构 / Steel structure

p314
桐庐莪山畲族乡先锋云夕图书馆
Tonglu Librairie Avant-Garde, Ruralisation Library

地点 Location
浙江省杭州市桐庐县莪山畲族乡戴家山村 /Daijiashan Village, Eshan She Nationality, Tonglu Town, Hangzhou City, Zhejiang Province

业主 Client
桐庐午夜文化创意有限公司 / Tonglu Midnight Cultural Creativity Co., Ltd.

合作单位 Cooperative Design
南京甲骨文空间设计有限公司 / ISO Workshop

设计团队 Project Team
[建筑] 张雷 戚威 刘玮 邵璇
[室内] 马海依
[景观] 陈隽隽

施工单位 Constructor
地方工匠

设计时间 Design Period
2015.05 - 2015.07

建造时间 Construction Period
2015.08 - 2015.10

基地面积 Site Area
600 m²

建筑面积 Floor Area
260 m²

层数 Storey
地上2层 /2 storeys

结构形式 Structure
砖石 + 土木 / Masonry + Rammed earth wood

p316
广元千佛崖摩崖造像保护建筑试验段工程
Experimental Protective Structure for Thousand Buddha Cliff

地点 Location
四川省广元市利州区嘉陵江东岸千佛崖风景区 /Qianfoya Scene Region, East Bank of Jialing River, Lizhou District, Guangyuan City, Sichuan Province

业主 Client
广元市千佛崖石刻艺术博物馆 /The Art Museum of Thousand Buddha Cliff Grottoes, Guangyuan

设计团队 Project Team
[建筑] 崔光海 安心 汪静 李京 揭小凤
[结构] 马智刚

施工单位 Constructor
重庆盛煌建筑工程有限公司

设计时间 Design Period
2011.03 - 2013.03

建造时间 Construction Period
2014.11 - 2015.05

基地面积 Site Area
470 m²

建筑面积 Floor Area
388.5 m²

层数 Storey
地上2层 / 2 storeys

结构形式 Structure
空间钢桁架结构 / Space steel truss structure

p318
摩梭家园
Homeland of Mosuo

地点 Location
四川省盐源县泸沽湖 / Lake Lugu , Yanyuan County,Sichuan Province

业主 Client
盐源县人民政府 / Yanyuan People's Government

设计团队 Project Team
[建筑] 张远平 郑欣 马俊 陈渊 陈贞妍

施工单位 Constructor
当地村民

设计时间 Design Stage
2014.03 - 2014.07

建造时间 Construction Stage
2015.03 - 2015.07

层数 Storey
地上3层 / 3 storeys

结构形式 Structure
木结构 / Timber Structure

p320
上海电子工业学校六号楼 / 学生浴室
Block 6 of Shanghai Electronic Industry School/Student Shower Block

地点 Location
上海市闵行区剑川路 910号 / No.910, Jianchuan Street, Minhang District, Shanghai City

业主 Client
华鑫置业·上海电子信息职业技术学院 /China Fortune Properties; Shanghai Technical Institute of Electronics and Information

设计团队 Project Team
[建筑] 冯路 赵青 姜宇
[结构] 吉峰

施工单位 Constructor
上海建工七建集团

设计时间 Design Stage
2011 - 2013

建造时间 Construction Stage
2012 - 2014

基地面积 Site Area
425 m²

建筑面积 Floor Area
503 m²

层数 Storey
地上2层 /2 storeys

结构形式 Structure
钢筋混凝土框架结构 /Reinforced concreate frame structure

p322
牛背山志愿者之家
Cattle Back Mountain Volunteer House

地点 Location
四川省甘孜藏族自治州泸定县蒲麦地村 /Pumaidi Village, Luding County, Canti Tibetan Autonomous Prefecture, SiChuan Province

业主 Client
蒲麦地志愿者 /Volunteer

设计团队 Project Team
[主持] 李道德
[建筑] 郑钰 陈昱 曹博 等

施工单位 Constructor
北京碧海怡景园林绿化有限公司

设计时间 Design Period
2014.06 - 2014.08

建造时间 Construction Period
2014.08 - 2014.11

基地面积 Site Area
150 m²

建筑面积 Floor Area
300 m²

层数 Storey
地上3层 / 3 storeys

结构形式 Structure
木结构 / Timber structure

p324
西河粮油博物馆及村民活动中心
Xihe Cereals and Oils Museum and Villagers' Activity Center

地点 Location
河南省信阳市新县西河村大湾 / Xihe Village Great Bay, Xin County, Xinyang City, Henan Province

业主 Client
西河村村民合作社 /Villager Cooperative of Xihe Village

设计团队 Project Team
[建筑] 何崴 陈龙 齐洪海 韩晓伟
[平面] 夏博洋
[展陈] 赵卓然

施工单位 Constructor
西河村村民合作社

设计时间 Design Period
2013.07 - 2014.09

建造时间 Construction Period
2014.06 - 2014.09

基地面积 Site Area
3 760 m²

建筑面积 Floor Area
1 532 m²

层数 Storey
地上1层 /1 storey

结构形式 Structure
砖混、木构、钢结构 /Brick structure, wood structure and steel structure

p326
箭厂胡同文创空间
Arrow Factory Media & Culture Creative Space

地点 Location
北京市东城区箭厂胡同38号 / No.38, Jianchang Lane, Dongcheng District, Beijing City

业主 Client
私人 / Private

设计团队 Project Team
[建筑] 王硕 张婧 吴亚萍
[灯光] 韩晓伟

设计时间 Design Period
2014.08 - 2014.10

建造时间 Construction Period
2014.10 - 2014.12

基地面积 Site Area
400 m²

建筑面积 Floor Area
400 m²

层数 Storey
地上2层 /2 storeys

结构形式 Structure
原有木桁架结构 /Original wood truss structure

p328
上海国际时尚中心
Shanghai International Fashion Center

地点 Location
上海市杨树浦杨树浦路 2888号 / No.2888, Yangshupu Street, Yangpu District, Shanghai City

业主 Client
上海纺织控股有限公司 /ShangTex Holding Co., Ltd.

设计团队 Project Team
[建筑] 邢同和 Pierre Chambron 袁静 周雯怡 石红梅 张驰 王岚 刘炳
[结构] 洪油然 骆正荣
[暖通] 郑兵 张云斌
[给排水] 周雪松
[电气] 刘前梅

施工单位 Constructor
上海申创建筑工程有限公司

设计时间 Design Period
2009.08 - 2012.05

建造时间 Construction Period
2010.01 - 2014.07

基地面积 Site Area
90 000 m²

建筑面积 Floor Area
100 000 m²

层数 Storey
地上6层 / 6 storeys

结构形式 Structure
钢框架结构 /Steel frame structure

p330
南京下关电厂码头遗址公园
Relics Park for the Coal Dock of Xiaguan Power Plant

地点 Location
江苏省南京市鼓楼区江边路1号 /No.1, Jiangbian Street, Gulou District, Nanjing City, Jiangsu Province

业主 Client
南京下关滨江开发建设投资有限公司 / Nanjing Xiaguan Binjiang Development Construction Investment Co., Ltd.

设计团队 Project Team
[建筑] 杨明 俞楠 于汶卉 顾鹏
[结构] 傅晋升
[强弱电] 刘向 王达威
[暖通] 景琳
[动力] 谈嘉辰

施工单位 Constructor
长江南京航道工程局

设计时间 Design Period
2012.10 - 2014.04

建造时间 Construction Period
2014.02 - 2014.10

基地面积 Site Area
9 167.3 m²

建筑面积 Floor Area
3 922 m²

层数 Storey
地上2层,地下1层 /2 storeys overground, 1 storey underground

结构形式 Structure
混凝土框架结构 (局部钢结构) / Concreate frame structure (Partly steel structure)

p332
天津拖拉机厂融创中心
Sunac Center of Tianjin Tractor Factory

地点 Location
天津市南开区红旗南路保泽路路口 /The Junction of Hongqi South Street and Baoze Street, Nankai District, Tianjin City

业主 Client
融创中国 /China Sunac

设计团队 Project Team
[建筑] 任治国 杨佩燊 刘振 申洪

[室内] 龚坤
施工单位 Constructor
天津市筑土建筑设计有限公司
设计时间 Design Period
2013.12 – 2014.04
建造时间 Construction Period
2014.05 – 2014.07
基地面积 Site Area
6 765 m²
建筑面积 Floor Area
3 664.8 m²
层数 Storey
地上 2层，局部 1层 /2 storeys, partly 1 storey
结构形式 Structure
排架结构 /Bent structure

---

p334
吉兆营清真寺翻建工程
Jizhaoying Mosque Renovation
地点 Location
江苏省南京市玄武区丹凤街吉兆营43号 /No.43, Jizhaoying, Danfeng Street, Nanjing City, Jiangsu Province
业主 Client
南京市伊斯兰教协会 /Nanjing Islamic Association
设计团队 Project Team
[建筑] 马晓东 韩冬青 高崧 孙鹏智
[结构] 韩重庆 王晨
[给排水] 贺海涛 王志东
[电气] 周桂祥 叶飞
施工单位 Constructor
中国江苏省国际经济技术合作集团有限公司
设计时间 Design Period
2009.02 – 2010.01
建造时间 Construction Period
2011.12 – 2014.01
基地面积 Site Area
661 m²
建筑面积 Floor Area
1 307.9 m²
层数 Storey
地上 4层 /4 storeys
结构形式 Structure
钢筋混凝土框架结构 /Reinforced concrete frame structure

---

p336
天仁合艺美术馆
T-Museum
地点 Location
浙江省杭州市上城区复兴街405号 /No.405, Fuxing Street, Shangcheng District, Hangzhou City, Zhejiang Province
业主 Client
天仁合艺美术馆 /T_Museum
设计团队 Project Team
[建筑] 王振飞 王鹿鸣 王凯
[结构] 郭昌锋
施工单位 Constructor
浙江舜业建设有限公司
设计时间 Design Period
2014.02 – 2014.03
建造时间 Construction Period
2014.03 – 2014.05
基地面积 Site Area
4 000 m²
建筑面积 Floor Area
2 450 m²
层数 Storey
地上 1层 /1 storey
结构形式 Structure
钢木结构 /Steel and wood structure

---

p338
回酒店
Hui Hotel
地点 Location
广东省深圳市福田区华强北华富路与红荔路交叉口东南侧 /The Intersection of Huafu Street and Hongli Street, Huaqiangbei, Futian District, Shenzhen City, Guangdong Province
业主 Client
回酒店.杨邦胜酒店设计集团 /Hui Hotel, YangBangsheng & Associates Group
合作单位 Cooperative Design
室内 / Interior: 杨邦胜酒店设计集团 / YangBangsheng & Associates Group
结构 / Structural: 同济人建筑设计 /Tongji Architects
机电 /E & M:筑道建筑工程设计 /Shenzhen Zhudao Architectural Engineering Design Co., Ltd.
幕墙 / Curtain Wall: 深圳市光华中空玻璃 / Shenzhen Guang Hua Insulating Glass Engineering Co.,Ltd.
设计团队 Project Team
[建筑] 刘晓都 王俊 姜轻舟 张震 李强 李嘉熹 梁广发 姚晓微 林怡琳 谢盛奋
设计时间 Design Period
2010 – 2011
建造时间 Construction Period
2011 – 2014
基地面积 Site Area
1 937 m²
建筑面积 Floor Area
10 915 m²
层数 Storey
地上 6层 /6 storeys
结构形式 Structure
框架结构 /Frame structure

---

p340
南京愚园修缮与重建
Restoration and Reconstruction of Nanjing Yu Garden
地点 Location
江苏省南京市秦淮区鸣羊街45号 /No.45 Mingyang Street, Qinhuai District, Nanjing City, Jiangsu Province
业主 Client
南京风光建设综合开发公司 /Nanjing Scenery Construction Comprehensive Development Company
设计团队 Project Team
[规划] 陈薇 王建国 是霏 杨俊
[建筑] 陈薇 高琛 胡石 戴薇薇 冯耀祖 等
[景观] 陈薇 朱舜 陶敏 杨冬辉 伍清辉 等
[修缮] 顾效 都荑
[结构] 梁沙河 孙逊
[水电设备] 赵元 罗振宁
施工单位 Constructor
苏州香山古建园林工程有限公司
设计时间 Design Period
2008.03 – 2010.12
建造时间 Construction Period
2009.03 – 2013.04
基地面积 Site Area
34 500 m²
建筑面积 Floor Area
4 120 m²
层数 Storey
地上 1-2层 /1-2 storeys
结构形式 Structure
木结构 +砖混结构 /Timber structure + Brick-concrete structure

居住建筑
Housing

---

p344
乌镇·雅园
Wuzhen Graceland
地点 Location
浙江省桐乡市乌镇白马墩村 /Baimadun village, Wuzhen Town, Tongxiang City, Zhejiang Province
业主 Client
浙江雅达置业有限公司 /Zhejiang YADA Property Co., Ltd.
设计团队 Project Team
[建筑] 蒋愈 孟骐 周勤 贺珉
[景观] 宋淑华
施工单位 Constructor
浙江恒力建设有限公司
设计时间 Design Period
2010.06 – 2018.10
建造时间 Construction Period
2011.03 – 2018.12
基地面积 Site Area
180 000 m²
建筑面积 Floor Area
600 000 m²
层数 Storey
住宅 /Apartments: 地上 5-18层，地下 1层 /5-18 storeys overground, 1 storey underground; 配套设施 / Supporting facilities: 地上 1-4层 / 1-4 storeys
结构形式 Structure
框架结构 /Frame structure

---

p346
泰康之家·燕园
Taikang Yanyuan Community
地点 Location
北京市昌平区南邵镇景荣街2号院 /No.2, Jingrong Street, Nanshao Town, Changping Distrct, Beijing City
业主 Client
泰康之家投资有限公司 /Taikang Community Investment Co., Ltd.
设计团队 Project Team
[建筑] 张广群 石华 王璐 杨帆 褚奕爽 等
[结构] 何鑫 逯晔 毛伟中 李阳 房梦茜
[设备] 袁煌 梁江 孙宗齐 刘立芳 马龙
[电气] 肖旎旎 陈婷 谭天博
施工单位 Constructor
中建一局集团建设发展有限公司
设计时间 Design Period
2013.09 – 2015.05
建造时间 Construction Period
2014.12 – 2017.08
基地面积 Site Area
143 000 m²
建筑面积 Floor Area
438 000 m²
层数 Storey
地上 15层，地下 2层 /15 storeys overground, 2 storeys underground
结构形式 Structure
框架剪力墙结构 /Frame-shear wall structure

---

p348
船长之家改造
Renovation of the Captain's House
地点 Location
福建省福州市连江县苔菉镇北茭村 /Beijiao Village, Tailu Town, Lianjiang Country, Fuzhou City, Fujian Province
业主 Client
私人 + 上海广播电视台东方卫星频道 /Private Client, Shanghai Dragon Television
合作单位 Cooperative Design
结构及设备 /Structural & MEP Engineering: 中国建筑科学研究院 /China Academy of Building Research
设计团队 Project Team
[主持] 董功
[项目] 刘晨
[驻场] 陈振强 赵亮亮
[建筑] 赵丹 江存裕 张钊
[建造] 孙栋平
[结构] 肖从真 杜义欣
施工单位 Constructor
福建昌达工程技术有限公司
设计时间 Design Period
2016.01 – 2016.08
建造时间 Construction Period
2016.05 – 2017.01
基地面积 Site Area
217 m²
建筑面积 Floor Area
470 m²
层数 Storey
地上 3层 /3 storeys
结构形式 Structure
混凝土承重墙 + 旧有砖混结构 / Concrete structure reinforcement+ Existing brick-concrete structure

---

p350
上海龙南佳苑
Shanghai Longnan Garden Social Housing Estate
地点 Location
上海市徐汇区龙水南路336弄 /No.336 Lane, South Longshui Street, Xuhui District, Shanghai City
业主 Client
上海汇成公共租赁住房建设有限公司 / Shanghai Huicheng Social Housing Construction Co., Ltd.
设计团队 Project Team
[高目] 张佳晨 徐文斌 黄巍 徐聪 易博文 等
[中星] 李婕 曹勇 张红革 黄庆
施工单位 Constructor
上海汇成建设工程有限公司
设计时间 Design Period
2012 – 2014
建造时间 Construction Period
2014 – 2017
基地面积 Site Area
48 112 m²
建筑面积 Floor Area
146 106 m²
层数 Storey
地上 17层，地下 2层 /17 storeys overground, 2 storeys underground
结构形式 Structure
框架剪力墙结构 /Frame-shear wall structure

---

p352
齐云山树屋
The Qiyun Mountain Tree House
地点 Location
安徽省黄冈市休宁县齐云山地质国家公园 /Qiyunshan National Nature Park, Xiuning County, Huanggang City, Anhui Province
业主 Client
自由人 /ZYJ
设计团队 Project Team
[主持] 相南
[建筑] 姚中 秦川 叶莹
设计时间 Design Period
2015.06 – 2015.11
建造时间 Construction Period
2016.12 – 2016.07
基地面积 Site Area
35 m²
建筑面积 Floor Area
120 m²
层数 Storey
地上 4层 /4 storeys
结构形式 Structure
钢结构 + 欧松板 /Steel structure with OSB

---

p354
随园嘉树养生中心
The Health & Longevity Center of the Suiyuan Jia Shu Project
地点 Location
浙江省杭州市余杭区良渚文化村 /Liangzhu Cultural Village, Yuhang District, Hangzhou City, Zhejiang Province
业主 Client
浙江万科南都房地产开发有限公司 /Zhejiang Vanke Nandu Real

p356
深深·深宅
Deep³ Courtyard
地点 Location
江苏省南通市 /Nantong City, Jiangsu Province
业主 Client
私人 / Private
设计团队 Project Team
[建筑] 马礼元 郭少珣 李硕 于庆元
[结构] 张明熠
[景观] 周方程
施工单位 Constructor
江苏通宁建设有限公司
设计时间 Design Period
2012.01 – 2013.06
建造时间 Construction Period
2013.01 – 2015.01
基地面积 Site Area
2 600 m²
建筑面积 Floor Area
650 m²
层数 Storey
地上 2层 / 2 storeys
结构形式 Structure
砖混结构 +钢结构屋顶 /Brick-concrete structure+ steel roof structure

p358
新青年公社
New Youth Commune
地点 Location
吉林省松花湖度假区 /Songhua Lake Resort, Jilin Province
业主 Client
万科松花湖度假区 /Vanke Songhua Lake Resort
设计团队 Project Team
[建筑] 王硕 张婧 曹世彪 兰添 吴亚萍 等
施工单位 Constructor
江苏宏景集团有限公司
设计时间 Design Period
2014.10 – 2015.04
建造时间 Construction Period
2015.05 – 2015.12
基地面积 Site Area
4 436 m²
建筑面积 Floor Area
10 080 m²
层数 Storey
地上 5层 /5 storeys
结构形式 Structure
混凝土框架结构 /Concrete frame structure

p360
浙江山地老宅
The Old Curtilage in the Mountains of Zhejiang
地点 Location
浙江省嵊州市 /Shengzhou City, Zhejiang Province
业主 Client
私人 / Private

Estate Development Co., Ltd.
设计团队 Project Team
[建筑] 陆臻 谢runs 王静伟 白玉龙
施工单位 Constructor
浙江常升建设集团有限公司
设计时间 Design Period
2013.05 – 2013.12
建造时间 Construction Period
2014.01 – 2015.03
基地面积 Site Area
64 000 m²
建筑面积 Floor Area
64 000 m²
层数 Storey
地上 5层, 地下 1层 / 5 storeys overground, 1 storey underground
结构形式 Structure
混凝土框架结构 / Concrete frame structure

p362
退台方院
Stepped Courtyards
地点 Location
福建省福州市长乐区湖南镇 / Hunan Town, Changle District, Fuzhou City, Fujian Province
业主 Client
福建天晴在线互动科技有限公司 /Fujian TQ Online Interactive Inc.
设计团队 Project Team
李虎 黄文菁 周亭婷 陈逸岚 汪剑怜 等
设计时间 Design Period
2012.02 – 2013.03
建造时间 Construction Period
2012.10 – 2014.08
基地面积 Site Area
44 570 m²
建筑面积 Floor Area
38 203 m²
层数 Storey
地上 7层 /7 storeys
结构形式 Structure
钢筋混凝土框架结构 /Reinforced concrete frame structure

p364
苏州阳山敬老院
Suzhou Yangshan Nursing Home
地点 Location
江苏省苏州市大阳山国家森林公园 /Yangshan National Forest Park, Suzhou City, Jiangsu Province
业主 Client
苏州国家高新技术产业开发区浒墅关分区管理委员会 /Suzhou SND Administration Committee Xushuguan Branch
设计团队 Project Team
[建筑] 张应鹏 黄志强 倪骏 王苏嘉 沈春华
[结构] 苗平洲 刘兰珣 吴玉英
[给排水] 张晓明 张琦
[暖通] 梁羽晴
施工单位 Constructor
江苏通力建设工程有限公司
设计时间 Design Period
2011.07 – 2012.08
建造时间 Construction Period
2012.02 – 2014.12
基地面积 Site Area
23 912.3 m²
建筑面积 Floor Area
29 843.13 m²
层数 Storey
地上 3层, 地下 1层 /3 storeys overground, 1 storey underground
结构形式 Structure
混凝土框架结构 /Concrete frame structure

p366
广州南湖山庄 C、D 区
Villa South Lake in Guangzhou-District C and D Residence
地点 Location

广东省广州市白云区同和路贤庄地段 1080号 / No.1080, Xianzhuang Zone, Tonghe Street, Baiyun District, Guangzhou City, Guangdong Province
业主 Client
广州市贤庄房地产开发有限公司 /Guangzhou Xianzhuang Real Estate Development Co., Ltd.
设计团队 Project Team
[建筑] 倪阳 林毅 张敏婷 邓心宇
[结构] 陈福熙
[设备] 高飞
[设备] 陈欣燕
施工单位 Constructor
中天建设集团有限公司、广东电白建设集团有限公司
设计时间 Design Period
2009.12 – 2011.01
建造时间 Construction Period
2011.03 – 2014.11
基地面积 Site Area
35 000 m²
建筑面积 Floor Area
69 500 m²
层数 Storey
C区: 地上 3-4层, 地下 1层 / 3-4 storeys overground, 1 storey underground;
D区: 地上 24-27层, 地下 2层 / 24-27 storeys overground, 2 storeys underground
结构形式 Structure
钢筋混凝土结构 / Reinforced concrete structure

p368
拙政别墅
The Humble Administrator's Villa
地点 Location
江苏省苏州市姑苏区百家巷 8号 / No.8, Baijia Street, Gusun District, Suzhou City, Jiangsu Province
业主 Client
苏州赞威置业有限公司 /Suzhou Tsan Wei Real Estate Co., Ltd.
设计团队 Project Team
[建筑] 俞挺 傅正伟 丁顺 王丽洋 陈滢滢 金丽丽
[结构] 沈新卫 张瑞红
[电气] 王宇
[给排水] 朱洪山
[暖通] 张云斌
施工单位 Constructor
南通建筑工程总承包有限公司
设计时间 Design Period
2009.10 – 2011.10
建造时间 Construction Period
2011.05 – 2014.06
基地面积 Site Area
34 908 m²
建筑面积 Floor Area
36 963 m²
层数 Storey
地上 2层, 地下 1层 /2 storeys overground, 1 storey underground
结构形式 Structure
钢筋混凝土框架结构 /Reinforced concrete frame structures

生产设施
Production Facilities

p372
唐山乡村有机农场
Tangshan Rural Organic Farm
地点 Location
河北省唐山市古冶区 /Guye District, Tangshan City, Hebei Province
业主 Client
中合丰景农业发展有限公司 /

Zhonghefengjing Agricultural Development Co., Ltd.
设计团队 Project Team
[建筑] 韩文强 李晓明 王汉 姜兆 黄涛
施工单位 Constructor
北京南森木结构工程有限公司
设计时间 Design Period
2015.06 – 2015.09
建造时间 Construction Period
2015.09 – 2016.04
基地面积 Site Area
6 000 m²
建筑面积 Floor Area
1 720 m²
层数 Storey
地上 1层 / 1 storey
结构形式 Structure
木结构 /Timber Structure

p374
松阳樟溪红糖工坊
Songyang Zhangxi Brown Sugar Workshop
地点 Location
浙江省丽水市松阳县兴村 /Xing Village, Songyang County, Lishui City, Zhejing Province
业主 Client
松阳县樟溪乡兴村村民委员会 /Villager Committee of Xing Village, Zhangxi Town, Songyang
设计团队 Project Team
徐甜甜 张龙潇 周洋 鲁勇 赵炜
施工单位 Constructor
松阳县樟溪乡兴村村民委员会
设计时间 Design Period
2015.06 – 2015.12
建造时间 Construction Period
2016.12
基地面积 Site Area
1 309 m²
建筑面积 Floor Area
1 234 m²
层数 Storey
地上 1层 /1 storey
结构形式 Structure
轻钢结构 /Steel Structure

p376
雅昌艺术中心
Artron Art Center
地点 Location
广东省深圳市南山区龙珠八路和广深高速公路交汇处 / The Intersection of Longzhu Eighth Street and Guangzhou-Shenzhen Expressway, Nanshan District, Shenzhen City, Guangdong Province
业主 Client
深圳雅昌彩色印刷有限公司 / Shenzhen Artron Colour Printing Co., Ltd.
合作单位 Cooperative Design
室内合作 / Interior: 极尚建筑装饰设计 /Wendell Burnette Architects; Artmost Design and Construction Co., Ltd.
结构 /Structure: 广州容柏生建筑结构设计事务所 /Guangzhou RBS Architecture Engineer Design Associates
机电 /Engineering: 天宇机电工程设计 /Shenzhen Tianyu Dynamo-electric Engineering Design Firm
幕墙 /Facade: 易科建筑幕墙技术有限公司 /ECO Building Facade Technologies Ltd
设计团队 Project Team
[主持] 孟岩
[项目经理] 周娅琳
[建筑] 饶恩辰 熊嘉伟 梁广发 吴春英 艾芸 孙艳花 黄志毅
[景观] 魏志姣 林挺 于晓兰 刘洁

陈丹平 黄陈航
[技术总监] 姚殿斌
设计时间 Design Period
2008 – 2013
建造时间 Construction Period
2010 – 2015
基地面积 Site Area
12 535 m²
建筑面积 Floor Area
41 504 m²
层数 Storey
地下 6层,地下 1层 / 6 storeys overground, 1 storey underground
结构形式 Structure
框架剪力墙结构 /Frame-shear wall structure

p378
松风翠山茶油厂
Song Feng Cui Camellia Oil Plant
地点 Location
江西省婺源县江湾镇中平村 / ZhongPing Village, Jiangwan Town, Wuyuan County, Jiangxi Province
业主 Client
松风翠有机农业发展有限公司 / Song Feng Cui Organic Agriculture Development Co., Ltd.
设计团队 Project Team
[建筑] 罗四维 卢珊 周伟 张俊 杨燕
[结构] 金姝 赵德良
施工单位 Constructor
徽印象古建筑公司
设计时间 Design Period
2012.02 – 2013.03
建造时间 Construction Period
2013.03 – 2014.12
基地面积 Site Area
18 500 m²
建筑面积 Floor Area
3 018 m²
层数 Storey
地上 1层 / 1 storey
结构形式 Structure
混合结构 / Mixed structure

p380
爱慕时尚工厂
Aimer Lingerie Factory
地点 Location
北京顺义区马坡镇秦武姚村 / Qinwuyao Village, Mapo Town, Shunyi District, Beijing City
业主 Client
爱慕集团 /Beijing AIMER Lingerie Co., Ltd.
设计团队 Project Team
[建筑设计 室内设计 景观设计 灯光设计 标识设计]
Crossboundaries: 蓝冰可 (Binke Lenhardt) 董灏 Cristina Portolés, Giacomo Butte
[施工图与机电设计]
BIAD秦禾设计所
[建筑] 侯新元 马跃
[结构] 周狄青 孙传波
[设备] 刘均 钱强
[电气] 白景录
施工单位 Construction
中北华宇建筑工程公司
设计时间 Design Period
2004 – 2011
建造时间 Construction Period
2011.02 – 2013.09
基地面积 Site Area
53 000 m²
建筑面积 Floor Area
53 000 m²
层数 Storey
地上 4层 / 4 storeys
结构形式 Structure
混凝土结构+钢结构 /Concrete structure + Steel structure

p382
武夷山竹筏育制场
Wuyishan Bamboo Raft Factory
地点 Location
福建省武夷山星村镇东南部 / Southeast of Xingcun Town, Wuyishan, Fujian Province
业主 Client
福建武夷山旅游发展股份有限公司 /Fujian Wuyishan Tourism Development Co., Ltd.
设计团队 Project Team
[建筑] 华黎 Elisabet Aguilar Palau 张婕 诸荔晶
[驻场] 赖尔逊
施工单位 Constructor
福建蓝海市政园林建筑有限公司
设计时间 Design Period
2011 – 2012
建造时间 Construction Period
2012 – 2013
基地面积 Site Area
14 629 m²
建筑面积 Floor Area
16 000 m²
层数 Storey
地上 1–2层 /1-2 storeys
结构形式 Structure
素混凝土结构+混凝土砌块外墙 / Fine concrete structure + Concrete brick as exterior wall

城市设计及其他
Urban Design Others

p386
常德老西门综合片区城市更新
Urban Renewal of Old West City Gate Region
地点 Location
湖南省常德市老西门 /Old West Citywall Gate Area Changde City, Hunan Province
业主 Client
常德市天源住建房地产开发有限公司 /TianYuan Construction Deviolopment Co., Ltd.
设计团队 Project Team
[建筑] 曲雷 何勍 王强 童佳明 宋亚涛 等
[结构] 李忠盛 郭伟 张慧 李廷平 等
[给排水] 李严 贺立军 何戬 等
[空调] 范向国 张英男 唐铁釜 等
[电气] 叶劲 刘涛 王开丰 等
[总配] 王雅萍
施工单位 Constructor
常德市天源工程有限公司
设计时间 Design Period
2011.05 至今
建造时间 Construction Period
2012.07至今
基地面积 Site Area
70 000 m²
建筑面积 Floor Area
200 000 m²
层数 Storey
地上 2-25层, 地下 1层 / 2-25 storeys overground, 1 storey underground

p388
驿道廊桥改造
Lounge Bridge Renovation
地点 Location
河北省阜平县八里庄村 /Bali Village, Fuping County, Hebei Province
业主 Client
八里庄村委会 /Committee of Bali Village
设计团队 Project Team
[建筑] 张东光 张意姝 刘文娟 张钰雯 李伟杰 等
施工单位 Constructor
河北泽田园林绿化工程有限公司
设计时间 Design Period
2016.05 – 2016.07
建造时间 Construction Period
2016.09 – 2017.01
基地面积 Site Area
320 m²
建筑面积 Floor Area
193 m²
层数 Storey
地上 1层 /1 storey
结构形式 Structure
轻木结构 / Light wood structure

p390
天府新区公安消防队站
Fire Station of Tianfu New District
地点 Location
四川省成都市天府新区万安镇东林社区四组 / Group Four, Donglin Community, Wan'an Town, Tianfu New District, Chengdu City, Sichuan Province
业主 Client
成都天投地产开发有限公司 / Chengdu Tianfu New Area Investment Group Co., Ltd.
设计团队 Project Team
[建筑] 刘艺 胡健 王子超 王俊涵 周娅
[结构] 张兴宇 罗实瀚
[电气] 郑庆军
[给排水] 王珂
[幕墙] 李铭
[暖通] 徐猛
[景观] 史笑微
[装饰] 蓝天
[造价] 陈帅锋
[绿建] 黄驰
施工单位 Constructor
四川中林建设有限公司
设计时间 Design Period
2014.05 – 2014.09
建造时间 Construction Period
2015.07 – 2016.07
基地面积 Site Area
1 280.47 m²
建筑面积 Floor Area
8 619.88 m²
层数 Storey
地上 4层 /4 storeys
结构形式 Structure
钢筋混凝土框架剪力墙结构 / Reinforced concrete frame-shear wall structure

p392
风雨桥
Wind and Rain Bridge
地点 Location
福建省龙岩市连城县朋口镇乡培田村 /Peitian Village, Xuanhe, Pengkou Town, Liancheng County, Longyan City, Fujian Province
业主 Client
培田村村民、香港大学 Gallant Ho 实验学习基金 /Villagers in Peitian Village, Gallait Ho Experiential Learning Fund of HongKong University
合作单位 Cooperative Design
地接团队 /Local Cooperator:
耕心乡村众创 /Gengxin Village Co-Creation
设计团队 Project Team
Donn Holohan 姜禾嘉 Man Ho Kwan 梁皓晴 刘畅 Elspeth Lee 香港大学建筑学生
施工单位 Constructor
龙岩培田鲁班古建筑工程有限公司、培田当地工厂
设计时间 Design Period
2015.10 – 2016.01
建造时间 Construction Period
2016.01 – 2016.05
基地面积 Site Area
20 m²
建筑面积 Floor Area
20 m²
层数 Storey
地上 1层 /1 storey
结构形式 Structure
榫卯结构 /Rivet and tenon structure

p394
隐庐莲舍
Lotus Tea House
地点 Location
江苏省宜兴市丁蜀镇莲花荡农场 / Lianhuadang farm, Dingshu Town, Yixing City, Jiangsu Province
业主 Client
丁蜀镇政府 /Dingshu government
设计团队 Project Team
[建筑] 唐芃 蔡陈翼
[结构] 杨波
[水电] 史海山
施工单位 Constructor
红太阳建设工程有限公司
设计时间 Design Period
2015.05 – 2015.11
建造时间 Construction Period
2015.11 – 2016.09
基地面积 Site Area
392 m²
建筑面积 Floor Area
392 m²
层数 Storey
地上 1层 /1 storey
结构形式 Structure
钢结构 /Steel structure

p396
六边体系
HEX-SYS
地点 Location
广东省广州市汉溪大道西 /West of Hanxi Avenue, Guangzhou City, Guangdong Province
业主 Client
广州万科 /Vanke Guangzhou
设计团队 Project Team
李虎 黄文菁 赵耀 Andrea Antonucci 陈逸岚 等
设计时间 Design Period
2014.03 – 2014.12
建造时间 Construction Period
2015.02 – 2015.08
基地面积 Site Area
5 680 m²
建筑面积 Floor Area
640 m²
层数 Storey
1层 /1 storey
结构形式 Structure
钢结构 /Steel structure

p398
衢州鹿鸣公园
Quzhou Luming Park
地点 Location
浙江省衢州市西区,石梁溪花园桥至白云桥段西岸 /Shiliang Brook Bridge to the West Part of Baiyun Bridge of Quzhou City, Zhejiang Province
业主 Client
衢州市西区开发建设管理委员会 /Development and Construction Committee of Quzhou West Area
设计团队 Project Team
[主持] 俞孔坚
[景观] 刘玉杰 高正敏 宁维晶
[建筑] 鲁晓静
施工单位 Constructor
宁波市园林工程有限公司
设计时间 Design Period
2013.01 – 2013.07
建造时间 Construction Period

p400
长沙滨江文化园
Riverside Cultural Park in Changsha

地点 Location
湖南省长沙市开福区新河三角洲 / Xinhe Delta, Kaifu District, Changsha City, Hunan Province
业主 Client
长沙市工务局 / Changsha City Public Works Bureau
设计团队 Project Team
[建筑] 陶郅 郭嘉 郭钦恩 陈子坚 陈坚 等
[结构] 孙文波 龚模松 王剑文
[给排水] 王学峰 陈欣燕 林方 梁志君 曾银波
[电气] 黄晓峰 俞洋 陈祖铭 李国有 伍尚仁
[空调] 王钊 张毅 郭宇
[景观] 陶郅 杜宇健 黄晓峰 陈天宁
施工单位 Constructor
中国建筑第五工程局有限公司
设计时间 Design Period
2005.12 – 2011.04
建造时间 Construction Period
2007.12 – 2015.12
基地面积 Site Area
124 000 m²
建筑面积 Floor Area
149 943 m²
层数 Storey
地上 5层, 地下 1层 / 5 storeys overground, 1 storey underground
结构形式 Structure
钢筋混凝土框架结构 / Concrete frame structure

p402
天空之桥
Sky Bridge

地点 Location
台湾云林县北港镇义民路与河堤道路交汇附近 / Junction of Hedi Road and Yimin Street., Beigang Township, Yunlin County, Taiwan Province
业主 Client
云林县政府 / Yunlin County Government
设计团队 Project Team
[建筑] 廖伟立
[机电] 冠升工程设计事务所
[结构] 富田构造设计事务所
施工单位 Constructor
佑镇营造
设计时间 Design Period
2012.01 – 2012.04
建造时间 Construction Period
2014.09 – 2015.06
基地面积 Site Area
305 m²
建筑面积 Floor Area
305 m²
层数 Storey
地上 1层 / 1 storey
结构形式 Structure
清水混凝土结构 / Bare concrete structure

p404
上海廊下郊野公园核心区景观
Veranda in Shanghai

地点 Location
上海市金山区漕廊公路 9133号 / No.9133, Caolang Road, Jinshan District, Shanghai City
业主 Client
上海市金山区廊下镇政府 / Government of Langxia Town Shanghai City
设计团队 Project Team
[规划] 张宁
[建筑] 李振国 李毅 袁珏
[景观] 王洪 唐颖 董小惠
施工单位 Constructor
金山当地施工单位
设计时间 Design Period
2013.12 – 2014.12
建造时间 Construction Period
2015.01 – 2015.05
基地面积 Site Area
11 536 m²
建筑面积 Floor Area
397 m²
层数 Storey
地上 1层 / 1 storey
结构形式 Structure
混凝土+钢结构 / Concrete structure + Steel structure

p406
华山绿工场
Green Factory

地点 Location
台湾台北市中正区八德路一段 1号华山大草原 / Huashan1914 Creative Park, No.1, Sec.1, Bade Road, Zhongzheng District, Taipei City, Taiwan Province
业主 Client
台北市都市更新处 / Taipei City Urban Regeneration Office
设计团队 Project Team
[建筑] 梁豫漳 蔡大仁 吴明杰 赵宇晨 洪正客 等
施工单位 Constructor
宜实工程
设计时间 Design Period
2013.08 – 2013.12
建造时间 Construction Period
2013.12 – 2014.02
基地面积 Site Area
1 035 m²
建筑面积 Floor Area
1 035 m²
层数 Storey
地上 1层 / 1 storey
结构形式 Structure
木结构 / Timber structure

p408
松鹤墓园接待中心
Reception Center for Songhe Cemetery

地点 Location
上海市嘉定区安亭镇嘉松北路 3458号 / No.3458, North Jiasong Street, Anting Town, Jiading District, Shanghai City
业主 Client
上海市嘉定区松鹤墓园 / Shanghai Songhe Cemetery
设计团队 Project Team
[建筑] 张斌 周蔚 袁怡 金燕琳 李佳 等
施工单位 Constructor
上海均泰建筑工程有限公司
设计时间 Design Period
2011.10 – 2014.10
建造时间 Construction Period
2013.06 – 2014.11
基地面积 Site Area
3 095 m²
建筑面积 Floor Area
4 936 m²
层数 Storey
地上 2层 / 2 storeys
结构形式 Structure
钢筋混凝土框架结构+局部钢结构 / Reinforced concrete frame structure + Partly steel frame structure

p410
格萨尔广场
Gesar Square

地点 Location
玉树藏族自治州玉树市结古大道与格萨路交汇处附近 / Intersection of Jiegu Avenue and Gesa Road, Yushu City Tibetan Autonomous Prefecture, Qinghai Province
业主 Client
玉树州三江源投资建设有限公司 / Yushu Sanjiangyuan Investment Construction Co., Ltd.
设计团队 Project Team
[建筑] 周恺 吴岳 章宁
[结构] 李悦谦
[给排水] 魏平
[暖通] 邵海
[电气] 王裕华
施工单位 Constructor
中铁二十一局集团第四工程有限公司
设计时间 Design Period
2011.05 – 2013.12
建造时间 Construction Period
2014.01 – 2014.09
基地面积 Site Area
69 300 m²
建筑面积 Floor Area
8 200 m²
层数 Storey
地上 2层, 地下 1层 / 2 storeys overground, 1 storey underground
结构形式 Structure
钢筋混凝土框架结构 / Reinforced concrete frame structure

p412
林建筑
Forest Building

地点 Location
北京市通州区大运河森林公园 / Grand Canal Forest Park, Tongzhou District, Beijing City
业主 Client
北京美景天成投资有限责任公司 / Fujian Wuyishan Tourism Development Co., Ltd.
设计团队 Project Team
[建筑] 华黎 赵刚 姜楠 赖尔逊 陈恺
施工单位 Constructor
凯康木结构
设计时间 Design Period
2011 – 2013
建造时间 Construction Period
2012 – 2014
基地面积 Site Area
14 629 m²
建筑面积 Floor Area
4 000 m² (一期 /1st Phase: 1830 m²)
层数 Storey
地上 1层 / 1 storey
结构形式 Structure
木结构 / Timber structure

p414
太阳公社竹构系列
Bamboo Design in Taiyang Farming Commune

地点 Location
浙江省杭州市临安区太阳镇 / Taiyang County, Lin an District, Hangzhou City, Zhejiang Province
业主 Client
太阳公社生态农场 / Taiyang Organic Farming Commune
[主持] 陈浩如
[建筑] 王春威 朱晓成 顾安婕
施工单位 Constructor
太阳公社竹工坊
设计时间 Design Period
2013
建造时间 Construction Period
2013 – 2014
基地面积 Site Area
2 000 m²
建筑面积 Floor Area
506 m²
层数 Storey
地上 1层 / 1 storey
结构形式 Structure
竹结构 / Bamboo structure

p416
苏仙岭景观瞭望台
Observation Deck in Suxianling

地点 Location
湖南省郴州市苏仙区苏仙岭风景名胜区 / Suxianling Scenic Resort Suxian District, Chenzhou City, Hunan Province
业主 Client
郴州市城市建设投资发展集团有限公司 / Chenzhou Urban Construction Investment Development Group Co., Ltd.
设计团队 Project Team
[建筑] 杨瑛 贺丽菱 李秀峰
[景观] 袁倩
[结构] 王巧清
施工单位 Constructor
汕头市东楚建筑工程有限公司
设计时间 Design Period
2011.01 – 2011.10
建造时间 Construction Period
2011.11 – 2014.05
基地面积 Site Area
458 m²
建筑面积 Floor Area
1 038 m²
层数 Storey
地上 3层 / 3 storeys
结构形式 Structure
空间钢结构 / Spatial steel structure

中国建筑设计
作品选 2013—2017

CHINESE ARCHITECTURE:
A Selection 2013–2017

中国建筑学会
指定国际交流图书

图书在版编目（CIP）数据

中国建筑设计作品选：2013—2017 / 中国建筑学会
《建筑学报》杂志社编著. -- 上海：同济大学出版社，2018.6
ISBN 978-7-5608-7894-2

Ⅰ.①中… Ⅱ.①中… Ⅲ.①建筑设计－作品集－中国
－现代 Ⅳ.① TU206

中国版本图书馆 CIP 数据核字（2018）第 119246 号

中国建筑设计作品选 2013—2017
中国建筑学会《建筑学报》杂志社 编著

ASC

出 版 人	华春荣
策 划	秦蕾 / 群岛工作室
责任编辑	秦蕾 李争
责任校对	徐春莲
书籍设计	黄晓飞
版式设计	曹楠＋李如珍
版 次	2018 年 6 月第 1 版
印 次	2018 年 6 月第 1 次印刷
印 刷	北京图文天地制版印刷有限公司
开 本	787mm×1092mm 1/16
印 张	29
字 数	724 000
书 号	978-7-5608-7894-2
定 价	268.00 元

出版发行	同济大学出版社
地 址	上海市杨浦区四平路 1239 号
邮政编码	200092
网 址	http://www.tongjipress.com.cn
经 销	全国各地新华书店

本书若有印刷质量问题，请向本社发行部调换。
版权所有 侵权必究

光明城联系方式：info@luminocity.cn

CHINESE ARCHITECTURE: A Selection 2013-2017
Edited by: China Architectural Society *Architectural Journal* press

ISBN 978-7-5608-7894-2
Publisher: Hua Chunrong
Initiated by: Qin Lei / Studio Archipelago
Editor: Qin Lei / Li Zheng
Proofreading: Xu Chunlian

Published in June 2018, by Tongji University Press,
1239, Siping Road, Shanghai, China, 200092.
www.tongjipress.com.cn
Contact us: info@luminocity.cn

All rights reserved
No part of this book may be reproduced in any
manner whatsoever without written permission from
the publisher, except in the context of reviews.